Trajectory Planning and Object Recognition for Robot Sensing and Control

Guest Editors

Gang Peng
Zhun Fan

MDPI

Basel • Beijing • Wuhan • Barcelona • Belgrade • Novi Sad • Cluj • Manchester

Guest Editors

Gang Peng
School of Artificial Intelligence and Automation
Huazhong University of Science and Technology
Wuhan
China

Zhun Fan
Shenzhen Institute for Advanced Study
University of Electronic Science and Technology of China
Shenzhen
China

Editorial Office
MDPI AG
Grosspeteranlage 5
4052 Basel, Switzerland

This is a reprint of the Special Issue, published open access by the journal *Sensors* (ISSN 1424-8220), freely accessible at: https://www.mdpi.com/journal/sensors/special_issues/T3W3214I64.

For citation purposes, cite each article independently as indicated on the article page online and as indicated below:

Lastname, A.A.; Lastname, B.B. Article Title. *Journal Name* **Year**, *Volume Number*, Page Range.

ISBN 978-3-7258-2863-0 (Hbk)
ISBN 978-3-7258-2864-7 (PDF)
https://doi.org/10.3390/books978-3-7258-2864-7

Contents

sensors

MDPI

Editorial

Recent Advances in Trajectory Planning and Object Recognition for Robot Sensing and Control

Gang Peng [1,*] and Zhun Fan [2]

1 School of Artificial Intelligence and Automation, Huazhong University of Science and Technology, Wuhan 430074, China
2 Shenzhen Institute for Advanced Study, University of Electronic Science and Technology of China, Shenzhen 518110, China; fanzhun@uestc.edu.cn
* Correspondence: penggang@hust.edu.cn

1. Trajectory Planning and Control

Nowadays, trajectory planning and control technology has been extensively utilized in industry, agriculture, service, construction, infrastructure inspection, military reconnaissance, disaster management, and other areas. However, the complex natural environment can significantly impact the accuracy of trajectory planning, and developing this technology entails overcoming substantial challenges. This Special Issue of the journal *Sensors* presents three articles on trajectory planning and control that are of significant importance for advancing this technology.

To address the issue of intricate environmental and dynamic limitations in real-world applications, the article [contribution 1] proposes a novel gradient potential field, termed the kino-gene regulatory network (K-GRN) potential field, which is based on a GRN model. By enhancing the gene regulation network model, the potential field becomes adaptable to the UAV's dynamic conditions and its surroundings, thereby extending the GRN into a kinodynamic GRN (K-GRN). The amalgamation of environmental insights and kinodynamic constraints within the potential field framework bolsters the adaptability and stability of the generated trajectories.

To improve the comprehensiveness of path planning, the article [contribution 2] proposes a framework for autonomous UAV-based landmine detection to determine the coverage route for scanning the target area. This method extracts the area of interest using segmentation based on deep learning and then constructs the coverage route plan for the aerial survey. Multiple coverage path patterns are used to identify the ideal UAV route. The effectiveness of the suggested framework is evaluated using several target areas of differing sizes and complexities.

In article [contribution 3], researchers design a stepped rotor BLSRM structure and combined particle swarm optimization (PSO) algorithm with finite element analysis to optimize the structural parameters of the motor. Subsequently, a performance analysis of the original and new motors was conducted using finite element analysis software, and the results showed that the stepped rotor BLSRM had an improved self-starting ability and significantly reduced torque fluctuation, verifying the effectiveness of the proposed motor structure and optimization method.

These three papers illustrate the continual development and innovation in trajectory planning and control technology.

2. Environment Perception and Target Detection

Environment perception and target detection, using deep learning, image processing, and other technologies, have a wide range of applications in medicine, sports, and other fields. In recent years, environmental perception and target detection technology has gained increasing attention and research. In this article, we introduce four new methods

Citation: Peng, G.; Fan, Z. Recent Advances in Trajectory Planning and Object Recognition for Robot Sensing and Control. *Sensors* **2024**, *24*, 4509. https://doi.org/10.3390/s24144509

Received: 11 June 2024
Accepted: 18 June 2024
Published: 12 July 2024

to enhance the accuracy of environment perception and target detection technology from different perspectives.

The article [contribution 4] proposes a novel human motion prediction method that utilizes dual-attention and multi-granularity temporal convolutional networks. By designing a dual-attention model for extracting inter-joint and intra-joint spatial features, provide richer information sources for motion prediction. Then, a multi-granularity temporal convolutional network is introduced to realize the discriminant fusion of different time granularities, flexibly capturing complex short-term and long-term temporal dependencies, and thereby further improving the model's performance. Article [contribution 5] proposes an evaluation model based on contrastive learning to evaluate the neurological function of patients with Alzheimer's disease. It introduces an objective dementia severity scale based on MRI using a contrastive learning framework to evaluate neurological function during Alzheimer's disease progression and can accurately discriminate between different stages of Alzheimer's disease progression. The article [contribution 6] focuses on an improved version of the H-GrabCut image segmentation algorithm to address the aforementioned issues and achieve background removal in complex pedestrian backgrounds. By incorporating the Bilateral-MSRCR image enhancement algorithm and introducing two-dimensional information entropy, UV component distance, and LBP texture features, it effectively resolves the resource limitations associated with nesting multiple deep learning algorithms while enabling more accurate distance measurement between robots and pedestrians. Furthermore, the article [contribution 7] designs an infrared small target detection method based on an attention mechanism. It proposes a top–down active attention module to obtain target knowledge–experience Gaussian shape features to improve the signal-to-clutter ratio gain and background suppression factor and the decision feedback equalization in order to find real targets to adapt to environmental changes in different scenarios. The proposed method can achieve better detection performance than conventional baseline methods and mathematical proofs are provided to validate the proposed method.

We believe that these research results will provide robust support for applying and promoting environment perception and target detection technology.

3. Multi-Sensor Fusion and Calibration

Multi-sensor fusion and calibration are critical for enhancing the accuracy and robustness of various robotic and sensing systems. This integration allows for the seamless combination of data from different sensors, enabling more reliable perception and detection capabilities. This article introduces three innovative methods aimed at improving the accuracy and robustness of multi-sensor fusion and calibration techniques.

In the realm of multi-sensor fusion and calibration, three notable contributions have been made. The paper [contribution 8] introduces the VILO algorithm, which addresses challenges faced by existing visual–inertial SLAM algorithms by tightly coupling vision, IMU, and 2D LiDAR observations, resulting in improved accuracy and robustness, particularly in environments with insufficient visual features. The paper [contribution 9] presents a novel tool frame calibration method tailored for ultrasonic testing systems, utilizing reflectors to enhance precision without additional measuring devices, thus overcoming limitations in underwater environments. Lastly, the paper [contribution 10] introduces a new method for hand–eye calibration error analysis and algorithm optimization, leveraging augmented reality markers to achieve higher precision calibration results. These advancements significantly enhance the accuracy and robustness of robotic systems across diverse applications, contributing to the ongoing development of environment perception and target detection technologies.

In conclusion, the advancements in multi-sensor fusion and calibration techniques, as demonstrated in articles [contributions 8–10], significantly improve the accuracy and robustness of robotic systems across diverse applications. These methodologies contribute to the ongoing development of environment perception and target detection technologies, facilitating their integration into real-world scenarios.

4. Autonomous Unmanned System

Autonomous unmanned systems represent a crucial frontier in modern technology, offering innovative solutions across various domains. These systems are instrumental in addressing complex challenges and enhancing efficiency in critical tasks, ranging from underwater exploration to disaster response and beyond.

The article [contribution 11] addresses challenges in underwater imaging, proposing a method based on the dark channel prior (DCP) to enhance visibility and eliminate color cast effectively. [Contribution 12] presents a nonlinear model predictive control (NMPC) scheme for underwater biomimetic vehicle-manipulator systems (UBVMS), enabling robust path following and dynamic obstacle avoidance. The article [contribution 13] proposes an autonomous optimal deployment method for mobile base stations in high-rise building fire environments, enhancing rescue operations through accurate positioning and collision avoidance. Furthermore, the article [contribution 14] introduces a novel recovery system for unmanned surface vehicles (USVs), incorporating physical modeling of disturbance forces and optimized genetic algorithms to enhance recovery efficiency and accuracy.

In conclusion, the advancements presented in refs. [contributions 11–14] contribute significantly to the development and effectiveness of autonomous unmanned systems. These methodologies address critical challenges and offer innovative solutions, enhancing capabilities and improving performance in real-world applications.

5. Conclusions

The theme of this Special Issue is trajectory planning and target recognition for robot sensing and control. This Special Issue selected 14 articles, which can be divided into the following four categories: trajectory planning and control [contribution 1–3], environmental perception and target detection [contributions 4–7], multi-sensor fusion and calibration [contributions 8–10], and autonomous unmanned systems [contributions 11–14]. The research objects cover UAV, human posture, underwater bionic aircraft, multi-sensor fusion, etc. The development of artificial intelligence algorithms provides a new solution for trajectory planning and target recognition of robot sensing and control and shows great potential in dealing with complex environments and dynamic constraints.

Conflicts of Interest: The authors declare no conflicts of interest.

List of Contributions

1. Hong, J.; Chen, D.; Li, W.; Fan, Z. Trajectory Planner for UAVs Based on Potential Field Obtained by a Kinodynamic Gene Regulation Network. *Sensors* **2023**, *23*, 7982.
2. Barnawi, A.; Kumar, K.; Kumar, N.; Thakur, N.; Alzahrani, B.; Almansour, A. Unmanned Ariel Vehicle (UAV) Path Planning for Area Segmentation in Intelligent Landmine Detection Systems. *Sensors* **2023**, *23*, 7264.
3. Gao, H.; Liu, Z.; Wang, X.; Li, D.; Zhang, T.; Yu, J.; Wang, J. A Novel Motor Structure with Extended Particle Swarm Optimization for Space Robot Control. *Sensors* **2023**, *23*, 4126.
4. Huang, B.; Li, X. Human Motion Prediction via Dual-Attention and Multi-Granularity Temporal Convolutional Networks. *Sensors* **2023**, *23*, 5653.
5. Zhang, Y.; Fan, W.; Chen, X.; Li, W.; on behalf of the for Alzheimer's Disease Neuroimaging Initiative. The Objective Dementia Severity Scale Based on MRI with Contrastive Learning: A Whole Brain Neuroimaging Perspective. *Sensors* **2023**, *23*, 6871.
6. Huang, X.; Wang, S.; Gao, X.; Luo, D.; Xu, W.; Pang, H.; Zhou, M. An H-GrabCut Image Segmentation Algorithm for Indoor Pedestrian Background Removal. *Sensors* **2023**, *23*, 7937.
7. Wang, X.; Lu, R.; Bi, H.; Li, Y. An Infrared Small Target Detection Method Based on Attention Mechanism. *Sensors* **2023**, *23*, 8608.
8. Peng, G.; Zhou, Y.; Hu, L.; Xiao, L.; Sun, Z.; Wu, Z.; Zhu, X. VILO SLAM: Tightly Coupled Binocular Vision–Inertia SLAM Combined with LiDAR. *Sensors* **2023**, *23*, 4588.
9. Zhang, H.; Wang, J.; Guo, C. Tool Frame Calibration for Robot-Assisted Ultrasonic Testing. *Sensors* **2023**, *23*, 8820.
10. Peng, G.; Ren, Z.; Gao, Q.; Fan, Z. Reprojection Error Analysis and Algorithm Optimization of Hand–Eye Calibration for Manipulator System. *Sensors* **2024**, *24*, 113.

11. Li, C.; Dong, X.; Wang, Y.; Wang, S. Enhancement and Optimization of Underwater Images and Videos Mapping. *Sensors* **2023**, *23*, 5708.
12. He, Y.; Dong, X.; Wang, Y.; Wang, S. Path Following and Collision Avoidance of a Ribbon-Fin Propelled Underwater Biomimetic Vehicle-Manipulator System. *Sensors* **2023**, *23*, 7061.
13. Li, K.; Huang, C.; Liang, J.; Zou, Y.; Xu, B.; Yao, Y.; Zhang, Y.; Liu, D. Research on Autonomous and Collaborative Deployment of Massive Mobile Base Stations in High-Rise Building Fire Field. *Sensors* **2023**, *23*, 7664.
14. Zhou, L.; Ye, X.; Huang, Z.; Xie, P.; Song, Z.; Tong, Y. An Improved Genetic Algorithm for the Recovery System of USVs Based on Stern Ramp Considering the Influence of Currents. *Sensors* **2023**, *23*, 8075.

sensors

MDPI

Article

A Novel Motor Structure with Extended Particle Swarm Optimization for Space Robot Control

Hongwei Gao [1,*], Zide Liu [1], Xuna Wang [1], Dongyu Li [1], Tian Zhang [2], Jiahui Yu [3] and Jianbin Wang [4,5,*]

[1] School of Automation and Electrical Engineering, Shenyang Ligong University, Shenyang 110000, China
[2] Beijing Bolean Intelligence Technology Co., Ltd., Beijing 100000, China; zzddddddz@163.com
[3] Department of Biomedical Engineering, Zhejiang University, Hangzhou 310027, China; jiahui.yu@zju.edu.cn
[4] Ocean College, Zhejiang University, Hangzhou 316021, China
[5] China Telecom Zhejiang Branch, Hangzhou 310014, China
* Correspondence: ghw1978@sohu.com (H.G.); 18905718365@189.cn (J.W.)

Abstract: This paper studies motor structures and optimization methods for space robots, proposing an optimized stepped rotor bearingless switched reluctance motor (BLSRM) to solve the poor self-starting ability and significant torque fluctuation issues in traditional BLSRMs. Firstly, the advantages and disadvantages of the 12/14 hybrid stator pole type BLSRM were analyzed, and a stepped rotor BLSRM structure was designed. Secondly, the particle swarm optimization (PSO) algorithm was improved and combined with finite element analysis for motor structure parameter optimization. Subsequently, a performance analysis of the original and new motors was conducted using finite element analysis software, and the results showed that the stepped rotor BLSRM had an improved self-starting ability and significantly reduced torque fluctuation, verifying the effectiveness of the proposed motor structure and optimization method.

Keywords: bearingless switched reluctance motor; finite element analysis; particle swarm optimization; self-starting; step rotor

Citation: Gao, H.; Liu, Z.; Wang, X.; Li, D.; Zhang, T.; Yu, J.; Wang, J. A Novel Motor Structure with Extended Particle Swarm Optimization for Space Robot Control. *Sensors* **2023**, *23*, 4126. https://doi.org/10.3390/s23084126

Academic Editors: Zhun Fan and Gang Peng

Received: 8 March 2023
Revised: 8 April 2023
Accepted: 17 April 2023
Published: 20 April 2023

1. Introduction

With the rapid development of modern industry and the leap in aerospace technology, aerospace tasks are becoming more complex. At the same time, it is becoming more and more common for artificial satellites to fail due to fuel exhaustion. Satellites can fail because of a single component, and the costs are high. Therefore, countries are paying more and more attention to on-orbit service technology [1–5]. On-orbit service refers to providing services for human-crewed or uncrewed spacecraft in space, including the repair and replacement of spacecraft components, spacecraft relocation, and liquid transfer. In the complex and harsh environment of space, space robots can assist astronauts in completing complex missions and even replace astronauts in conducting some dangerous extravehicular operations. Space robots can assist in completing space operations and ensure the safety of astronauts and the successful completion of missions. Therefore, the use of space robots for on-orbit services [6] has attracted more and more attention and is becoming one of the main trends in the development of aerospace technology.

Space robots [7,8] providing extravehicular services can complete extensive tasks, including capturing and maintaining small satellites, target handling, and on-orbit assembly. The robots also put forward higher requirements for the performance of motor drives, and researchers are paying more and more attention to small, lightweight motors with high power density. The increase in motor speed not only helps to increase the power density of the motor, but also reduces the size and weight of the equipment, as shown in Figure 1. Therefore, high-speed and ultra-high-speed motors are increasingly being used in aerospace, flywheel energy storage, and other fields.

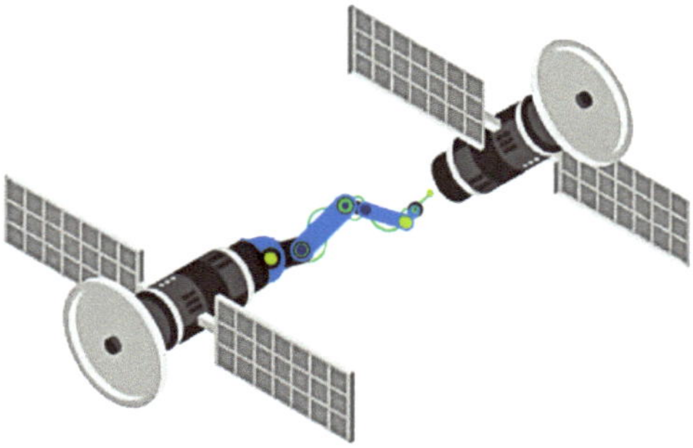

Figure 1. Concept image of a robotic arm for on-orbit service. The robotic arm is used to assist in the combination of the two satellites.

The bearingless switched reluctance motor (BLSRM) [9–14] is a new type of motor that combines drive and suspension. It combines bearingless technology and switched reluctance motors (SRM). SRM does not require permanent magnets or electromagnets, but instead uses the magnetic reluctance difference between the iron core and the stator windings to generate torque. The rotor and stator of the BLSRM are not supported by bearings, but by magnetic levitation technology. The bearingless motor has no mechanical wear and a long service life.

It has the advantages of having a compact structure, enormous output power, and no need for lubrication. In addition, it has the benefits of a simple structure and the robust fault tolerance of SRM. Compared with traditional servo motors, the BLSRM does not require vulnerable components such as brushes or slip rings, making it suitable for special requirements in the space environment, such as radiation resistance and high–low-temperature endurance. Its compact structure allows for a higher power output in the same space. The manufacturing cost is relatively low as it does not require a feedback control system or precise sensors. Its simple structure usually results in higher efficiency compared to servo motors, especially under partial load or low-speed operation. Due to its good fault tolerance, flexible control, and adaptability in harsh environments, it is widely used in aerospace.

However, due to the structure of the BLSRM motor itself, there are disadvantages, such as the significant torque fluctuation and existence of a torque dead zone. To solve these problems, many experts in the aerospace field and other areas have carried out research on the motor structure and performance. Many domestic and foreign universities and scientific research institutions have introduced bearingless switched reluctance motors [15–20]. Yue Qi proposed a novel hybrid rotor type 12/14 BSRM by keeping the air gap constant and modifying the aligned area between the torque and rotor poles to solve this problem. Fangxu Li changed a BLSRM with a 12/12 structure to reduce the axial space and produce higher torque for the purpose of improving performance. Zhenyao Xu proposed a stepped rotor structure and optimally designed the torque pole arc and rotor pole shape to solve the self-starting problems and reduce the torque ripple. Single-neuron-based rotor suspension control of a 12/14 hybrid pole BLSRM with autonomous rotation and suspension poles was presented. This hybrid pole structure produced suspension force, linearly concerning the rotor position. Wenjuan Hao proposed a bearingless linear switched reluctance motor to simplify the levitation force control based on a structural decoupling design for the mover. Shaofei Tang presented a 12/14 bearingless switched reluctance motor (BLSRM) with a small permanent magnet. However, due to the dead torque zone, BLSRM structures cannot self-start at certain rotor positions.

Traditional motors, such as induction motors and permanent magnet synchronous motors, usually have rotors with iron cores. In contrast, the rotor of a BLSRM is coreless. Due to the absence of a core for support, when the rotor is in certain positions, the magnetic field on the rotor does not overlap with that on the stator, which makes it impossible to generate enough magnetic force to start the motor. Additionally, there are significant torque fluctuations during the start-up and acceleration processes. Therefore, we optimized the motor by focusing on two main aspects: we improved the stepped structure to eliminate the dead zone, and we solved the problem of the significant torque fluctuations during start-up.

The particle swarm algorithm [21,22] is also known as the bird swarm foraging algorithm or particle swarm optimization algorithm (particle swarm optimization, PSO). PSO has many advantages, such as a simple expression, easy implementation, few setting parameters, and a fast convergence speed. Since being proposed, it has attracted academic attention. At present, its application field has been extended from its initial function [23], neural network training [24], to practical engineering problems, such as optimal control [25,26] and structural optimization [27,28], and has achieved good results. In addition, PSO can handle multi-dimensional, nonlinear, and non-convex optimization problems, and is able to search for a global optimal solution. However, the problem of motor parameter optimization is usually a high-dimensional problem. In order to reduce torque ripple and improve the motor's starting ability, in this paper, we consider torque pole arc, rotor pole arc, and step structure geometry parameters as optimization variables. In addition, PSO has been widely used in the design and optimization of motors. PSO is a heuristic optimization algorithm that optimizes problems by simulating the behavior of birds searching for food. However, the particle swarm optimization algorithm easily reverts to finding the local optimal solution, and is not suitable for high dimension problems..

In this study, we conducted the initial design and optimization of the BLSRM structure based on its structure and working principles. With the objective of addressing the issues of self-starting and oscillation by designing a stepped rotor, an optimization method for motor parameters was implemented by integrating finite element analysis with the PSO algorithm. The initial motor was optimized using this method to resolve the aforementioned problems. The main contributions of this paper are summarized as follows.

(1) Aiming to solve the problems of the poor self-starting ability and significant torque fluctuation of the 12/14 hybrid stator pole bearingless switched reluctance motor, we propose two critical parameters, namely, step height and step angle, and conducted an experimental analysis on the starting torque, average torque, and torque fluctuations under different conditions in order to design a stepped rotor bearingless switched reluctance motor structure.

(2) To solve the problems of the need to rely on experience, the time- and labor-consuming nature, the high degree of specialization, and the difficulty of obtaining optimal the structural parameters of the current motor design, we combined the finite element analysis method and particle swarm algorithm to establish a systematic motor structure optimization method.

(3) We improved the particle swarm optimization algorithm and enhanced its global optimization capability. Based on this method, we optimized the design of the stepped rotor bearingless switched reluctance motor.

(4) The finite element analysis results demonstrated the following: Compared with the original motor, when the excitation current was 2A, the initial design of the stepped rotor bearingless switched reluctance motor increased by 0.11 N·min in the starting torque and decreased by 1.9% in torque ripple. After optimization, the torque ripple was further reduced by 9.2%. In contrast, the starting torque remained unchanged, proving that the stepped rotor bearingless switched reluctance motor structure was adequate and that the proposed motor optimization method was feasible.

The remainder of this paper is organized as follows: Section 1 reviews the related work of the BLSRM. In Section 2, we review the structure and working principle of the stepped rotor BLSRM and introduce the initial design proposal for the stepped rotor

BLSRM. Section 3 introduces the standard particle swarm optimization algorithm (PSO) and the optimized particle swarm algorithm, considering the characteristics of the motor optimization problem. Section 4 combines the improved particle swarm algorithm with the finite element method to optimize the initially designed motor structure. Section 5 analyzes the motor's performance using finite element analysis software to verify the motor structure's effectiveness and compare it with the original 12/14 hybrid stator pole type BLSRM. Finally, Section 6 concludes the paper with a summary and discussion.

2. Motor Design for Space Robots

Due to the particularity of the working environment and tasks of space robots, there is a high requirement regarding the performance of the motor drive. Therefore, the design of the motor structure has a significant influence on the space robot. The traditional 12/14 hybrid stator magnetic pole BLSRM requires a better self-starting ability and significant torque fluctuation, which cannot directly meet the hardware requirements of space robots; thus, we redesigned the magnetic torque pole and rotor magnetic pole arc and adopted a stepped rotor based on its structure. As the number of poles of the stepped rotor increased during motor start-up, the magnetic field generated by the rotor was also gradually strengthened. Due to the gradual increase in the number of poles of the stepped rotor, the magnetic field of the rotor was able to interact more fully with the magnetic field of the stator, thereby improving the starting torque and torque stability of the motor. In this way, we designed a stepped rotor BLSRM structure, as shown in Figure 2.

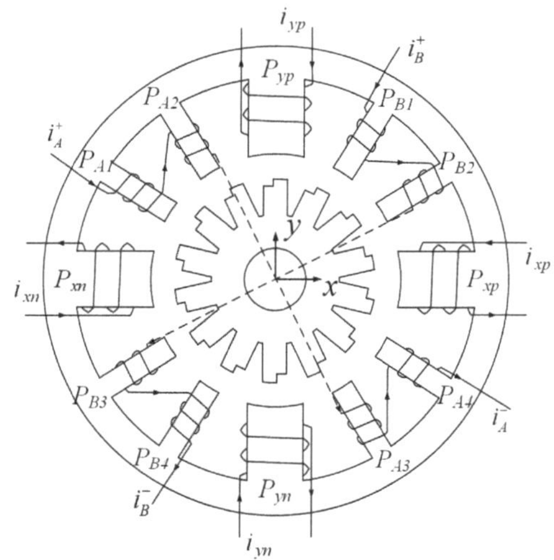

Figure 2. Structure of the stepped rotor BLSRM.

2.1. Structure and Working Principle of the BLSRM

The rotor pole arc of the stepped rotor BLSRM was slightly larger than the torque pole arc so that, when the motor was started, no matter where the rotor was located, there would be a torque winding phase that would provide positive torque to the rotor.

In addition, the rotor's stepped structure increased the motor inductance's rising interval. According to the electromagnetic torque calculation, i.e., Formula (1), it was found that if the effective length of the torque winding conduction became more extensive, thereby increasing the torque during commutation, then the torque fluctuation would decrease.

$$T = \frac{1}{2} i^2 \frac{dL}{d\theta} \tag{1}$$

where T is the electromagnetic torque, i is the torque winding current, L is the inductance, and θ is the rotor position.

There are two types of poles, namely, suspension poles and torque poles, on the stator of a step rotor BLSRM. The stepped rotor BLSRM has two types of poles, a suspension force pole and a torque pole. The torque winding, composed of four coils in phases A and B, provides torque to rotate the rotor. P_{xp}, P_{xn}, P_{yp}, and P_{yn} correspond to the positive and negative directions of the X-axis and the positive and negative directions of the Y-axis. We can satisfy the rotor's requirement to be suspended by turning on one or two windings.

2.2. Parameter Conditions

To compare the performance with the original 12/14 hybrid stator pole BLSRM and to verify the effectiveness of the stepped rotor BLSRM structure, we kept the main dimensions of the motor the same as those of the original motor, including the stator and rotor outer diameter, air gap length, and iron core length. In the design of the stepped rotor BLSRM, the rotor pole arc β_r and the torque pole arc β_{st} are critical parameters that directly affect the motor torque fluctuation and self-starting ability. Moreover, this paper has two unique and vital parameters: step height, h_{sr}, and step angle, β_{rs}, as shown in Figure 3.

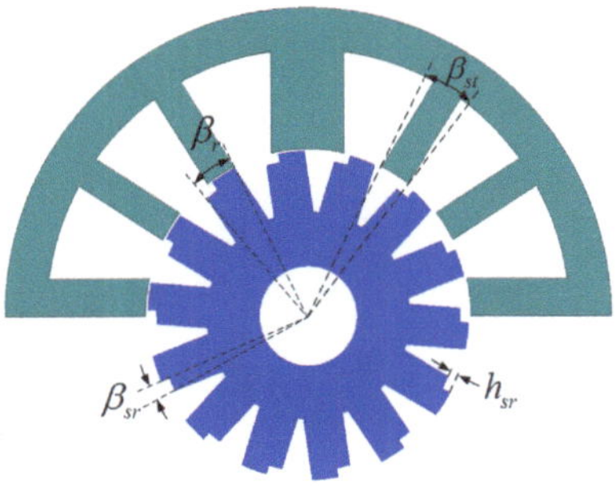

Figure 3. Important parameters of the stepped rotor BLSRM.

To ensure that the motor has a specific self-starting ability at any rotor position, it is required that the rising periods of the inductance curves of the two adjacent phases have an inevitable overlap; that is, the rotor pole arc and torque pole arc are required to satisfy:

$$\beta_r > \beta_{st} \tag{2}$$

To make the motor's inductance as small as possible when the rotor is not aligned, and to generate positive torque at any rotor position, the rotor pole arc, β_r, and the torque pole arc, β_{rs}, must meet the following necessary conditions:

$$\begin{cases} \beta_r + \beta_{st} \le \frac{2\pi}{N_r} = \frac{360^\circ}{14} = 25.7^\circ \\ \beta_r > \frac{2\pi}{qN_r} = \frac{360^\circ}{2\times 14} = 12.85^\circ \end{cases} \tag{3}$$

where q is the number of phases of the motor and N_r is the number of poles of the rotor pole. This can be expressed by the following triangle, ABC, in Figure 4.

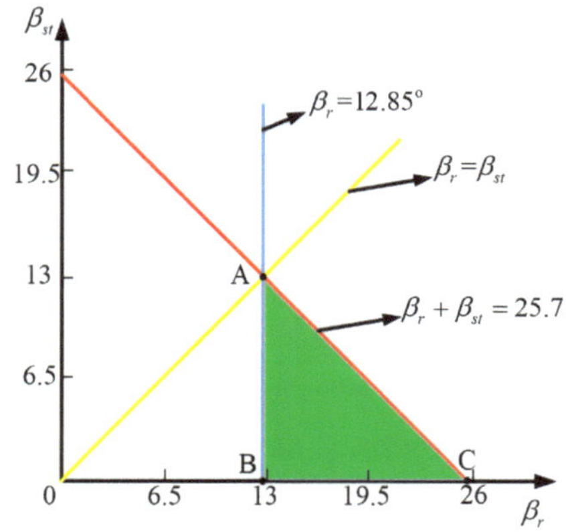

Figure 4. Feasible range of the rotor pole arc and torque pole arc.

The stepped structure has two parameters, namely, step angle, β_{rs}, and step height, h_{rs}. The design should be based on the principle of affecting other performance areas as little as possible to determine the range of variation, so β_{rs} and h_{rs} should meet the following conditions:

$$\begin{cases} 1^\circ < \beta_{rs} < 7^\circ \\ 0.1\ \text{mm} < h_{rs} < 0.5\ \text{mm} \end{cases} \tag{4}$$

To ensure self-starting capability, the difference between the torque pole and rotor pole arc should be as slight as possible. According to the constraint relationship between the torque pole and rotor pole arc shown in Figure 4, and based on the design principle that the average torque of the motor should be as large as possible, we chose 12°/13.5°, 11.5°/14°, and 11°/14.5°. During the optimization of the motor structure, three combinations of uniformly distributed parameters of torque poles and rotor pole arcs were compared, and the average torque of one cycle was calculated.

2.2.1. Design of Step Angle

The step angle is an essential parameter of the step rotor structure. It is a part of the rotor pole arc, and the stepped rotor structure can increase the rise interval of the motor inductance, allowing the motor to produce greater electromagnetic torque during start-up as well as low-speed operation, thereby improving the motor's output torque. According to the relationship between the torque pole fox and the rotor pole arc, the design of the step angle can be divided into the following situations:

$$\begin{cases} \beta_{rs} > \beta_r - \beta_{st} \\ \beta_{rs} = \beta_r - \beta_{st} \\ \beta_{rs} < \beta_r - \beta_{st} \end{cases} \tag{5}$$

According to these three situations and the feasible range of the rotor pole arc and torque pole arc shown in Figure 4, based on the determination of the main dimensions of the motor, the step height was fixed at 0.2 mm. The torque was not changed under different conditions.

Finite element analysis of the magnetic field was carried out for the combination of the three step angles of the pole and rotor pole arc. The average torque, starting torque, and torque fluctuation were calculated for the different step angles. Figures 5–7 show the different torque characteristic curves for different step angles.

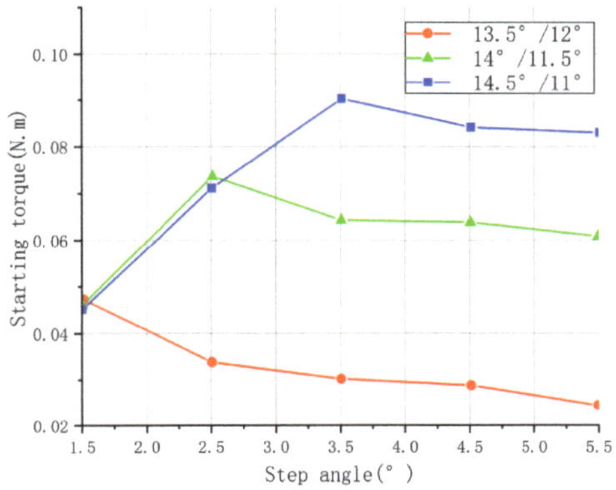

Figure 5. Starting torque of different step angles for three parameter combinations.

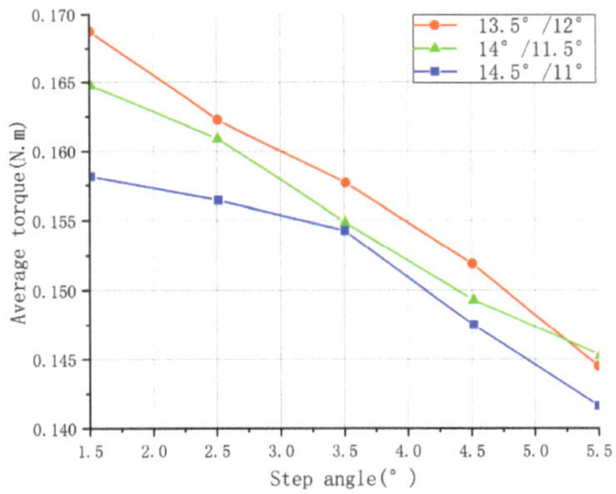

Figure 6. Average torque of different step angles for three parameter combinations.

As can be seen in Figure 5, when the step angle was less than $(\beta_r - \beta_{st})$, the starting torque was proportional to the step angle; when the step angle was greater than $(\beta_r - \beta_{st})$, the starting torque was inversely proportional to the step angle. Regardless of the torque and rotor pole arc, the motor's starting torque was the largest when the step angle was equal to $(\beta_r - \beta_{st})$.

The larger the step angle, the larger the angle of each step taken by the motor, and the fewer pulses required for each step. Therefore, the motor's rotation speed increased in the same amount of time, but the torque output for each step decreased. From the average torque curve shown in Figure 6, the average torque decreased with the increase in the step angle.

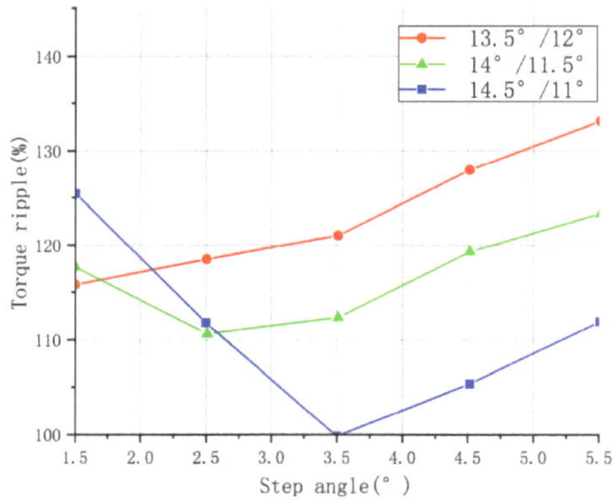

Figure 7. Torque fluctuation of different step angles for three parameter combinations.

The torque ripple curve shown in Figure 7 demonstrates that the torque ripple also adhered to the following laws with the increase in the step angle. When the number of magnetic poles was fixed, and if the stepping angle of the rotor was small, the magnetic field change of the motor was relatively smooth, so the torque fluctuation of the motor was small. When the stepping angle was large, the magnetic field change of the motor was relatively abrupt, possibly causing a larger torque fluctuation in the motor. When the step angle was less than ($\beta_r - \beta_{st}$), the torque ripple was inversely proportional to the step angle; when the angle was greater than ($\beta_r - \beta_{st}$), the torque ripple was proportional to the step angle.

To summarize, the starting torque and torque fluctuation were optimal when the step angle was equal to the difference between the rotor pole arc and the torque pole arc, so the size of the step angle was determined by the torque and the pole–pole arc of the rotor. That is,

$$\beta_{rs} = \beta_r - \beta_{st} \tag{6}$$

2.2.2. Design of Step Height

The direct effect of the stepped structure was to increase the air gap between the rotor pole and the torque pole. This reduced the intensity of eddy currents, decreased energy losses, and improved motor efficiency. If the step angle determined the range of the stepped structure, then the step height determined the strength of its effect. The proper step height was able to improve the positioning accuracy of the motor and reduce the probability of step loss occurring. At the same time, it also reduced the noise and vibration of the motor. When the torque winding current was constant, the motor torque was determined by the rate of change in the inductance. The step height directly affected the change in inductance, thus affecting the torque characteristics. Therefore, the torque characteristics were analyzed at different step heights.

Figure 8 shows the starting torque curves for the structures of the three parameter combinations at different step heights. As the height of the stepped rotor increased, the inductance of the motor increased, as did the rise time of the inductance. This led to a decrease in the response speed of the motor, which, in turn, led to a slower generation of electromagnetic torque and a decrease in the starting torque of the motor. It can be seen in the figure that with the increase in the step height, the starting torque of the three structures gradually decreased. The starting torque of the torque pole arc–rotor pole arc was 11°/14.5° as a whole, larger than that of the other two structures: 11.5°/14° took second place, and 12°/13.5° was the smallest.

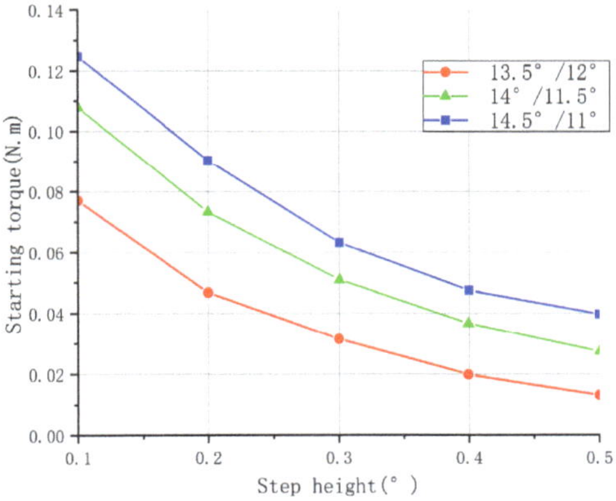

Figure 8. Starting torque of different step heights for three parameter combinations.

Figure 9 shows the average torque curves for the structures of the three parameter combinations at different step heights. When the step height of the salient rotor of the stepped motor increased, the starting torque of the motor decreased, but the motor's average torque was also affected. When the step height was relatively small, the motor's torque curve was smaller, resulting in a higher starting torque, but a lower average torque. However, when the step height increased, the torque curve increased, resulting in a lower starting torque, but a higher average torque. According to the graph, the optimal performance was achieved when the step height was 14/11.5, with a step size of 0.1.

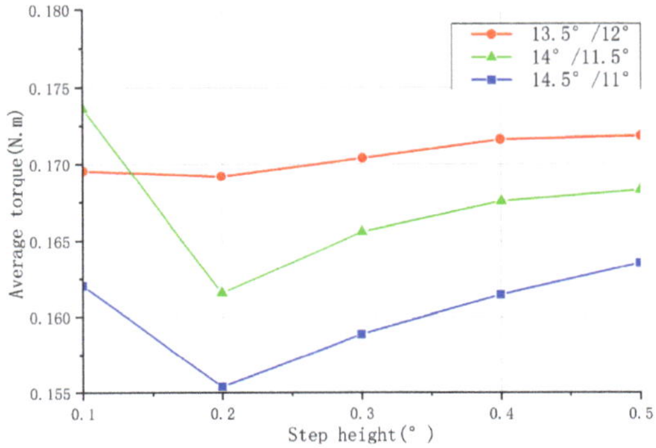

Figure 9. Average torque of different step heights for three parameter combinations.

Figure 10 shows the torque fluctuation curves for the three structures at different step heights. When the step height of the salient pole rotor increased, the required magnetic flux change for each step also increased accordingly. This led to a more uniform distribution of magnetic flux between the rotor and stator, reducing the rapid changes in magnetic flux and thus decreasing the torque ripple. This effect was more pronounced when the step height increased to an appropriate level. Figures 8–10 show that when the rotor pole arc–torque pole arc was 13.5°/12°, the average torque of the motor was higher than that of the other two combinations. Its starting torque was higher than that of the two other combinations, and the torque ripple was more prominent, completely deviating from the original design intention of increasing the starting torque and reducing the torque ripple; thus, this combination was unsuitable. The 14°/11.50 and 14.5/11° structures had two rotor pole arc–torque pole arc combinations, each with 5 step heights, for a total of 10 combinations. The rotor pole arc and torque pole arc were visible. When the step height was 14°, 11.5°, or 0.1, the starting torque was the most significant, the torque fluctuation was the slightest, and the average torque was the highest.

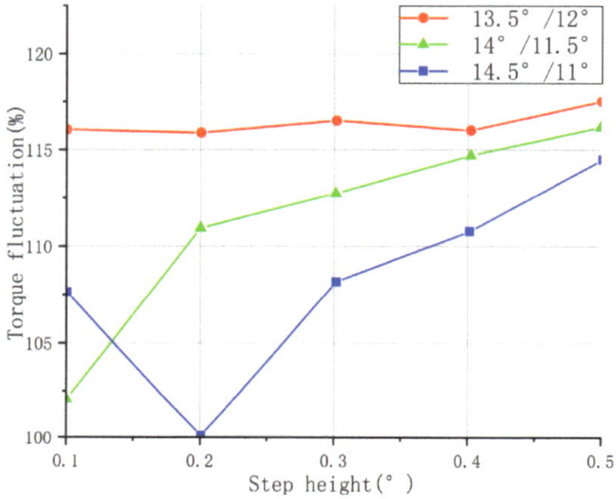

Figure 10. Torque fluctuation at different step heights for three parameter combinations.

In summary, the rotor pole arc, torque pole arc, step angle, and step height of the motor were determined as follows:

$$\begin{cases} \beta_r = 14^\circ \\ \beta_{st} = 11.5^\circ \\ \beta_{rs} = 2.5^\circ \\ h_{rs} = 0.1 \text{ mm} \end{cases} \tag{7}$$

According to the motor drive requirements of the space robot, based on the above analysis, we redesigned the motor structure of the stepped rotor BLSRM, thereby improving the performance of the space robot. Then, we optimized the PSO algorithm and used finite element electromagnetic field analysis software to determine the structural parameters that significantly affected the torque and the structural parameters of the step angle on the rotor. Moreover, we assessed the performance of the motor and further completed the preliminary design of the stepped rotor BLSRM structure.

3. Improved PSO

In PSO, the n-dimensional problem to be optimized corresponds to an n-dimensional solution space. The potential solution for the problem to be optimized is regarded as the mass-free and volume-free moving particles in that space, described by the position and velocity such that each particle has different characteristics, thus forming a particle population. During PSO, an evolutionary search is carried out through the cooperation and competition of each particle in the population.

PSO achieves the purpose of optimization by completing an evolutionary search through the cooperation and competition of each particle in the group. The key to this algorithm is to update the particle position and velocity, that is, to approach the optimal global solution quickly. Each particle updates itself by tracking the optimal solutions for individual and group extreme values. For the optimal particle of the group, the powerful personal matter and the extreme group value are relevant. Therefore, when the particle is updated, the learning ability of the self-cognition and social cognition parts is zero, and there is no autonomous evolution mechanism. Therefore, it is very likely that it will fall into local extreme points and be unable to escape, so the PSO algorithm can mature quickly and early.

We proposed eliminating the inferior and bolstering the superior value to solve this problem. The main idea is that during iterative updates, we sort the entire particle swarm according to the fitness value and eliminate the 2^n particles with lower fitness values in this generation, according to the dimension, n, of the problem. Taking the particle with the most significant group fitness value as the center, we divide the solution space into 2^n regions, and 2^n new particles with different directions and uniform distributions are generated in these new regions.

When there is a new particle whose fitness value is larger than that of the original optimal particle of the group, the group extremum is updated; that is, it is successful in helping the group's optimal particles to escape the optimal local solution. The key to PSO is updating the particle position and speed, that is, quickly growing close to the optimal global solution. This paper addresses the problems of the standard PSO and combines them with the optimization of the structural parameters of the motor to improve the standard PSO.

In this paper, the adaptive inertial weight method is adopted; that is, the inertial weight changes with the change in the particle's fitness value. The expression is as follows:

$$\omega = \begin{cases} \omega_{min} + \frac{(\omega_{max} - \omega_{min}) * (f - f_{min})}{(f_{avg} - f_{min})}, f \leq f_{avg} \\ \omega_{max}, f > f_{avg} \end{cases} \tag{8}$$

where f represents the current fitness value of the particle, f_{min} and f_{avg} represent the minimum fitness value and the average fitness value of all the current particles, and ω_{max} and ω_{min} represent the maximum and minimum values, respectively, usually taken as $\omega_{max} = 0.9$ and $\omega_{min} = 0.4$.

The learning factor c_1 decreases with the increase in the number of iterations, and c_2 increases with the increase in the number of iterations, meeting the requirements, and generally follows $c_1 + c_2 \leq 4$. The change formula of the learning factor is

$$\begin{cases} c_1 = c_{1,ini} + (c_{1,fin} - c_{1,ini})\frac{t}{t_{max}} \\ c_2 = c_{2,ini} + (c_{2,fin} - c_{2,ini})\frac{t}{t_{max}} \end{cases} \tag{9}$$

where $c_{1,ini}$ and $c_{2,ini}$, respectively, represent the initial values of c_1 and c_2, and $c_{1,fin}$ and $c_{2,fin}$, respectively, represent the final values of c_1 and c_2. In this paper, $c_{1,ini} = 2.5$, $c_{2,ini} = 0.5$, $c_{1,fin} = 0.5$, and $c_{2,fin} = 2.5$.

The formula for generating new particles is as follows:

$$\begin{cases} V_i(t) = \left(1 - t/t_{max}()_{d,lgbest}\right) \\ X_i(t) = X_{gbest} \end{cases} \tag{10}$$

where t and t_{max} are the current number of iterations and the maximum number of iterations, respectively, and $1 - t/t_{max}$ is a coefficient that linearly decreases as the number of iterations increases. It can make the optimal particle search step of the group more prominent and increase the probability of escaping from the local optimum in the early stage of optimization, as well as making the search step size of the optimal particle of the group smaller in the later stage of optimization, which is beneficial to improving the accuracy of the algorithm. V_{gbest} and X_{gbest}, respectively, are the speed and position of the optimal particle of the group, and $Direc_{t,l}$ is the direction of the first particle in the first cutting area, where d, l meet the following conditions:

$$\begin{cases} 2 \leq d \leq 2^n \\ 1 \leq l \leq ESC \end{cases} \tag{11}$$

where ESC is the escape capability coefficient, which represents the number of new particles generated in each direction, meaning that the ESC is an integer greater than or equal to 1. In dealing with a complex multipolar value problem, there is a large difference in fitness value between the optimal local value and the optimal global value, which produces a new particle in each direction; thus, the escape ability coefficient ESC, which represents the number of new particles generated in each direction, is set as an integer greater than or equal to 1. When it cannot escape from the local optima, the coefficient can be appropriately increased to further enhance the global search capability of the algorithm. We should ensure that Formula (10) is satisfied when adjusting the escape ability coefficient. Figure 11 shows the Improved PSO flow chart.

$$\begin{cases} ESC \geq 1 \\ 2^n \cdot ESC < sizepop \end{cases} \tag{12}$$

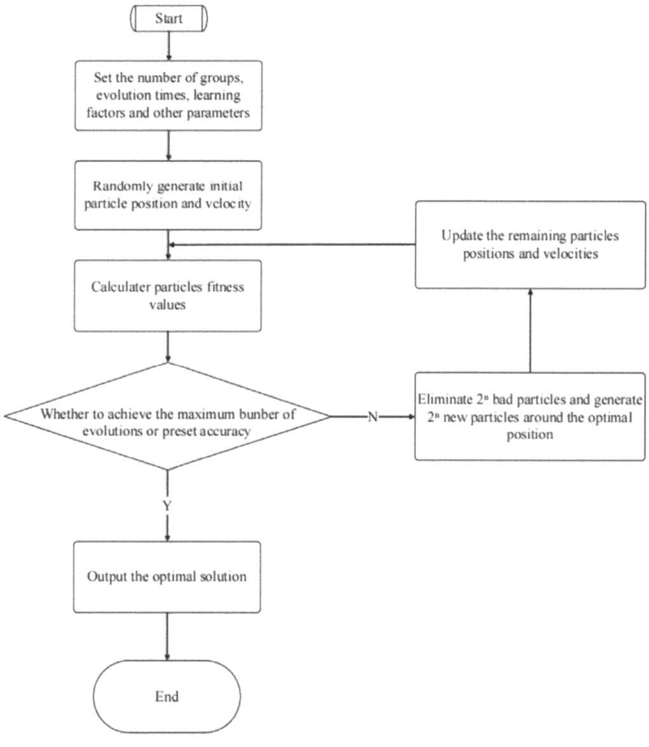

Figure 11. Improved PSO flow chart, showing the basic process of the improved PSO.

It can be seen that the elimination of the particles with lower fitness values and the procedure of generating new particles are merged into the update iterative process, so the algorithm steps do not become complicated, and the algorithm can help the particles to escape from the optimal local domain while eliminating particles with relatively poor fitness values. The particles also improve the global convergence speed of the algorithm.

This section described the process of improving the particle swarm optimization algorithm and enhancing its global optimization ability. Next, to solve the common problems of the current motor design, such as the need to rely on experience, the time-consuming and labor-intensive nature, the high degree of specialization, and difficulty of obtaining optimal structural parameters, the finite element analysis method and particle swarm algorithm were combined to establish a systematic motor structure optimization. The technique was used to optimize the stepped rotor bearingless switched reluctance motor design.

4. Optimized Design of Stepped Rotor BLSRM

4.1. The Selection of Optimization Variables

To reduce the torque fluctuation, improve the motor's starting ability, and ensure the motor's output torque, the torque pole arc β_{st} of the motor, the rotor pole arc β_r, and stepped structure geometry parameters were considered as optimization variables first. However, as β_{rs} was part of the rotor pole β_r, the step angle was interrelated and could not be used as an optimization variable simultaneously. To maximize the effect of the stepped structure, we chose the torque pole arc β_{st}, rotor pole arc β_r, and step height h_{rs} as the optimization variables. The step angle should satisfy the following condition:

$$\beta_{rs} = \beta_r - \beta_{st} \tag{13}$$

4.2. Constraint Setting

To ensure that the motor has a specific self-starting ability when the torque winding current is constant at 2A, the starting torque T_s met the following condition:

$$T_s \geq 0.05\ \text{N·m} \tag{14}$$

Increasing the motor's starting torque risked causing the output torque to decrease. Therefore, during the optimization process, the average torque T_{ave} of the motor during a running cycle met the following condition:

$$T_{ave} \geq 0.9T_{ave0} \tag{15}$$

where T_{ave0} is the average torque of the original 12/14 BLSRM in an operating cycle.

4.3. Determination of the Objective Function

To solve the problems of the low starting torque and significant torque fluctuation of the 12/14 BLSRM, the starting torque of the motor was restrained by Formula (12) to enhance the self-starting ability. The average torque was calculated using Formula (13) to guarantee the output torque of the motor. Therefore, selecting torque fluctuation as the only optimization target was a necessary step, ensuring the optimization effect and simplifying the optimization problem.

The objective function is defined as

$$F(x) = T_{ripple} \tag{16}$$

The calculation of the torque fluctuation was as follows:

$$T_{ripple} = \frac{T_{max} - T_{min}}{T_{ave}} \times 100\% \tag{17}$$

4.4. Method for Calculating Fitness Value Based on Finite Element

The accuracy of the optimization target calculation is paramount when using PSO to solve practical problems. However, BLSRM is a complex system with nonlinear, strong coupling. When deriving its mathematical model, it is easy to ignore edge effects, magnetic saturation, and magnetic leakage. Moreover, the rotor's particular stepped structure complicates this paper's mathematical model. Therefore, an optimization target calculation method based on finite element electromagnetic field analysis is more accurate than a calculation method based on mathematical models.

The fitness function was used to evaluate the fitness of each particle. The fitness function was designed based on the objective function and torque ripple, and its purpose was to map the value of the objective function to the fitness value for comparison and selection by the optimization algorithm. Generally, the smaller the value of the fitness function, the more suitable the particle is for the optimization problem. We used the FEMM 2D electromagnetic field analysis software to model and analyze each particle. First, we established the finite element model, shown in Figure 12, according to the position information of the particle. Then, we carried out an analysis to obtain the maximum, minimum, and average torques. Following this, we calculated the torque ripple, that is, the fitness value.

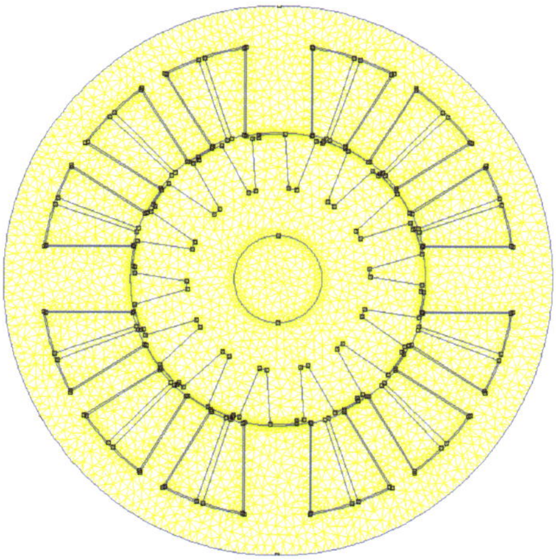

Figure 12. Finite element model of the stepped rotor BLSRM.

4.5. Algorithm Implementation

Figure 13 shows the block diagram of the optimized design system of the stepped rotor BLSRM.

We used Matlab to prepare the improved particle swarm algorithm program for the optimization design process. We connected it to the 2D finite element analysis software FEMM. We repeatedly called the FEMM software to complete the model building and electromagnetic field analysis of the finite element of each particle in the optimization process. Moreover, we obtained performance indicators such as the starting torque, average torque, and torque fluctuation to calculate the fitness value of each particle.

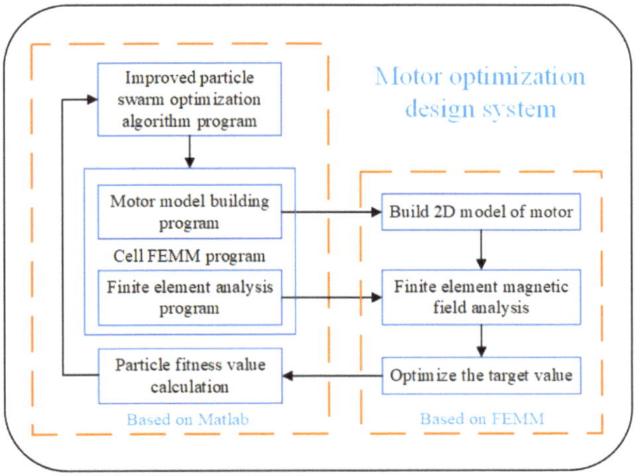

Figure 13. Block diagram of the stepped rotor BLSRM optimization design system.

The optimization process was as follows:

Optimization process

Begin

Step 1: Set the parameters such as the inertia weight, learning factor, and escape ability coefficient;

Step 2: Randomly generate an initial group;

Step 3: Call the FEMM finite element magnetic field analysis software to model and analyze each particle and return the optimization target value;

Step 4: Calculate the fitness value of the particle according to the returned optimization target value;

Step 5: Compare the initial individual extreme importance and the initial group extreme value after calculating the fitness values of all the particles in the group;

Step 6: Perform iteration, sort the entire particle swarm by fitness value, eliminate the 2^n particles with the lower fitness values in the generation according to the number of variables, generate new 2^n particles at the particle positions with the maximum fitness values, and give them different directions and uniformly distributed speeds;

Step 7: Update the positions and velocities of the remaining particles;

Step 8: Update the individual extremum values and the group extremum values;

Step 9: If the stop condition is met (the preset accuracy or the number of iterations is reached), stop the search, and the result will be the output; otherwise, return to step 6 to continue the search.

End

Figure 14 shows the change curve of the fitness value of the optimal particle in the stepped rotor BLSRM parameter optimization process as the number of iterations increased. It can be seen that the most optimal solution was found when the 28th iteration was reached.

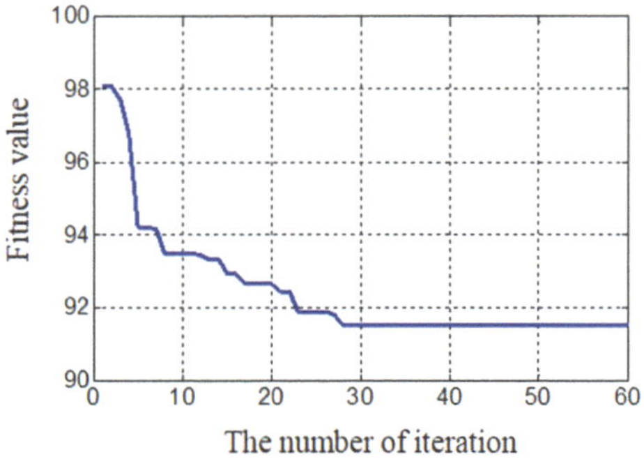

Figure 14. Convergence curve of the objective function.

5. Analysis and Verification of Results

To improve the motor drive performance of the space robots, we used finite element electromagnetic field analysis software to analyze various aspects of the stepped rotor BLSRM in order to verify the effectiveness of the improved PSO algorithm and the motor structure.

5.1. Optimization Algorithm Test

To verify the effectiveness of the improved PSO, single-peak and multi-peak test functions were used to test and compare the PSO before and after the improvement. Single-peak test functions are used to evaluate the global search ability and convergence speed of optimization algorithms, and they are a type of function with only one extreme point. Multi-peak test functions, on the other hand, contain multiple optimal solutions and are more suitable for comparing the global search abilities of two algorithms. Figure 15 compares the convergence process of the objective functions of the two algorithms. Table 1 shows the comparison of the optimization results.

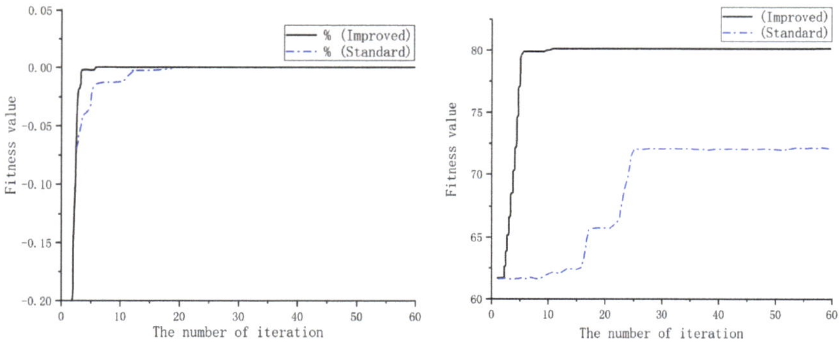

(**a**) Unimodal test function (**b**) Multimodal test function

Figure 15. Comparison of convergence processes.

Table 1. Comparison of PSO optimization results before and after improvement.

	Standard PSO	Improved PSO	The Optimal Value
Unimodal function	-3.43×10^{-5}	-8.53×10^{-26}	0
Multimodal function	72.64	80.70658	80.70658

It can be seen in Figure 15a and Table 1 that both the improved PSO and the standard PSO converged to a position close to 0 under the unimodal test function. The improved PSO achieved the design requirements after 5 iterations, while the standard PSO reached them after about 20 iterations. Therefore, the improved PSO improved the search efficiency and accuracy of the algorithm. Under the multimodal test function of Figure 15b, it can be seen that the objective process of the standard PSO converged to a local optimal solution position after about 25 iterations, rather than the optimal global solution. In contrast, the optimization result of the improved PSO accurately converged to the optimal global position after five iterations. For multi-extreme optimization problems, the standard PSO was able to easily stay at the local optimal solution position, unable to escape, and was affected by the parameter settings, initial populations, and other factors. However, the improved PSO ran the local optimal solution position and searched for the optimal global solution. Through the improved PSO algorithm, the optimized values of the rotor pole arc, torque pole arc, step angle, and step height of the motor could be calculated.

5.2. Characteristic Analysis of BLSRM

We use finite element electromagnetic field analysis software to analyze various characteristics of the stepped rotor BLSRM.

5.2.1. Inductance Characteristics

Figure 16a,b show the torque winding inductance characteristic curves of the original 12/14-pole BLSRM and the stepped rotor BLSRM, respectively, at different currents. It can be seen in the figure that when the wind was fixed, the inductance characteristics of the two BLSRM torque windings were similar on the whole, and they both changed significantly with the change in the rotor position. The overlap area of the rotor poles was related, and the overlap area varied with the rotor position angle. From a local point of view, the inductance rise period of the stepped rotor-type BLSRM occupied a more extensive range; it did not remain at the maximum inductance position, but continued to change because the rotor pole arc was slightly larger than the torque pole arc.

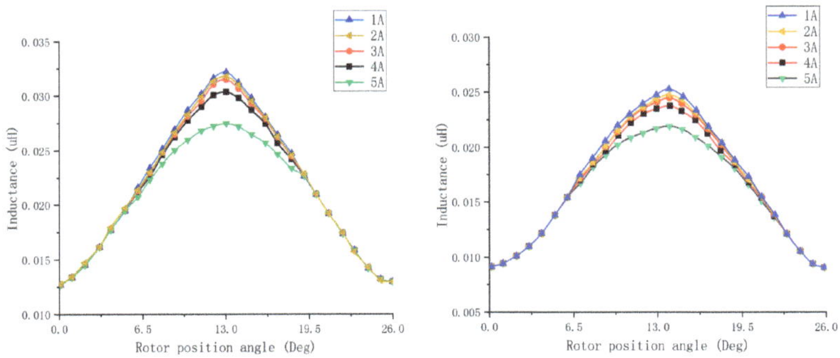

(**a**) Original 12/14 BLSRM (**b**) Stepped rotor BLSRM

Figure 16. Inductance characteristic curve of torque windings.

5.2.2. Torque Characteristics

Figure 17 shows the torque characteristics of the torque windings at different currents for the original 12/14 pole type BLSRM and the stepped rotor BLSRM. According to the previous analysis of the inductance characteristics, the inductance varied with the rotor position angle. When the current is constant, the motor torque should be proportional to the rate of change in the inductance. It can be seen in the figure that the motor torque changed significantly with the change in the rotor position angle. The positive torque interval corresponded to the inductance rise phase, and the negative torque interval corresponded to the inductance fall phase. The stepped rotor BLSRM produced slightly less torque than the original BLSRM because the new motor had a more minor torque pole arc than the original, resulting in a slightly smaller inductance at the aligned position, and, thus, a slightly smaller slope in the inductance rise phase. However, since the inductance rise interval of the stepped rotor BLSRM was longer than that of the original BLSRM, the range of positive torque was more comprehensive, and the average torque generated by the new motor in one operating cycle did not decrease significantly.

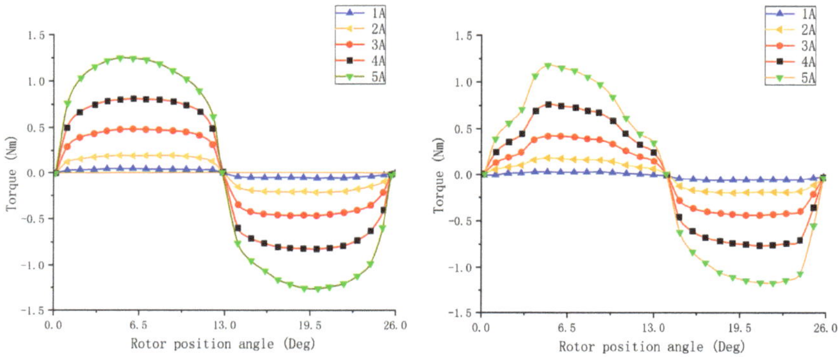

(**a**) Original 12/14 BLSRM (**b**) Stepped rotor BLSRM

Figure 17. Torque characteristic curve of torque windings.

5.2.3. Starting Torque and Torque Ripple

Figure 18 compares the original 12/14 type BLSRM with the stepped rotor BLSRM before and after optimization. Figure 19 shows a torque ripple comparison of the two. It can be seen that the starting torque was significantly improved compared to the 12/14 BLSRM due to the initial design of the stepped rotor BLSRM. Moreover, the torque ripple of the motor was significantly reduced, and the advantage of the starting torque remained. Compared with the original 12/14 type BLSRM, the optimized stepped rotor BLSRM had the ability to self-start, significantly reducing the torque fluctuation. Figure 18 shows that when the torque winding current was 1A, the starting torque of the new motor was slightly larger than that of the original motor. With the increase in the current, the starting torque advantage of the new motor became more pronounced, proving that the self-starting ability of the new motor was enhanced. Figure 19 shows that, in the case of different torque winding currents, the torque fluctuation of the improved motor was always significantly smaller than that of the original motor.

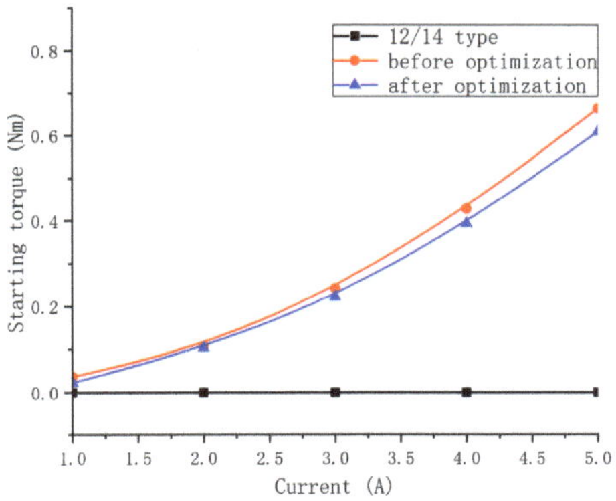

Figure 18. Comparison of starting torques of three motor structures.

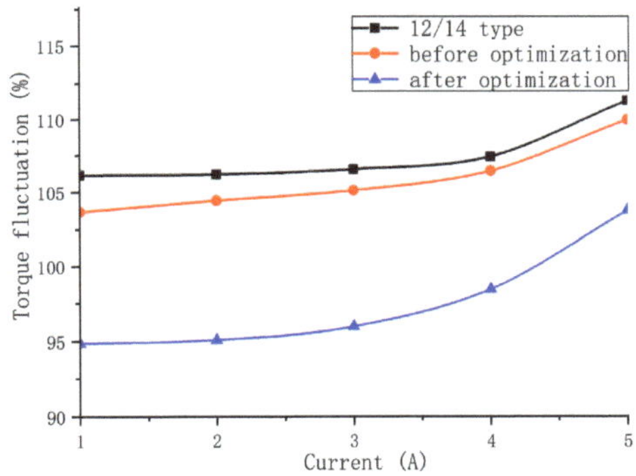

Figure 19. Comparison of torque fluctuations of three motor structures.

5.2.4. Structural Parameter Comparison

Table 2 shows the optimization results, initial design, and structural parameters of the original 12/14 BLSRM and the stepped rotor BLSRM. It can be seen that the optimized motor rotor pole arc, torque pole arc, step angle, and step height are all different from the original values.

5.3. Dynamic Simulation

To further verify the effectiveness of the stepped rotor BLSRM structure, in this paper, we used Matlab to build a motor system simulation model and to perform dynamic simulations.

This section focuses on the simulation of the motor from start-up to the point at which a predetermined speed was reached, assuming that the rotor had been steadily suspended in the central position, that the initial torque load was 0.1 N.m, and that there was no load other than the gravity of the radial rotor and the shaft itself.

Table 2. Structural parameters of stepped rotor BLSRM before and after optimization.

Parameter	Type 12/14	Stepped Type Initial Value	Stepped Type Optimization Value
Phase	2	2	2
Suspension force pole	4	4	4
Torque pole	8	8	8
Number of rotor poles	14	14	14
Suspension force pole arc (deg)	25.7	25.7	25.7
Torque pole arc (deg)	12.85	11.5	10.3
Rotor pole arc (deg)	12.85	14	13.61
Air gap length (mm)	0.3	0.3	0.3
Stator outer diameter (mm)	112	112	112
Stator yoke height (mm)	7.7	7.7	7.7
Rotor outer diameter (mm)	59.6	59.6	59.6
Rotor yoke height (mm)	9.7	9.7	9.7
Core length (mm)	40	40	40
Shaft diameter (mm)	18	18	18
Stepped angle (deg)	-	0.1	0.22
Stepped height (mm)	-	2.5	3.31

Figures 20 and 21 show the torque and speed variation curves. It can be seen that the alternating conduction of the two-phase torque windings of the A phase and the B phase were able to generate continuous torque to accelerate the rotor. During the starting stage, the torque ripple of the stepped rotor structure motor was small, and the speed increased steadily. After reaching the predetermined speed of 3000 rpm, the torque remained almost constant. This demonstrates that, under appropriate control strategies, the stepped rotor BLSRM structure improves its self-starting capability, has a larger starting torque value, and reduces torque ripple, showing a good dynamic performance.

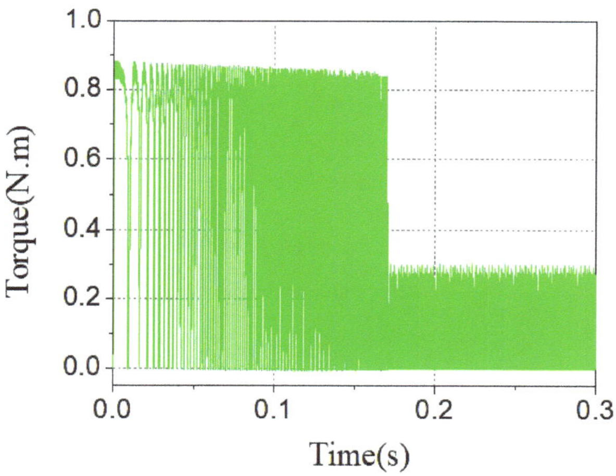

Figure 20. Torque variation during the speed regulation process.

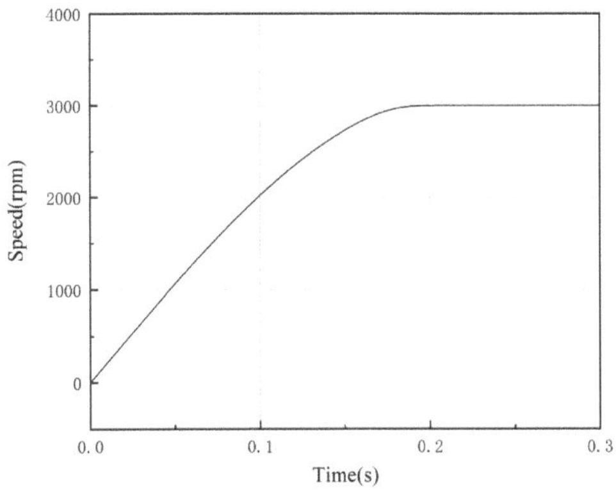

Figure 21. Speed variation during the speed regulation process.

6. Conclusions

Firstly, aiming at aerospace tasks and considering the application of space robots, we designed a stepped rotor BLSRM structure to optimize the motor drive performance of a space robot based on the 12/14 hybrid stator pole type BLSRM. We also analyzed the influence of its motor structure and parameter conditions on the motor's performance. Using an excitation current of 2 A, a torque pole arc of 11.5°, a rotor pole arc of 14°, a step angle of 0.1°, and a step height of 2.5 mm, the improved switched reluctance motor without bearings showed an increase of 0.11 N·m in starting torque and a reduction of 1.9% in torque ripple compared to the original motor design. Secondly, aiming to solve the prematurity problem of the standard PSO, we proposed a method by which the fittest particles would help the best to escape in order to improve the algorithm. According to the stepped rotor BLSRM structure characteristics, we combined the finite element method to calculate the fitness values and formed a motor structure optimization method based on the improved PSO. We used it to complete the optimization design of the stepped rotor BLSRM structure. After optimization, it was found that the torque pole arc was 10.3°, the rotor pole arc was 13.61°, the step angle was 0.22°, and the step height was 3.31 mm. Compared with the original initial motor, the torque ripple was further reduced by 9.2%. The starting torque remained unchanged. Finally, the effectiveness and feasibility of the optimization algorithm and motor structure were verified through algorithm testing and finite element analysis. The results show the following:

(1) Compared with the standard PSO, the proposed PSO had a faster convergence speed, higher accuracy, and more robust global optimization capability;
(2) Compared with the original 12/14 type BLSRM, the torque ripple of the stepped rotor type BLSRM w reduced, and the starting torque is significantly increased. Furthermore, the optimized motor reduces the torque ripple under the condition of ensuring self-starting ability. This proves that the stepped rotor BLSRM structure is adequate and feasible.

Author Contributions: Conceptualization, H.G. and J.Y.; methodology, Z.L. and X.W.; software, Z.L. and X.W.; validation, D.L., T.Z. and J.W.; formal analysis, Z.L. and X.W.; investigation, D.L. and T.Z.; resources, J.Y.; data curation, Z.L., X.W. and D.L.; writing—original draft preparation, Z.L. and X.W.; writing—review and editing, Z.L. and H.G.; visualization, Z.L. and X.W.; supervision, H.G. and J.W.; project administration, J.Y.; funding acquisition, H.G. and J.W. All authors have read and agreed to the published version of the manuscript.

Funding: This work was supported by the Liaoning Province Higher Education Innovative Talents Program #1 under Grant No. LR2019058, and the Liaoning Province Joint Open Fund for Key Scientific and Technological Innovation Bases #2 under Grant No. 2021-KF-12-05.

Data Availability Statement: The data used to support the findings of this study are available from the corresponding author upon request.

Acknowledgments: The authors would like to acknowledge the support from the following projects: Liaoning Province Higher Education Innovative Talents Program Support Project (Grant No. LR2019058), Liaoning Province Joint Open Fund for Key Scientific and Technological Innovation Bases (Grant No. 2021-KF-12-05), China Postdoctoral Science Foundation (Grant No. 2022M712756), and NSFC-Zhejiang Joint Fund for Industrialization and Informatization (Grant No. U1809211).

Conflicts of Interest: The authors declare no conflict of interest.

Abbreviations

BLSRM	Bearingless Switched Reluctance Motor
SRM	Switched Reluctance Motors
PSO	Particle Swarm Optimization
β_r	Rotor Pole Arc
β_{st}	Torque Pole Arc
h_{sr}	Step Height
β_{rs}	Step Angle

References

1. Li, W.J.; Cheng, D.Y.; Liu, X.G.; Wang, Y.B.; Shi, W.H.; Tang, Z.X.; Gao, F.; Zeng, F.M.; Chai, H.Y.; Luo, W.B.; et al. On-orbit service (OOS) of spacecraft: A review of engineering developments. *Prog. Aerosp. Sci.* **2019**, *108*, 32–120. [CrossRef]
2. Li, Y.; Sun, X.; Liu, X.; Wu, J.; Liu, Q. Deployment of On-Orbit Service Vehicles Using a Fuzzy Adaptive Particle Swarm Optimization Algorithm. *Model. Simul. Eng.* **2021**, *2021*, 6644339. [CrossRef]
3. Han, D.; Dong, G.Q.; Huang, P.E.; Ma, Z.Q. Capture and detumbling control for active debris removal by a dual-arm space robot. *Chin. J. Aeronaut.* **2021**, *35*, 342–353. [CrossRef]
4. Hatty, I. Viability of On-Orbit Servicing Spacecraft to Prolong the Operational Life of Satellites. *J. Space Saf. Eng.* **2022**, *9*, 263–268. [CrossRef]
5. Truong, V.; Greco, T.; Kassam, I.; D'Silva, O.; Tan, E. A cost effective methodology for building flight spares for robotic life extension on the international space station. *Acta Astronaut.* **2019**, *162*, 405–408. [CrossRef]
6. Ding, X.L.; Wang, Y.C.; Wang, Y.B.; Xu, K. A review of structures, verification, and calibration technologies of space robotic systems for on-orbit servicing. *Sci. China Technol. Sci.* **2021**, *64*, 462–480. [CrossRef]
7. Shi, L.; Xiao, X.; Shan, M.; Wang, X. Force control of a space robot in on-orbit servicing operations. *Acta Astronaut.* **2022**, *193*, 469–482. [CrossRef]
8. Yang, J.; Peng, H.; Zhang, J.; Wu, Z. Dynamic modeling and beating phenomenon analysis of space robots with continuum manipulators. *Chin. J. Aeronaut.* **2022**, *35*, 226–241. [CrossRef]
9. Ahn, J.W.; Lukman, G.F. Switched reluctance motor: Research trends and overview. *CES Trans. Electr. Mach. Syst.* **2018**, *2*, 339–347. [CrossRef]
10. Yang, Y.F.; Sun, L.; Zhu, X.B.; Hu, D. Comparative performance analysis of bearingless switched reluctance motor. *AIP Adv.* **2021**, *11*, 025307. [CrossRef]
11. Kumar, V.M.; Vinoth Kumar, K.; Saravanakumar, R. Switched Reluctance Motor Converter Topologies: A Review. *Innov. Electr. Electron. Eng.* **2020**, *626*, 55–63. [CrossRef]
12. Wang, Z.; Cao, X.; Deng, Z.; Li, K. Modeling and Characteristic Investigation of Axial Reluctance Force for Bearingless Switched Reluctance Motor. *IEEE Trans. Ind. Appl.* **2021**, *57*, 5215–5226. [CrossRef]
13. Degano, M.; Murataliyev, M.; Di Nardo, M. Reluctance motors (including switched reluctance motors). In *Encyclopedia of Electrical and Electronic Power Engineering*; García, J., Ed.; Elsevier: Amsterdam, The Netherlands, 2023; pp. 319–328. ISBN 978-0-12-823211-8.
14. Qi, Y.; Xu, Z.; Zhang, Y. Characteristics Analysis of a Novel Hybrid Rotor Type 12/14 Bearingless Switched Reluctance Motor. In Proceedings of the 2021 24th International Conference on Electrical Machines and Systems (ICEMS), Gyeongju, Republic of Korea, 31 October–3 November 2021; pp. 339–343.
15. Li, F.; Wang, H.; Zhou, G.X.; Pang, M.; Wei, X.; Zhang, M. Design and Experimental Investigation of Bearingless Switched Reluctance Motor with Permanent Magnet into Stator Yoke. In Proceedings of the 2018 21st International Conference on Electrical Machines and Systems (ICEMS), Jeju Island, Republic of Korea, 7–10 October 2018; pp. 1968–1973.
16. Xu, Z.; Lee, D.H.; Ahn, J.W. Stepped rotor type 12/14 bearingless SRM for self-starting and torque ripple reduction. In Proceedings of the 2015 IEEE 13th International Conference on Industrial Informatics (INDIN), Cambridge, UK, 22–24 July 2015; pp. 414–419.

17. Xu, Z.; Zhang, F.; Lee, D.H.; Ahn, J.W. Design and analysis of novel 12/14 hybrid pole type bearingless switched reluctance motor with short flux path. *J. Electr. Eng. Technol.* **2012**, *7*, 705–713. [CrossRef]
18. Sobhan, P.V.S.; Kumar, G.V.N.; Rao, P.V.R. Rotor autocentering and speed control of hybrid bearingless SRM using single-neuron adaptive PID controller. *Int. J. Eng.* **2018**, *7*, 5. [CrossRef]
19. Hao, W.; Wang, Y. A brief analysis of a bearingless linear switched reluctance motor with mover structural decoupling design. *IEEJ Trans. Electr. Electron. Eng.* **2019**, *14*, 652–653. [CrossRef]
20. Tang, S.; Wang, H.; Xue, B.; Liang, J. Levitation control of novel bearingless switched reluctance motor with biased permanent magnet. In Proceedings of the 2015 18th International Conference on Electrical Machines and Systems (ICEMS), Pattaya, Thailand, 25–28 October 2015; pp. 1360–1365. [CrossRef]
21. Jain, N.K.; Nangia, U.; Jain, J. A review of particle swarm optimization. *J. Inst. Eng. (India) Ser. B* **2018**, *99*, 407–411. [CrossRef]
22. Chopard, B.; Tomassini, M. *An Introduction to Metaheuristics for Optimization*, 1st ed.; Springer: Berlin/Heidelberg, Germany, 2018; Chapter 6; pp. 97–102. ISBN 978-3-319-93073-2.
23. Houssein, E.H.; Gad, A.G.; Hussain, K.; Suganthan, P.N. Major advances in particle swarm optimization: Theory, analysis, and application. *Swarm Evol. Comput.* **2021**, *63*, 100868. [CrossRef]
24. de Almeida, B.S.G.; Leite, V.C. A powerful technique for solving engineering problems. In *Particle Swarm Optimization*; Del Ser, J., Villar, E., Osaba, E., Eds.; IntechOpen: Rijeka, Croatia, 2019; pp. 1–21. ISBN 978-1-78984-537-2.
25. Hu, Z.; Yang, G. Simulation control model of synchronous motor based on PSO algorithm optimization in power system. *Energy Rep.* **2022**, *8*, 1044–1054. [CrossRef]
26. Zhu, S.P.; Keshtegar, B.; Seghier, M.E.A.B.; Zio, E.; Taylan, O. Hybrid and enhanced PSO: Novel first order reliability method-based hybrid intelligent approaches. *Comput. Methods Appl. Mech. Eng.* **2022**, *393*, 114730. [CrossRef]
27. Sahu, A.; Mohanty, K.B.; Mishra, R.N. Design of MPC-PSO based torque regulator for DTC-SVM induction motor drive. In Proceedings of the 2021 1st International Conference on Power Electronics and Energy (ICPEE), Bhubaneswar, India, 2–3 January 2021; pp. 1–6. [CrossRef]
28. Scalcon, F.P.; Vieira, R.P.; Gründling, H.A. PSO-Based Fast Mechanical Parameters Estimation of Switched Reluctance Motor Drives. *J. Control Autom. Electr. Syst.* **2022**, *33*, 1286–1293. [CrossRef]

sensors

MDPI

Article

VILO SLAM: Tightly Coupled Binocular Vision–Inertia SLAM Combined with LiDAR

Gang Peng [1,2], Yicheng Zhou [1,2,*], Lu Hu [1,2], Li Xiao [1,2], Zhigang Sun [1,2], Zhangang Wu [3] and Xukang Zhu [3]

1 School of Artificial Intelligence and Automation, Huazhong University of Science and Technology, Wuhan 430074, China
2 Key Laboratory of Image Processing and Intelligent Control, Ministry of Education, Wuhan 430074, China
3 Shantui Construction Machinery Co., Ltd., Jining 272073, China
* Correspondence: m202173171@hust.edu.cn

Abstract: For the existing visual–inertial SLAM algorithm, when the robot is moving at a constant speed or purely rotating and encounters scenes with insufficient visual features, problems of low accuracy and poor robustness arise. Aiming to solve the problems of low accuracy and robustness of the visual inertial SLAM algorithm, a tightly coupled vision-IMU-2D lidar odometry (VILO) algorithm is proposed. Firstly, low-cost 2D lidar observations and visual–inertial observations are fused in a tightly coupled manner. Secondly, the low-cost 2D lidar odometry model is used to derive the Jacobian matrix of the lidar residual with respect to the state variable to be estimated, and the residual constraint equation of the vision-IMU-2D lidar is constructed. Thirdly, the nonlinear solution method is used to obtain the optimal robot pose, which solves the problem of how to fuse 2D lidar observations with visual–inertial information in a tightly coupled manner. The results show that the algorithm still has reliable pose-estimation accuracy and robustness in many special environments, and the position error and yaw angle error are greatly reduced. Our research improves the accuracy and robustness of the multi-sensor fusion SLAM algorithm.

Keywords: multi-sensor fusion; pose estimation; lidar; visual inertial system

Citation: Peng, G.; Zhou, Y.; Hu, L.; Xiao, L.; Sun, Z.; Wu, Z.; Zhu, X. VILO SLAM: Tightly Coupled Binocular Vision–Inertia SLAM Combined with LiDAR. *Sensors* **2023**, *23*, 4588. https://doi.org/10.3390/s23104588

Academic Editor: Simon X. Yang

Received: 19 March 2023
Revised: 14 April 2023
Accepted: 27 April 2023
Published: 9 May 2023

1. Introduction

There are many excellent visual simultaneous localization and mapping (SLAM) systems, such as LSD-SLAM [1], DSO [2], and SVO [3]. These algorithms are based on pure vision. When there are no visual features, the accuracy and robustness of the pose estimation decreases rapidly, and the algorithm may fail. Therefore, in the follow-up development of visual SLAM, in order to overcome the shortcomings of pure vision, the strategy of multi-sensor fusion is adopted.

Because the inertial measurement unit (IMU) sensor-based inertial navigation algorithm has better short-term tracking performance, it can provide short-term accurate and reliable poses in the case of loss of visual features (light changes, missing texture features, and fast motion). In addition, the pure visual SLAM algorithm can provide speed constraints for the IMU-based inertial navigation algorithm, which can avoid the rapid spread of IMU measurement speed errors to a certain extent. Therefore, researchers have adopted the idea of loosely coupled fusion and used filters to fuse vision and IMU information to construct visual inertial SLAM algorithms. For example, the visual inertial SLAM algorithm in the literature [4–7] regards the IMU-based inertial navigation algorithm as the state prediction equation of the extended Kalman filter (EKF), and it uses the result of pure visual pose estimation as the measurement update equation of the EKF to realize the loosely coupled fusion of vision and IMU. Loosely coupled fusion means that the IMU and the camera estimate motion separately and then fuse their pose estimation results. In the loosely coupled fusion idea, because the visual features are invisible in the state optimizer, they cannot be adjusted by the information of the IMU. When encountering the positions of visual feature points with poor accuracy, the accuracy of the previous pure visual estimation decreases, and finally, the accuracy of the estimation of the whole state optimizer decreases. Tightly coupled fusion optimizes the camera pose and the position of the visual feature point together as the optimization variable of the entire state optimizer, which can optimize the position of the visual feature while optimizing the pose. This fusion strategy can effectively improve the disadvantages of loose coupling.

Although the addition of the IMU improves the accuracy and robustness of pure visual SLAM algorithms, the performance of visual–inertial SLAM systems can still be adversely affected by issues such as low lighting conditions, varying depth of field, and occlusion in complex environments, regardless of tightly coupled fusion or loosely coupled fusion. Moreover, due to the limitations of IMU precision, visual–inertial systems often struggle to maintain stability during long-term operation in weakly textured and unstructured scenes. Therefore, it is necessary to add high-precision ranging sensors, such as lidar. Lidar has the characteristics of strong anti-interference and high precision. When the visual features are lost, the lidar odometer can provide speed constraints for the IMU odometer and suppress the rapid spread of IMU measurement errors. At the same time, the motion distortion of lidar data can also be effectively corrected by the IMU, thereby improving the accuracy and robustness of the SLAM system.

In this paper, we propose a tightly coupled visual–inertial–lidar odometry framework, called VILO SLAM, by incorporating a low-cost 2D lidar into a visual–inertial SLAM system, to improve the positioning accuracy and robustness in complex environments. We combine the advantages of vision, IMU, and low-cost 2D lidar to make up for the insufficient information of each and build a tightly coupled state estimation problem based on nonlinear optimization. The main contributions of our work can be summarized as follows:

- Tightly coupling low-cost 2D lidar observations with stereo vision and inertial observations improves the accuracy and robustness of pose estimation in traditional visual–inertial SLAM algorithms in scenarios where visual features are lost due to darkness, strong light, or lack of texture.
- A lidar residual factor is constructed using the 2D lidar odometry model, and the Jacobian matrix of lidar residuals with respect to the state variables to be estimated is derived.
- The residual constraint equation of vision-IMU-LiDAR is constructed, and optimal robot pose estimation is obtained using nonlinear optimization, which solves the problem of fusing 2D lidar observations with binocular visual–inertial information in a tightly coupled manner.

2. Related Work

The most commonly used strategy in visual–inertial odometry is the tightly coupled approach, where vision and inertial measurements are combined into the same state vector. This method leverages visual measurements (such as feature points) and inertial measurements (such as accelerations and angular velocities) to construct error terms that include residuals from both sources. By optimizing the visual state variables and inertial measurement state variables simultaneously, the system achieves tightly coupled estimation. For example, Leutenegger et al. used a tightly coupled strategy for IMU measurement and integrated it into the key frame-based beam adjustment optimization visual SLAM. They then designed a tightly coupled SLAM algorithm called OKVIS based on a monocular camera and IMU [8]. In the SLAM process of this algorithm, there is a problem of repeated calculation of IMU points, which reduces the real-time performance of the entire algorithm. To solve this problem, the Christian Forster team designed the IMU pre-integration algorithm [9] and tightly coupled it with visual feature points to realize the visual inertial SLAM algorithm. Campos et al. fused IMU pre-integration and visual measurement to launch the ORB-SLAM3 system [10]. Xiao et al. designed tightly coupled real-time visual–inertial odometry based on the sliding-window method [11]. Yuan et al. proposed a multi-sensor fusion-state estimator based on a feature point optical-flow-tracking monocular vision, IMU, and wheel odometer measurement [12]. Hashim and Eltoukhy utilized available measurements obtained from group velocity vectors, feature measurements, and an inertial measurement unit, and proposed a computationally cheap geometric nonlinear SLAM filter algorithm that could account for the unknown bias inevitably present in velocity measurements [13]. Qin's and Li's teams at the Hong Kong University of Science and Technology adopted a sliding-window mechanism to construct the constraint equations of the IMU pre-integration error and the visual reprojection error. Additionally, they used a non-linear optimization method to propose the VINS-Fusion [14] SLAM algorithm, compatible with both monocular and binocular vision inertia. The algorithm also supports the expansion of fusion global sensor GPS to further improve the accuracy and robustness of the algorithm.

When visual features are lost, the visual–inertial SLAM algorithm degenerates into a classic inertial navigation algorithm. If the visual features cannot be recovered for a long time and the robot is in a state of uniform motion, resulting in no excitation of the IMU, the pose deviation estimated by the inertial navigation algorithm increases rapidly over time. Therefore, some researchers merge lidar, which has a strong anti-interference ability, with vision or IMU to further improve the accuracy and robustness of the visual SLAM algorithm. For example, in [15], Tixiao Shan constructed the residual constraint expressions of LiDAR and IMU pre-integration, computed robot poses through factor graph optimization, and proposed the LIO-SAM algorithm. In [16], a laser–inertial–vision tightly coupled SLAM framework is constructed, which improves the accuracy of robot pose estimation. In [17], the EKF is used to achieve loosely coupled fusion between the pure binocular vision-based pose-estimation results and the 2D lidar-based pose-estimation results, and a loosely coupled SLAM algorithm based on binocular vision-2D lidar is designed. However, the loose coupling of 2D lidar information and visual information has the problem of accuracy in specific scenes. As with the previously discussed visual-IMU loosely coupled fusion, if the position accuracy of the visual feature points is low, the accuracy of the entire fusion algorithm decreases.

Thus far, for unknown and complex indoor and outdoor environments, there are few studies on SLAM technology based on vision, inertia, and 2D lidar for tightly coupled optimization of multi-sensor fusion, and there is no mature solution similar to VINS-Fusion and ORB-SLAM.

3. Multi-Sensor Pose Estimation Based on Tightly Coupled Optimization

An overview of the proposed system is shown in Figure 1. The system receives sensor data from a 2D LiDAR, IMU, and a stereo camera. We seek to estimate the robot's state and trajectory using these observations. The state estimation problem can be formulated as a maximum a posteriori (MAP) problem. Under the assumption of a Gaussian noise model, solving this MAP inference is equivalent to solving a nonlinear least-squares problem. In the system, we construct residual constraints of the 2D LiDAR, IMU, and camera, and perform optimization using Google's Ceres nonlinear optimization library. Finally, the system outputs the pose of the robot in the world coordinate system at 10 Hz.

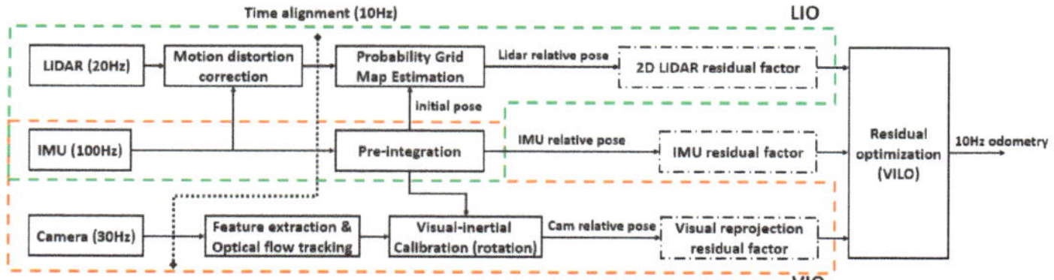

Figure 1. The system structure of VILO.

Due to the large amount of camera data, the camera-IMU frame should be reduced to 10 Hz considering the real-time performance of the robot pose solution. In addition, different sensor data acquisition frequencies are not consistent. If the time synchronization between sensors is not correct, the data provided by them may not be collected at the same time, which may lead to certain noise between data and affect the accuracy and robustness of the system. Therefore, the camera, IMU, and lidar need to be aligned in time, as shown in Figure 2.

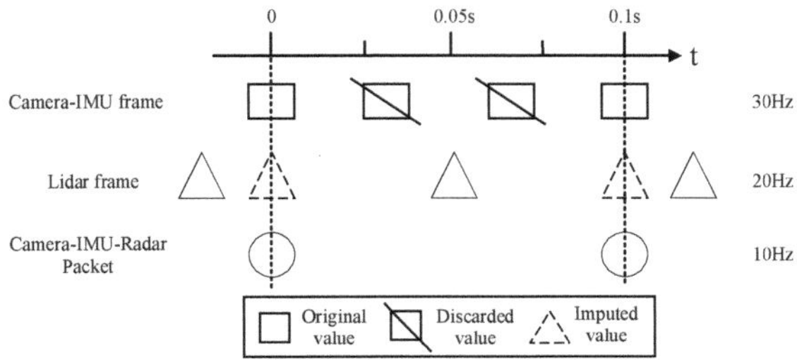

Figure 2. Camera-lidar time alignment diagram.

First, the camera frame is used as the alignment mark, and the lidar data frame is interpolated at the 10 Hz time position of the camera-IMU frame. The interpolation of lidar data frames needs to be completed in combination with IMU. The specific operation can be divided into two steps. The first step is to record the time stamp of the nearest lidar frame before the camera-IMU frame timestamp and to calculate the time difference between the two. The second step is to integrate the original IMU data according to the time difference, calculate the pose change and transform the lidar data frame to the camera-IMU frame time. Then the data of the camera-IMU-lidar are collected into a data structure for subsequent processing.

The multi-sensor tightly coupled pose estimation method in this section is designed based on the binocular-IMU local pose estimator in the VINS-Fusion algorithm. Its innovation lies in the fusion of 2D lidar information with binocular vision-IMU pose estimation through tight coupling to form the binocular vision-IMU-2D lidar odometry (VILO) front-end pose estimator. It also lies in the addition of lidar observations to further restrict the pose to improve the accuracy and robustness of the SLAM front-end pose estimation in indoor and outdoor environments. The state variable χ to be solved in VILO is shown in the following formula:

$$\begin{aligned} \chi_w &= [x_0, x_1, \cdots x_n, \lambda_0, \lambda_1, \cdots \lambda_m] \\ x_k &= \left[P_{b_k}^w, R_{b_k}^w, v_{b_k}^w, b_{a_k}, b_{g_k}\right], k \in [0, n] \end{aligned} \tag{1}$$

w represents the world coordinate system. In the state variable χ, x_k is the IMU state when the k-th image is captured. It contains the position, speed, and direction of the IMU in the world coordinate system, as well as the accelerometer bias and gyroscope bias in the IMU body coordinate system. Additionally, n is the total number of key frames, m is the total number of feature points in the sliding window, and λ_i is the inverse depth of the i-th observed feature point.

In VILO's local pose estimation process, the observation set used to constrain the variable to be estimated is defined by the following formula:

$$Z = [Z_{c_i}, b_{i,i+1}, L_{i,i+1}] \quad (i, i+1) \in \kappa \tag{2}$$

In the formula, Z_{c_i} represents the visual measurement, $b_{i,i+1}$ is the body coordinate system when the i-th and i+1-th images are acquired, and $L_{i,i+1}$ represents the measured value of the lidar odometry from the i-th key frame to the i+1-th key frame. κ is the key frame set in the sliding window.

A factor graph (Figure 3) is constructed using the residual items in VILO as constraint factors and the variables to be optimized as nodes. In Figure 3, the circled node represents the inverse depth of the visual feature point observed by the camera, and the non-solid rectangular boxes are the visual residual factor, IMU residual factor, and lidar residual factor respectively. The variables to be estimated are restricted by the three residual factors. The construction of the IMU residual factor and the 2D lidar residual factor is only related to adjacent frames. The construction of the visual residual factor depends on the common-view relationship of feature points on adjacent frames. If the number of common-view feature points is scarce, the visual residual factor is not successfully constructed.

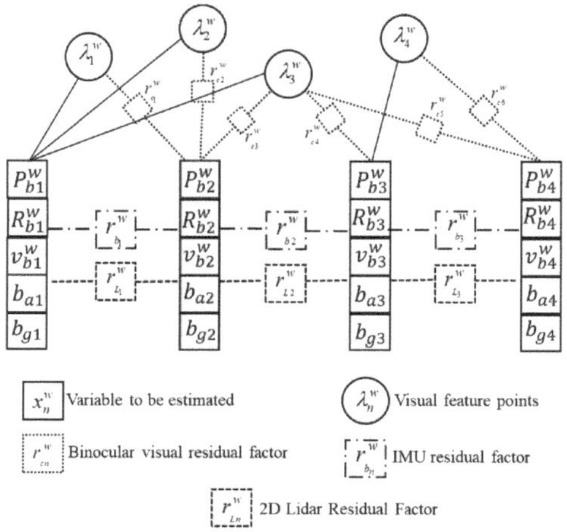

Figure 3. Binocular vision-IMU-2D lidar pose-estimation factor graph.

3.1. Nonlinear Least-Squares Model Based on Vision-IMU-Lidar

According to the residual factor of vision-IMU-lidar, and using the Mahalanobis distance to indicate the degree of deviation between the residual of each sensor and the covariance matrix, the VILO nonlinear least squares problem can be constructed as follows:

$$
\begin{aligned}
\chi_w{}^* = \arg\min_{\chi_w} & \left(\left\| r_p - H_p \chi \right\|^2 + \sum_{(i,j)\in\kappa} \left\| r_b(\{\hat{a}_t, \hat{\omega}_t\}_{t_i \le t \le t_j}, x_i, x_j) \right\|^2_{\Sigma_{b_{i,j}}} \right) \\
& + \sum_{i\in\kappa} \sum_{m\in\gamma_i} \rho(\left\| r_c(\hat{z}_i^m, x_i) \right\|^2_{\Sigma_{c_{i,m}}}) \\
& + \sum_{(i,j)\in\kappa} \rho \left(\left\| r_L(z_t^L - h_t^L(\chi)_{t_i \le t \le t_{i+1}}, x_i, x_{i+1}) \right\|^2_{\Sigma_{b_{i,i+1}}} \right)
\end{aligned} \tag{3}
$$

In (3), $\|r\|^2_\Sigma$ is the Mahalanobis distance of the residual r when the covariance matrix is Σ. The Mahalanobis distance is defined as $\|r\|^2_\Sigma = r^T \Sigma^{-1} r$. The first term is the prior information of the marginalization of the key frames in the sliding window, r_b is the residual item of the IMU, r_c is the binocular vision residual item, r_L is the residual error of the 2D lidar odometry, and κ is the key frame set in the sliding window. In order to improve the robustness, the Huber loss function ρ [18] is used for the visual residual r_c and the 2D laser odometry residual r_L, where ρ is

$$\rho(s) = \begin{cases} 1 & s \geq 1 \\ 2\sqrt{s} - 1 & s < 1 \end{cases} \tag{4}$$

3.1.1. Visual Reprojection Residual Constraint

The pose of the key frame is defined as the pose $\left(P_b^w, R_b^w\right)$ of the IMU coordinate system b, relative to the world coordinate system w. A certain visual feature point k in binocular vision is observed for the first time in key frame i, and its 3D space position coordinate is a function of the pose $\left(P_{b_i}^w, R_{b_i}^w\right)$ of the current key frame and the inverse depth value λ_k of the feature point. Next, the feature point k will continue to be tracked in the subsequent key frame j. Then, we can formulate the visual reprojection residual representation of the feature point k, which is a function of the key frame i and j poses $\left(P_{b_i}^w, R_{b_i}^w\right)$ and $\left(P_{b_j}^w, R_{b_j}^w\right)$, and the inverse depth λ_k of the feature point, defined as the following formula:

$$r_c\left(z_j^k, \chi\right) = r_{c_{jk}}\left(P_{b_i}^w, R_{b_i}^w, P_{b_j}^w, R_{b_j}^w, \lambda_k\right) \tag{5}$$

The advantage of defining the visual reprojection error to the unit ball is that the algorithm can support a wide-angle-lens model with severe distortion. It is also applicable to general camera models. Any pixel in the camera can be mapped to a ray on the unit ball. Therefore, the visual residual r_c based on the unit ball is:

$$\begin{aligned} r_{c_{jk}}(P_{b_i}^w, R_{b_i}^w, P_{b_j}^w, R_{b_j}^w, \lambda_k) &= \begin{bmatrix} b_1 & b_2 \end{bmatrix}^T \cdot \left(\hat{P}_{c_j}^k - \frac{P_{c_j}^k}{\left\| P_{c_j}^k \right\|} \right) \\ \hat{P}_{c_j}^k &= K_c^{-1}(\hat{z}_j^k) = K_c^{-1} \begin{pmatrix} \hat{u}_{c_j}^k \\ \hat{v}_{c_j}^k \end{pmatrix} \\ P_{c_j}^k &= R_b^c(R_w^{b_j}(R_{b_i}^w(R_c^b \frac{1}{\lambda_k} K_c^{-1} \begin{pmatrix} \hat{u}_{c_i}^k \\ \hat{v}_{c_i}^k \end{pmatrix} + P_c^b) + P_{b_i}^w) + P_w^{b_j}) + P_b^c \end{aligned} \tag{6}$$

In (6), K_c^{-1} is the inverse transformation of the camera projection matrix used to convert 2D pixel coordinates to 3D coordinates in the camera coordinate system; $\hat{P}_{c_j}^k$ represents the 3D space coordinates of the k-th feature point in the key frame j camera coordinate system, and is the observation value; and $P_{c_j}^k$ represents the 3D space coordinates of the k-th feature point in the key frame j camera coordinate system, and is the prediction value, which is transformed from key frame i to j through the pose $(P_{b_i}^w, R_{b_i}^w)$ of the i-th key frame and the pose $(P_{b_j}^w, R_{b_j}^w)$ of the j-th key frame. $\begin{bmatrix} b_1 & b_2 \end{bmatrix}$ represents the two orthogonal basis vectors of the tangent plane of the unit sphere, used to map the 3D camera coordinate difference to the unit sphere; and (P_b^c, R_b^c) is the pose from the IMU to the camera, obtained through pre-calibration.

3.1.2. IMU Residual Constraints

The IMU sensor can measure the acceleration and angular velocity information of the robot. The amount of measurement information is small, and the measurement frequency of IMU is high. Therefore, when the initial pose of the robot is known, the IMU can be used to continuously integrate time to obtain the speed and pose of the robot relative to the reference coordinate in real time. However, the IMU measurement carries a lot of noise, or the robot moves at a constant speed. If only IMU is used to estimate the pose, the accuracy and reliability of the pose estimation are low. Therefore, the integral information from the IMU can be used for data complementary fusion with the information of sensors, such as vision sensors, to improve the accuracy and robustness of the pose estimation.

In the visual inertial odometry (VIO), the IMU integration information between key frames is added as a residual constraint to the entire BA (bundle adjustment) optimization. According to the IMU pre-integration model, the IMU residual is as follows:

$$\begin{aligned} r_b(z_{b_{i+1}}^{b_i},\chi) &= r_b(\{\hat{a}_t,\hat{\omega}_t\}_{t_i\le t\le t_{i+1}},x_i,x_{i+1}) \\ &= \begin{bmatrix} \delta\alpha_{b_{i+1}}^{b_i} \\ \delta\beta_{b_{i+1}}^{b_i} \\ \delta\theta_{b_{i+1}}^{b_i} \\ \delta b_{a b_{i+1}}^{b_i} \\ \delta b_{g b_{i+1}}^{b_i} \end{bmatrix} = \begin{bmatrix} R_w^{b_i}(P_{b_{i+1}}^w - P_{b_i}^w + \frac{1}{2}g^w\Delta t_i^2 - v_{b_i}^w\Delta t_i) - \hat{\alpha}_{b_{i+1}}^{b_i} \\ R_w^{b_i}(v_{b_{i+1}}^w + g^w\Delta t_k - v_b^w) - \hat{\beta}_{b_{i+1}}^{b_i} \\ 2\left[(\hat{\gamma}_{b_{i+1}}^{b_i})^{-1}\otimes R_{b_i}^{w-1}\otimes R_{b_{i+1}}^w\right]_{xyz} \\ b_{ab_{i+1}} - b_{ab_i} \\ b_{gb_{i+1}} - b_{gb_i} \end{bmatrix} \end{aligned} \tag{7}$$

Equation (7) subtracts the measured value of the IMU pre-integration from the predicted value to obtain the IMU predicted score residual item. Among the variables in the equation, $\delta\alpha_{b_{i+1}}^{b_i}$ is the pre-integrated three-dimensional coordinate position residual of the IMU, $\delta\beta_{b_{i+1}}^{b_i}$ is the pre-integrated velocity residual of the IMU, and $\delta\theta_{b_{i+1}}^{b_i}$ is the IMU pre-integrated pose residual, that is, the rotation error of the IMU three axes. $\delta b_{a b_{i+1}}^{b_i}$ and $\delta b_{g b_{i+1}}^{b_i}$ are the zero-point error residual items of the accelerometer and gyroscope, respectively, in the IMU.

3.1.2.1. 2D Lidar Residual Error Constraints

Usually, 2D lidar is used in the field of ground mobile robots. The lidar information is passed through the Bayesian conditional probability $p(z_t|x_t,m)$ model of the radar to establish the maximum likelihood estimation problem. Then, the maximum likelihood estimation is converted into a scanning matching problem based on a laser point cloud, and a nonlinear least-squares problem that minimizes the error of scanning matching probability is constructed. Finally, a nonlinear optimization method is used to solve the pose of the lidar relative to the reference coordinate system, which is called the 2D lidar odometry algorithm. Because the pose estimated by the odometry algorithm has the advantage of a continuous and accurate scale, it can be tightly coupled and fused with the pose estimated by the VIO algorithm to improve the accuracy and robustness of the visual SLAM algorithm in indoor and outdoor scenes.

2D lidar Odometry Algorithm

On the basis of obtaining lidar data, the pose-estimation problem of lidar in the map can be transformed into a nonlinear least-squares problem as shown in (8) [19]:

$$\xi^* = \arg\min_{\xi}[1 - M(\mathrm{S}_i(\xi))]^2 \tag{8}$$

ξ is the pose of lidar relative to grid map coordinate system. $\mathrm{S}_i(\xi)$ represents the coordinates of the end point of the i-th scanning ray of the laser in the grid map coordinate system when lidar is in the pose variable ξ. $M(\mathrm{S}_i(\xi))$ is the probability that the grid map is an obstacle at the given coordinate position $\mathrm{S}_i(\xi)$. The purpose of (8) is to find the ξ variable that minimizes the objective function $[1 - M(\mathrm{S}_i(\xi))]^2$. $\mathrm{S}_i(\xi)$ is represented as follows:

$$\mathrm{S}_i(\xi) = \begin{pmatrix} \cos(\theta) & -\sin(\theta) \\ \sin(\theta) & \cos(\theta) \end{pmatrix}\begin{pmatrix} s_{i,x} \\ s_{i,y} \end{pmatrix} + \begin{pmatrix} p_x \\ p_y \end{pmatrix} \tag{9}$$

In (9), $\xi = (p_x, p_y, \theta)^{\mathrm{T}}$, where p_x and p_y represent the coordinates of the origin of the laser sensor coordinate system relative to the reference coordinate system, and θ is the heading angle of the robot. The purpose of (9) is to transform the observation point of lidar to the grid map coordinate system.

Therefore, the core of the scan matching algorithm is that at the beginning it gives an estimated value ξ of lidar with a smaller deviation from the actual pose, and then finds a pose increment $\Delta\xi$ according to the laser data acquired at the current time t so that the end point of the laser scan and the probability grid map are optimally matched at the current time t, that is, (10) is established:

$$\Delta\xi^{*}\arg\min_{\Delta\xi}\sum_{i=1}^{n}[1-M(S_i(\xi+\Delta\xi))]^2 \rightarrow 0 \tag{10}$$

In order to solve the optimal increment $\Delta\xi$ that minimizes (10) at the current time t, the Gauss–Newton iteration method is used to perform the first-order Taylor expansion on the nonlinear function $M(S_i(\xi+\Delta\xi))$ to obtain (11):

$$\Delta\xi^{*}\arg\min_{\Delta\xi}\sum_{i=1}^{n}\left[1-M(S_i(\xi))-\nabla M(S_i(\xi))\frac{\partial_{S_i(\xi)}}{\partial_{\xi}}\Delta\xi\right]^2 \rightarrow 0 \tag{11}$$

In (11), i is the i-th laser beam of the laser sensor at time t. Sum the n beams of laser light, and obtain the partial derivative of $\Delta\xi$ from the objective function in (11) so that the partial derivative is 0. Finally, the pose increment $\Delta\xi$ is as follows:

$$\Delta\xi = \mathrm{H}^{-1}\sum_{i=1}^{n}\left[\nabla M(S_i(\xi))\frac{\partial S_i(\xi)}{\partial\xi}\right]^{\mathrm{T}}[1-M(S_i(\xi))] \tag{12}$$

In (12), $M(S_i(\xi))$ represents the probability value of an obstacle at the coordinate $S_i(\xi)=(x_i,y_i)$. According to equation (9), (x_i,y_i) may not be an integer pair. To express $\nabla M(S_i(\xi))$, assume that $P_{i,i}=(x_i,y_i)$, $P_{01}=(x_0,y_1)=(floor(x_i),Ceiling(y_i))$, then we can write the approximation of $\nabla M(S_i(\xi))$, as shown in (13). The Hessian matrix H is as shown in (14), and the partial derivative of $S_i(\xi)$ to ξ is as shown in (15):

$$\begin{cases}\frac{\partial M(P_{i,i})}{\partial x} & \approx \frac{y_i-y_0}{y_1-y_0}(M(P_{11})-M(P_{01})) \\ & +\frac{y_1-y_i}{y_1-y_0}(M(P_{10})-M(P_{00})) \\ \frac{\partial M(P_{i,i})}{\partial y} & \approx \frac{x_i-x_0}{x_1-x_0}(M(P_{11})-M(P_{10})) \\ & +\frac{x_1-x_i}{x_1-x_0}(M(P_{01})-M(P_{00}))\end{cases} \tag{13}$$

$$\mathrm{H}=\sum_{i=1}^{n}\left[\nabla M(S_i(\xi))\frac{\partial S_i(\xi)}{\partial\xi}\right]^{\mathrm{T}}\left[\nabla M(S_i(\xi))\frac{\partial S_i(\xi)}{\partial\xi}\right] \tag{14}$$

$$\frac{\partial \mathrm{S_i}(\xi)}{\partial\xi}=\begin{pmatrix}1 & 0 & -\sin(\theta)s_{i,x} & -\cos(\theta)s_{i,y} \\ 0 & 1 & \cos(\theta)s_{i,x} & -\sin(\theta)s_{i,y}\end{pmatrix} \tag{15}$$

The core of the 2D lidar odometry algorithm is to use the first-order Taylor series expansion of the Gauss–Newton iteration method to approximately replace the nonlinear least squares problem. The objective function $1-M(\mathrm{S_i}(\xi))$ is approximated near the lidar pose ξ, and multiple optimization iterations are performed until the increment $\Delta\xi$ is small enough, at which time iteration is stopped, the optimal variable ξ at the current time t is obtained, and (11), the formula of the residual sum of squares of the original model, is minimized.

The above analysis shows that the operating speed of the 2D lidar odometry algorithm is quite fast. Using the incremental formula (12), only a few simple iterative calculations are required to obtain the laser pose $\xi_t=(p_x,p_y,\theta)^{\mathrm{T}}$ at time t.

3.1.2.3. Residual Items Based on 2D Lidar Odometry Model

The residual factor r_L of the 2D lidar pose can be expressed by the relative displacement $\Delta\hat{P}^{L_i}_{L_{i+1}}$ between the *i*-th frame and the *i*+1-th frame observed by the lidar odometry, and the relative displacement $\Delta P^{L_i}_{L_{i+1}}$ of the last two key frames before and after the state variable to be optimized. $\Delta\hat{P}^{L_i}_{L_{i+1}}$ and $\Delta P^{L_i}_{L_{i+1}}$ are, respectively, called the observed value and the predicted value. The specific form is described in the following formula:

$$r_L(z^L_t - h^L_t(\chi)_{t_i \le t \le t_{i+1}}, x_i, x_{i+1}) = z^L_t - h^L_t(x_i, x_{i+1}) = \Delta\hat{P}^{L_i}_{L_{i+1}} - \Delta P^{L_i}_{L_{i+1}} \tag{16}$$

In (16), L represents the lidar coordinate system. $x_i = (P^w_{b_i}, R^w_{b_i})$ represents the pose of the IMU relative to the world coordinate system at the *i*-th frame.$h^L_t(\chi)_{t_i \le t \le t_{i+1}}$ represents the relative pose estimation between the *i*-th frame and the *i*+1-th frame. The residual term does not include the error term about the pose but rather the error term about the position. The reason why the error term of the pose is discarded is that the IMU can provide a higher-precision integration of the pose angle. Therefore, pose is not used as a residual constraint here. The specific expansion form of $h^L_t(x_{t-1}, x_t)$ (that is, $\Delta P^{L_i}_{L_{i+1}}$) is shown in the following formula:

$$\begin{aligned} h^L_t(x_i, x_{i+1}) &= \Delta P^{L_i}_{L_{i+1}} = R^{L_i}_w (P^w_{L_{i+1}} - P^w_{L_i}) \\ &= R^{L_i}_w (P^w_{b_{i+1}} + R^w_{b_{i+1}} P^b_L - P^w_{b_i} - R^w_{b_i} P^b_L) \\ &= R^{L_i}_w (P^w_{b_{i+1}} - P^w_{b_i}) + R^{L_i}_w R^w_{b_{i+1}} P^b_L - R^{L_i}_w R^w_{b_i} P^b_L) \\ &= R^L_b R^{b_i}_w (P^w_{b_{i+1}} - P^w_{b_i}) + R^L_b R^{b_i}_w R^w_{b_{i+1}} P^b_L - R^L_b P^b_L \end{aligned} \tag{17}$$

Finally the residual term of the relative displacement $\Delta\hat{P}^{L_i}_{L_{i+1}}$ between the position and pose of the variable to be optimized $x_i = (P^w_{b_i}, R^w_{b_i})$, $x_{i+1} = (P^w_{b_{i+1}}, R^w_{b_{i+1}})$ and the lidar odometry frames *i* and *i*+1 are obtained as follows:

$$r_L = \Delta\hat{P}^{L_i}_{L_{i+1}} - R^L_b R^{b_i}_w (P^w_{b_{i+1}} - P^w_{b_i}) - R^L_b R^{b_i}_w R^w_{b_{i+1}} P^b_L + R^L_b P^b_L \tag{18}$$

In (18), $\Delta P^{L_i}_{L_{i+1}}$ can be obtained by lidar odometry, and (R^b_L, P^b_L) is the pose of the lidar coordinate system relative to the IMU coordinate system, which can be obtained by pre-calibration.

Because the lidar odometry can only observe the pose yaw angle θ, the residual model can only constrain the pose yaw angle in the state variable χ to be optimized. Then the state variable χ is specifically $\chi = (P^w_{b_i}, \theta^w_{b_i}, P^w_{b_{i+1}}, \theta^w_{b_{i+1}})$, and the Jacobian matrix of the residual term r_L with respect to the state variable χ to be estimated is as follows:

$$\begin{aligned} J[0]^{3\times 10} &= \begin{bmatrix} \frac{\partial r_L}{\partial P^w_{b_i}} & \frac{\partial r_L}{\partial \theta^w_{b_i}} \end{bmatrix} = \begin{bmatrix} -R^L_b R^{b_i}_w & -R^L_b \left[R^{b_i}_w (P^w_{b_{i+1}} - P^w_{b_i} + R^{b_{i+1}}_w P^b_L) \right]^\wedge \end{bmatrix} \\ J[1]^{3\times 10} &= \begin{bmatrix} \frac{\partial r_L}{\partial P^w_{b_{i+1}}} & \frac{\partial r_L}{\partial \theta^w_{b_{i+1}}} \end{bmatrix} = \begin{bmatrix} R^L_b R^{b_i}_w & -R^L_b R^{b_i}_w R^{b_{i+1}}_w (P^b_L)^\wedge \end{bmatrix} \end{aligned} \tag{19}$$

In (19), $\wedge$ is the transformation of the vector to the antisymmetric matrix. $J[0]^{3\times 10}$ and $J[1]^{3\times 10}$ are the first-order partial derivatives of the lidar residual r_L with respect to the pose $(P^w_{b_i}, \theta^w_{b_i})$ and $(P^w_{b_{i+1}}, \theta^w_{b_{i+1}})$ of the *i*-th and *i*+1-th key frame, also known as the Jacobian matrix.

The covariance matrix Σ_L of the residual term of the lidar odometry is the covariance matrix of the pose estimation in the lidar odometry model. The covariance matrix can be obtained by the laser scanning matching algorithm.

Thus far, the least squares problem based on binocular-IMU-lidar has been constructed, and the visual residual r_c, IMU residual r_b, and 2D lidar odometry residual r_L have been analyzed. Furthermore, the Jacobian matrix analytical formula of the residual r_L of the lidar odometry with respect to the state variable χ_w to be estimated has been derived. Therefore, Google's Ceres nonlinear optimization library can be used to solve the least squares model. After three sensor measurements are jointly constrained and continuous iterative optimization is carried out, the pose estimation can be solved with high accuracy and robustness in indoor and outdoor environments.

3.2. *Pseudo-Code Description*

The pseudo code for VILO residual calculation and optimization is shown in Algorithm 1. Firstly, the SLAM system and variables are initialized. Secondly, the data from the 2D lidar, IMU, and camera are received and processed in a loop. It is important to note that motion distortion correction is performed on the 2D lidar scan data using the IMU raw data before time-aligning all the sensor data. After obtaining the time-aligned data packets, the IMU pre-integration residual, visual reprojection residual, and 2D lidar odometry residual are calculated. Finally, the final pose is solved using the Ceres nonlinear optimization library.

Algorithm 1 VILO Residual Calculation and Optimization

Input: IMU data, camera images, and 2D LiDAR point cloud data
Output: The pose of robot $x = [P_b^w, R_b^w]$ in the world coordinate system
// Step 1. Initialize system and variables
$x_i = \left[P_{b_i}^w, R_{b_i}^w, v_{b_i}^w, b_{a_i}, b_{g_i}\right]$ //Initialize state variables according to Equation (1);
$Z = [Z_{c_i}, b_{i,i+1}, L_{i,i+1}]$ //Initialize observation set according to Equation (2);
// Step 2. Receive data and perform pose estimation and optimization
while received IMU, camera image and 2D LiDAR point cloud **do**
  Undistorted pcd = Motion distortion correction (imu data, point cloud)
  Z = sync_process(imudata, image, Undistortedpcd)//Time alignment
  if empty(Z) **then**
    continue;
  end if
  r_b = processIMU(Z, x_i)//IMU residual according to Equation (7)
  F = featureTracker(Z) //Extract features (pixel coordinates) and optical flow tracking
  r_c = processImage(F, x_i) //Visual reprojection residual according to Equations (5) and (6)
  $\Delta\xi$ = probabilityGridMap(Z, x_i) //LiDAR pose estimation according to Equations (8)–(15)
  r_L = processLiDAR($\Delta\xi, x_i$) //Construct 2D lidar odometry residual according to Equation (18)
  x_{i+1} = optimization(r_b, r_c, r_L) //Ceres optimization, get output pose
end while

4. Experiment

In order to verify the accuracy and effectiveness of the tightly coupled pose estimation algorithm based on binocular VILO proposed in this paper, some extreme experimental environments need to be selected, such as insufficient light or darkness, lack of texture characteristics (indoor white walls), and frequent movement of dynamic obstacles (people). Therefore, this section focuses on the corridor environment of the indoor experimental building with extreme conditions for algorithm comparison and verification experiments. In this paper, we compare the performance of VILO (ours), VINS Fusion and ORB-SLAM2.

In this pose-estimation experiment of the three types of algorithms, it is necessary to use the cumulative error of each algorithm to measure the excellence of each algorithm. Therefore, the visual loop detection function is not turned on during the operation of the three types of algorithms because it would eliminate the cumulative pose error. All the methods are executed on a computing device equipped with an Intel i7-8700 CPU using the robot operating system (ROS) in Ubuntu Linux. The sensor mounting platforms are shown in Figure 4.

Figure 4. Sensor mounting platforms. PC is used for data acquisition of robot sensors, pose estimation and mapping. Embedded controller is used for robot motion control.

4.1. Motion Trajectory Comparison Experiment

Experimental configuration description: The algorithm verification environment is a corridor environment on the first floor of the experimental building with a width of 3 m, a long length, corners, and a relatively empty hall environment. The area on the first floor is about 250×100 m^2. The scene of this experiment is shown in Figure 5, where (a) is the satellite map of the experimental building, in which you can clearly see the outline of the corridor, and (b–d) show the scene inside the corridor. The robot is controlled to traverse every scene in the corridor as much as possible, the linear velocity of movement is maintained at 0.5 m/s, and the angular velocity is maintained at 0.5 rad/s. A large number of fixed marking points are arranged inside the corridor to obtain the real position of the robot in order to analyze the positioning error of the robot.

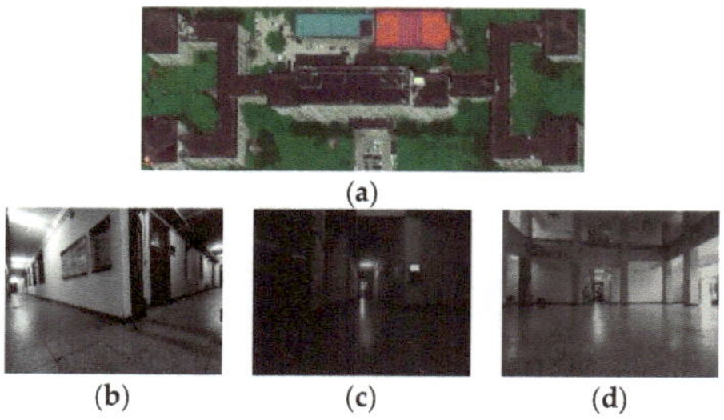

Figure 5. Indoor corridor environment. (**a**) Corridor outline; (**b**) Corner of the corridor; (**c**) Insufficient light inside the corridor; (**d**) Corridor hall scene.

The extreme situations in the experiment are as follows:

(1) Parts of the corridor walls are completely white without obvious visual features;
(2) In the corridor hall, the walls are covered with tiles with high reflectivity, which affects the camera observation data;
(3) The similarity of some corridor scenes is relatively high, and there are no special markings, which affects the accuracy of 2D lidar odometry;
(4) On some floors of the corridor, there are cracks, uneven heights, and large vibrations when the robot moves, causing large fluctuations in the measured values of each sensor;
(5) During the experiment, there were many pedestrians in the field of view of the camera and 2D lidar.

The trajectories of each algorithm in an experiment are shown in Figure 6, including the standard trajectory of the corridor environment, the trajectory of the algorithm in this paper (binocular-IMU-2D lidar), the trajectory of the VINS-Fusion algorithm (binocular-IMU), and the trajectory of the ORB-SLAM2 algorithm (pure binocular vision).

(a) (b) (c) (d)

Figure 6. The trajectory of each algorithm (top view). (**a**) Standard trajectory; (**b**) This paper; (**c**) VINS-Fusion; (**d**) ORB-SLAM2.

The motion trajectory of the pose estimation algorithm in this paper, shown in Figure 6b, is highly consistent with the real corridor contour, demonstrating that it effectively restores the real robot motion trajectory, and the overall error is small. The trajectory of the VINS-Fusion algorithm in Figure 6c does not closely match the contour of the corridor. In the scene where some visual features of the corridor are missing, the pose-estimation drifts, causing the recorded trajectory to curve or be lost when compared to the actual trajectory of the robot. Especially in the lower part of the trajectory in Figure 6c, when the robot returns, it encounters the loss of visual features, and the robot also moves at a constant speed, meaning the IMU is not stimulated, which causes the VINS-Fusion algorithm to fail, and the algorithm exits early. The ORB-SLAM2 algorithm in Figure 6d has the largest deviation from the real contour of the corridor and thus the worst performance among the three algorithms. The reason for this is that it only relies on a single visual sensor. In the case of few visual features or high visual noise (more pedestrians), the pose-estimation error increases rapidly, resulting in a large deviation between the estimated motion trajectory and the true trajectory. Like the VINS-Fusion algorithm, in the lower part of the trajectory, when the robot returns, it encounters complete loss of visual features, and the algorithm also fails.

Because the algorithm proposed in this paper integrates lidar information, it overcomes the problem of the original binocular vision-IMU algorithm being overly dependent on visual information. In the case of loss of visual features and lack of IMU excitation, reliable and high-precision pose estimation can still be achieved. Because the lidar measurement value has strong anti-interference ability and can adapt to many extreme environments, after vision is restored, the robot can still combine the lidar-estimated pose for joint optimization to obtain a more precise pose.

4.2. Pose Estimation Accuracy Verification Experiment

The robot was manually controlled to repeat two to three closed-loop motions along the corridor environment, referring to the robot moving from the same starting position and then returning to the starting position. For each closed-loop motion, it was ensured that the ending pose at the starting point was consistent with the starting pose (the position error was less than 0.1 cm, and the heading angle error was less than 5°). Furthermore, rosbag, the message recording tool in ROS, was used to record binocular vision, IMU, and 2D lidar data in the same bag file.

Later, the recorded data set was applied to the ORB-SLAM2 algorithm, the VINS-Fusion algorithm, and the binocular vision inertial pose estimation algorithm of tightly coupled fusion 2D lidar proposed in this paper to conduct a positioning-accuracy-comparison verification experiment. The relative pose between the start point and the end point of the motion obtained by each pose-estimation algorithm was used as the pose error to evaluate the effect of the pose estimation.

The robot in this paper was a mobile robot on the ground, so only the three-dimensional pose (p_x, p_y, θ) of the robot coordinate system relative to the global coordinate system, where p_x, p_y is the offset of the X and Y axes, and θ is the yaw angle of the robot, needed to be recorded during the experiment. In order to more clearly illustrate the effectiveness of the pose estimation algorithm proposed in this paper, three-pose estimation experiments were performed on each of the three algorithms. First, the initial pose-estimation value of each algorithm, that is, the pose of the robot's starting point, was recorded. Starting from the same starting point, the robot moved along the same path and returned to the starting point again, at which time the pose of the robot was recorded. By calculating the average value of the three motions of each experimental parameter of each algorithm and comparing the average value with the initial pose, the pose-estimation error of each algorithm was obtained.

The results of this pose-estimation experiment are shown in Table 1. We computed the mean absolute error (MAE) for each experiment, which represents the average absolute difference between the predicted and true values and is presented in the table as the absolute value of the "Average" minus "Initial pose" parameter in units of meters. This parameter can measure the accuracy of the pose estimation. To demonstrate the accuracy differences between algorithms more clearly, we separately displayed the MAE errors of the X axes, Y axes, and yaw angle for each algorithm, which better measures the differences between the algorithms.

Table 1. Corridor environment pose-estimation experiment results.

Algorithm	Parameters	Initial Pose	1st Finish Pose	2nd Finish Pose	3rd Finish Pose	Average	MAE
ORB-SLAM2	X axis offset (m)	0.032	6.56	5.36	7.28	6.40	6.368
	Y axis offset (m)	−0.028	−5.32	−5.98	−6.46	−5.92	5.892
	yaw angle (degrees)	0.351	8.563	7.253	10.26	8.692	8.341
VINS-Fusion	X axis offset (m)	−0.021	5.221	4.336	4.758	4.772	4.793
	Y axis offset (m)	0.033	−4.532	−5.142	−4.349	−4.674	4.707
	yaw angle (degrees)	0.283	5.286	5.463	5.852	5.534	5.251
Our method	X axis offset (m)	−0.011	−3.326	−3.635	−3.867	**−3.609**	**3.598**
	Y axis offset (m)	0.016	−3.855	−4.126	−3.732	**−3.904**	**3.920**
	yaw angle (degrees)	−0.203	4.563	4.068	3.659	**4.10**	**4.303**

From the analysis in Table 1, it can be seen that the ORB-SLAM2 algorithm has a large error in the pose estimation during the entire closed-loop motion process. The position estimation deviations of the X axis and Y axis both exceed 5 m, and the yaw angle error is also large. The pose-estimation error of the VINS-Fusion algorithm is much lower than that the ORB-SLAM2 algorithm. Because the VINS-Fusion algorithm uses IMU to make up for the lack of visual information, it performs well if the robot does not move at a constant speed, but this is not in line with reality. In real situations, the robot often moves at a constant speed, so the IMU information is invalid, and the VINS-Fusion algorithm degenerates to a pure visual pose-estimation algorithm. In the case of loss of visual features or large interference (more pedestrians), the reliability of the VINS-Fusion algorithm performance decreases. In the method of this paper, the average pose error estimation is the smallest. The position estimation error drops below 4 m, and the yaw angle error is much smaller than that of the VINS-Fusion algorithm. This is because 2D lidar odometry compensates for the degradation and instability issues of VINS-Fusion. When a robot moving at a constant speed enters a feature-sparse scene, the visual-inertial odometry is severely affected, while lidar odometry works normally. By tightly coupling low-cost 2D lidar, camera, and IMU, the accuracy and robustness of the pose estimation are effectively improved.

Overall, in terms of accuracy, after the robot moves about 758.62 m, the algorithm in this paper reduces the Euclidean distance of the position error by 38.7% and the yaw angle error by 48.4% compared with ORB-SLAM2. Compared with VINS-Fusion, the Euclidean distance of the position error is reduced by 20.8%, and the yaw angle error is reduced by 18.1%. In terms of robustness, after many experiments, the algorithm in this paper has not failed in many extreme situations (visual occlusion, uniform motion, and more pedestrians), so the algorithm is robust. In terms of real-time performance, the algorithm in this paper can achieve 10 Hz, which can meet the needs of tasks that do not require high real-time performance. Therefore, the tightly coupled fusion VILO pose estimation method proposed in this paper can effectively improve the accuracy and robustness of the robot pose estimation and has strong feasibility. It can provide reliable pose estimation for other modules of visual SLAM, such as environment mapping.

4.3. Comparative Experiment on Dense Mapping of Spatial Environment Based on Three Types of Pose-Estimation Algorithms

In order to further highlight the performance of the tightly coupled fusion 2D lidar pose-estimation algorithm proposed in this paper in real space environment mapping, a 3D dense mapping algorithm based on super-pixel bins [20] is applied to ORB-SLAM2, VINS-Fusion, and the pose-estimation algorithm proposed in this paper to conduct 3D dense-mapping-quality comparison experiments.

The 3D dense maps in Figure 7 are all represented by 3D dense point cloud maps. In Figure 7a, the middle image is an actual scene with a lack of visual features in a corridor, and the right image is a 3D dense point cloud image established at this scene, corresponding to the map position indicated by the circle in the left image. The loss of visual features causes the pose tracking of the ORB_SLAM2 algorithm to fail. As a result, the mapping algorithm fails to build the map at this location, the algorithm exits early, and ultimately the entire environment mapping is incomplete.

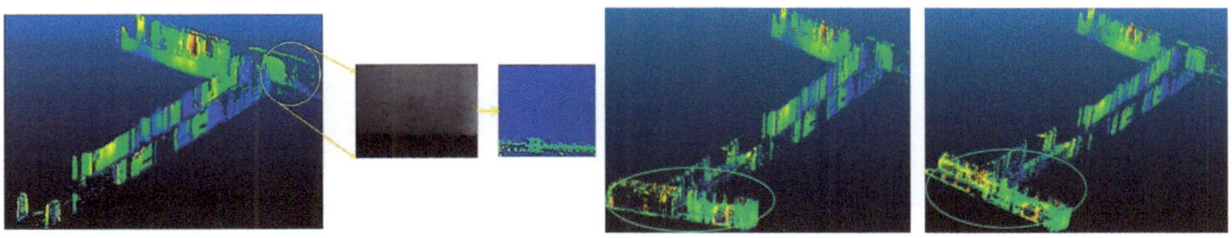

(a) (b) (c)

Figure 7. 3D mapping comparison experiment based on three types of pose-estimation algorithms. (**a**) Based on ORB_SLAM2; (**b**) Based on VINS_Fusion; (**c**) Based on algorithm of this paper.

In Figure 7b, because the algorithm fuses binocular vision and IMU sensor information, when the visual features are lost, the algorithm can still perform short-term pose tracking based on IMU information. Therefore, compared with the mapping effect of ORB_SLAM2, based on this algorithm, the spatial 3D environment map can be completely established without visual loss. However, when the robot is moving at a constant speed, the acceleration of the IMU is not stimulated, which leads to a decrease in the short-term tracking accuracy of the algorithm and an increase in the cumulative error, which makes the established 3D map deform. The 3D dense map indicated by the ellipse in Figure 7b corresponds to a time when the robot was moving at a constant speed. The map shows that the lower part of the corridor has been deformed.

Figure 7c shows the dense 3D environment map established by the pose-estimation algorithm proposed in this paper. The map has no obvious distortion, indicating that the algorithm can ensure the accuracy and robustness of pose estimation under extreme conditions such as loss of visual features and uniform motion of the robot, and it truly restores the actual environment. According to the comparison of actual environment mapping experiments, the pose-estimation algorithm proposed in this paper can effectively overcome the robot-pose-estimation problem in some extreme environments and has good robustness.

5. Summary

Aiming to improve the accuracy and robustness of the visual–inertial SLAM algorithm, we designed and implemented VILO, which includes a binocular visual–inertial system and low-cost 2D lidar. With the help of 2D lidar odometry, the 3D pose of the 2D lidar relative to the local reference coordinate system is obtained before sensor data fusion. Then, a residual constraint equation can be constructed, which allows the low-cost 2D lidar information to be used to optimize with the binocular visual–inertial state estimator based on tightly coupled optimization, and the optimal robot pose is solved through nonlinear optimization. The experimental results show that the VILO SLAM algorithm can still have high pose accuracy and robustness in multiple extreme environments. We hope that our work can provide a feasible idea and scheme for SLAM of a mobile robot.

Author Contributions: Methodology, G.P. and L.H.; software, Y.Z.; validation, Y.Z. and L.H.; investigation, L.X. and Z.S.; data curation, X.Z.; project administration, Z.W. All authors have read and agreed to the published version of the manuscript.

Funding: This research was funded by Hubei Province Core Technology for Bridging Development Gaps Project (HBSNYT202213), Hubei Province Unveiling Science and Technology Project (2021BEC008), Hubei Province Major Science and Technology Project (2020ACA006), Hubei Province Natural Science Foundation of China (No. 2019CFB526), and Shandong Province Key Research and Development Project of China (2019JZZY010443).

Institutional Review Board Statement: Not applicable.

Informed Consent Statement: Not applicable.

Data Availability Statement: Data sharing not applicable.

Conflicts of Interest: The authors declare no conflict of interest.

References

1. Engel, J.; Schöps, T.; Cremers, D. LSD-SLAM: Large-scale direct monocular SLAM. In Proceedings of the Computer Vision–ECCV 2014: 13th European Conference, Part II 13, Zurich, Switzerland, 6–12 September 2014; Springer International Publishing: Berlin/Heidelberg, Germany, 2014.
2. Engel, J.; Koltun, V.; Cremers, D. Direct sparse odometry. *IEEE Transact. Pattern anal. Machine Intel.* **2017**, *40*, 611–625. [CrossRef]
3. Forster, C.; Pizzoli, M.; Scaramuzza, D. SVO: Fast semi-direct monocular visual odometry. In Proceedings of the 2014 IEEE International Conference on Robotics and Automation (ICRA), Hong Kong, China, 31 May–7 June 2014; pp. 15–22.
4. Lahemer, E.S.; Rad, A. An adaptive augmented vision-based ellipsoidal SLAM for indoor environments. *Sensors* **2019**, *19*, 2795. [CrossRef] [PubMed]
5. Lynen, S.; Achtelik, M.W.; Weiss, S.; Chli, M.; Siegwart, R. A robust and modular multi-sensor fusion approach applied to mav navigation. In Proceedings of the IEEE/RSJ International Conference on Intelligent Robots and Systems, Tokyo, Japan, 3–7 November 2013; pp. 3923–3929.
6. López, E.; García, S.; Barea, R.; Bergasa, L.M.; Molinos, E.J.; Arroyo, R.; Romera, E.; Pardo, S. A multi-sensorial simultaneous localization and mapping (SLAM) system for low-cost micro aerial vehicles in GPS-denied environments. *Sensors* **2017**, *17*, 802. [CrossRef] [PubMed]
7. Yan, Y.; Zhang, B.; Zhou, J.; Zhang, Y.; Liu, X. Real-Time Localization and Mapping Utilizing Multi-Sensor Fusion and Visual–IMU–Wheel Odometry for Agricultural Robots in Unstructured, Dynamic and GPS-Denied Greenhouse Environments. *Agronomy* **2022**, *12*, 1740. [CrossRef]
8. Leutenegger, S.; Lynen, S.; Bosse, M.; Siegwart, R.; Furgale, P. Keyframe-based visual–inertial odometry using nonlinear optimization. *Int. J. Robot. Res.* **2015**, *34*, 314–334. [CrossRef]
9. Forster, C.; Carlone, L.; Dellaert, F.; Scaramuzzza, D. IMU preintegration on manifold for efficient visual-inertial maximum-a-posteriori estimation. In *Robotics Science and Systems*; Georgia Institute of Technology: Roma, Italy, 2015; pp. 1–20.
10. Campos, C.; Elvira, R.; Rodríguez, J.J.G.; Montiel, J.M.M.; Tardos, J.D. ORB-SLAM3: An Accurate Open-Source Library for Visual, Visual-Inertial and Multi-Map SLAM. *IEEE Transact. Robot.* **2021**, *37*, 1874–1890. [CrossRef]
11. Xiao, J.; Xiong, D.; Yu, Q.; Huang, K.; Lu, H.; Zeng, Z. A Real-Time Sliding-Window-Based Visual-Inertial Odometry for MAVs. *IEEE Transact. Ind. Inform.* **2020**, *16*, 4049–4058. [CrossRef]

12. Wang, Y.; Li, Z.; Su, C.-Y. Multisensor-Based Navigation and Control of a Mobile Service Robot. *IEEE Trans. Syst. Man Cybern. Syst.* **2019**, *49*, 1–11.
13. Hashim, H.A.; Eltoukhy, A.E.E. Nonlinear Filter for Simultaneous Localization and Mapping on a Matrix Lie Group Using IMU and Feature Measurements. *IEEE Transact. Syst. Man Cybernet. Syst.* **2022**, *52*, 2098–2109. [CrossRef]
14. Qin, T.; Cao, S.; Pan, J.; Shen, S. A general optimization-based framework for global pose estimation with multiple sensors. *arXiv* **2019**, arXiv:1901.03642.
15. Tixiao, S.; Englot, B.; Meyers, D.; WAmg, W.; Ratti, C.; Rus, D. Lio-sam: Tightly-coupled lidar inertial odometry via smoothing and mapping. In Proceedings of the 2020 IEEE/RSJ International Conference on Intelligent Robots and Systems (IROS), Las Vegas, NV, USA, 10 February 2020.
16. Jiarong, L.; Zhang, F. R3LIVE: A Robust, Real-time, RGB-colored, LiDAR-Inertial-Visual tightly-coupled state Estimation and mapping package. In Proceedings of the 22 International Conference on Robotics and Automation (ICRA), Philadelphia, PA, USA, 23–27 May 2022.
17. Ming, J. Binocular Vision Positioning and 2D Lidar Mapping of Indoor Environment Based on Point-Line FEATURE. Master's Thesis, Southwest Jiaotong University, Chengdu, China, 2018; p. 91.
18. Qin, T.; Li, P.; Shen, S. VINS-Mono: A robust and versatile monocular visual-inertial state estimator. *IEEE Trans. Robot* **2018**, *34*, 1004–1020. [CrossRef]
19. Hess, W.; Kohler, D.; Rapp, H.; Andor, D. Real-time loop closure in 2D LIDAR SLAM. In Proceedings of the 2016 IEEE International Conference on Robotics and Automation (ICRA), Stockholm, Sweden, 9 June 2016.
20. Wang, K.; Gao, F.; Shen, S. Real-time Scalable Dense Surfel Mapping. In Proceedings of the 2019 International Conference on Robotics and Automation (ICRA), Montreal, QC, Canada, 16–21 May 2019; pp. 6919–6925.

sensors

MDPI

Article

Human Motion Prediction via Dual-Attention and Multi-Granularity Temporal Convolutional Networks

Biaozhang Huang [1,2] and Xinde Li [1,2,*]

[1] Key Laboratory Measurement and Control of CSE Ministry of Education, School of Automation, Southeast University, Nanjing 210002, China; bzhuang@seu.edu.cn

[2] Nanjing Center for Applied Mathematics, Nanjing 211135, China

* Correspondence: xindeli@seu.edu.cn

Abstract: Intelligent devices, which significantly improve the quality of life and work efficiency, are now widely integrated into people's daily lives and work. A precise understanding and analysis of human motion is essential for achieving harmonious coexistence and efficient interaction between intelligent devices and humans. However, existing human motion prediction methods often fail to fully exploit the dynamic spatial correlations and temporal dependencies inherent in motion sequence data, which leads to unsatisfactory prediction results. To address this issue, we proposed a novel human motion prediction method that utilizes dual-attention and multi-granularity temporal convolutional networks (DA-MgTCNs). Firstly, we designed a unique dual-attention (DA) model that combines joint attention and channel attention to extract spatial features from both joint and 3D coordinate dimensions. Next, we designed a multi-granularity temporal convolutional networks (MgTCNs) model with varying receptive fields to flexibly capture complex temporal dependencies. Finally, the experimental results from two benchmark datasets, Human3.6M and CMU-Mocap, demonstrated that our proposed method significantly outperformed other methods in both short-term and long-term prediction, thereby verifying the effectiveness of our algorithm.

Keywords: human motion prediction; multi-granularity; attention mechanism; temporal convolutional networks

Citation: Huang, B.; Li, X. Human Motion Prediction via Dual-Attention and Multi-Granularity Temporal Convolutional Networks. *Sensors* **2023**, *23*, 5653. https://doi.org/10.3390/s23125653

Academic Editor: Sašo Blažič

Received: 17 May 2023
Revised: 12 June 2023
Accepted: 14 June 2023
Published: 16 June 2023

1. Introduction

With the rapid development of artificial intelligence technology, an increasing number of intelligent devices are being applied in industrial production and daily human life. Human motion prediction, a key technology for enhancing device intelligence, aims to capture the intrinsic temporal evolution within historical human motion sequences to generate predictions for future motion. Human motion prediction has been widely applied in fields such as autonomous driving [1], human–computer interaction [2,3], human emotion recognition [4], and human behavior analysis [5–7]. However, due to the high dimensionality, joint spatial collaboration, hierarchical human structure, and strong temporality characteristics of human motion, capturing temporal dynamic information and spatial dependency features for precise human motion prediction remains a challenging research hotspot.

Human motion prediction is a typical task in the computer vision field. Traditional human motion prediction algorithms, such as hidden Markov models (HMMs) [8], Gaussian process dynamic models (GPDMs) [9], and restricted Boltzmann machines [10], as shown in Figure 1, often require extensive prior knowledge and assumptions, making it challenging to capture the complexity and diversity of human motion and so restricting their application impact.

As more and more large-scale motion capture datasets become available, an increasing number of deep learning models have been designed and have demonstrated excellent performance, such as convolutional neural networks (CNNs) [11], graph neural networks

(GNNs) [12–14], temporal modules such as recurrent neural networks (RNNs) [15–20], temporal convolutional networks (TCNs) [21,22], and attention mechanisms [23,24]. Although these deep learning models have exhibited effectiveness in human motion prediction, there are still limitations in two aspects:

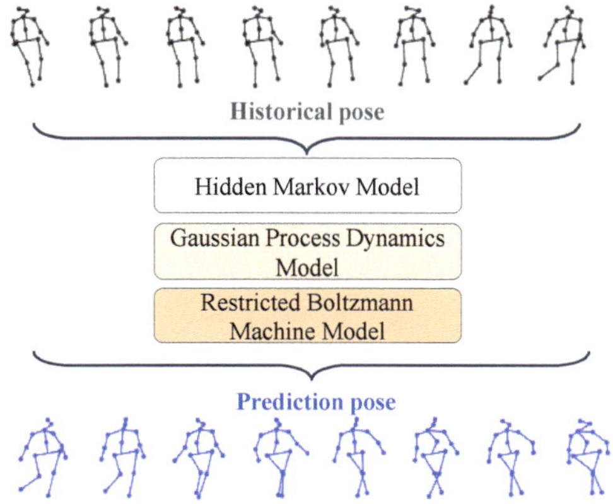

Figure 1. The process of human motion prediction. The top layer represents the historical pose data; the middle layer shows the classical methods used; and the bottom layer shows the output result, i.e., the predicted pose.

(a) Spatial relationship modeling: In most previous studies, spatial joint graphs were designed based on the human physical structure, typically utilizing graph neural networks (GNNs) [25] to capture spatial correlations. However, GNNs are limited by the local and linear aggregation of node features and may not effectively capture the global and nonlinear dynamics of human motion. The introduction of adaptive graphs aimed to overcome these limitations, but they still have drawbacks, such as overlooking the correlation between critical 3D coordinate information, which results in a loss of relevant internal data feature information.

(b) Simultaneously capturing complex short-term and long-term temporal dependencies: Most research has employed temporal learning components to capture temporal correlations. RNNs are a classic approach, but they face gradient vanishing or exploding issues when learning long time sequences. More advanced models such as LSTM and GRU mitigate the issue of vanishing gradients to a certain degree, but pose challenges in training and lack a parallel computation capability. Self-attention mechanisms [26,27] attempt to capture temporal dependencies but still struggle to effectively model long-range dependencies. TCNs [22,28] capture long-term dependencies through fixed kernel sizes, adopting an independent module framework that can only capture single dependency relationships from a temporal scale perspective. Fixed receptive fields limit their ability to adaptively learn multi-scale temporal dependencies.

To tackle these challenges in human motion prediction, a novel method based on dual attention and multi-granularity temporal convolutional networks (DA-MgTCNs) was proposed. This approach effectively captures spatial correlations and multi-scale temporal dependencies. Specifically, joint attention and channel attention were combined to design a dual-attention structure for extracting spatial features and capturing information on spatial correlations between and within human joints. TCNs were employed to model long-term temporal dependencies, and the concept of multi-granularity was introduced into the TCN to further enhance performance. The multi-granularity TCN (MgTCN) employed convolution kernels of varying scales in its convolution operations across multiple branches, enabling it to effectively capture multi-scale temporal dependencies in a flexible manner.

The MgTCN module was comprised of a combination of multi-granularity causal convolutions, dilated convolutions, and residual connections. Each branch of the module was composed of multiple causal convolution layers with varying dilation factors. This design enabled the adaptive selection of different receptive fields based on varying motion styles and joint trajectory features for short-term and long-term human motion prediction.

The main contributions of this paper are as follows:

(1) We designed a dual-attention model for extracting inter-joint and intra-joint spatial features, more effectively mining spatial relationships between joints and different motion styles, providing richer information sources for motion prediction.

(2) We introduced a multi-granularity temporal convolutional network (MgTCN) that employed multi-channel TCNs with different receptive fields for learning, thus achieving discriminative fusion at different time granularities, flexibly capturing complex short-term and long-term temporal dependencies, and thereby further improving the model's performance.

(3) We conducted extensive experiments on the Human3.6M and CMU-MoCap datasets, demonstrating that our method outperformed most state-of-the-art approaches in short-term and long-term prediction, verifying the effectiveness of the proposed algorithm.

The remainder of this paper is organized as follows: Section 2 reviews related work. Section 3 details the proposed methodology. In Section 4, we describe experiments conducted on two large-scale datasets, comparing the performance of the proposed method with baselines. Section 5 provides a summary and conclusion, as well as a discussion of future work.

2. Related Work

In this section, we review the literature relevant to our dual-attention multi-granularity temporal convolutional networks (DA-MgTCNs) model, focusing on existing methods for human motion prediction, temporal convolutional networks (TCNs), multi-granularity (Mg) convolutions, and attention mechanisms.

2.1. Human Motion Prediction

The development of human motion prediction has evolved through several phases. Traditional methods primarily rely on statistical approaches, such as hidden Markov models (HMMs) [8], Gaussian processes (GPs) [9], and restricted Boltzmann machines [10], to learn underlying patterns and structures from data in order to predict future human motion [15]. Although these methods have achieved some success in certain scenarios, they still face challenges in capturing complex spatial and temporal dependencies, computational efficiency, and scalability.

With the rapid development of deep learning, researchers have started to apply it to human motion prediction tasks. Recurrent neural networks (RNNs) have been widely adopted for the temporal information modeling of human motion. Some representative works include Fragkiadaki et al. [15]'s RNN model, Martinez et al. [17]'s RNN-based joint angle prediction model, and Li et al. [11]'s convolutional recurrent neural network (CRNN) model. Although RNNs have achieved high accuracy in human motion prediction, they may lead to error accumulation and eventually converge to a statical average pose due to the continuous computation of time series.

To address this issue, researchers have improved RNNs. Chiu et al. [29] used LSTM units to model the underlying structure of human motion hierarchically, but this method did not adequately capture the spatial structure of the human body. Martinez et al. [17] introduced a residual structure using GRUs to model the velocity of human motion sequences, focusing on short-term temporal modeling but ignoring long-term dependencies and spatial structure. Jain et al. [16] combined LSTM and fully connected (FC) layers in a structural RNN model to encode high-level spatio-temporal structures in human motion sequences. Guo et al. [30] employed FC layers and GRUs to model local structures and capture long-term temporal dependencies, but they did not account for the interactions between different limbs. These RNN-based models faced challenges in capturing long-term dependencies and error accumulation.

2.2. Temporal Convolutional Networks

The temporal convolutional network (TCN) was developed to address these issues. The fundamental TCN architecture includes causal convolution, dilated convolution, and residual blocks [31]. Compared to RNNs and LSTMs, TCNs offer the advantages of parallel computation and larger receptive fields. Recent research has shown that 1D convolution can effectively represent time-series data [31–33], achieving significant success in various sequence learning tasks, such as machine translation [34], speech synthesis [35], video analysis [36], and semantic segmentation [37]. The contextual size of the network may be easily increased by stacking numerous one-dimensional convolutional layers, and creating hierarchical feature representations for input sequences enables the efficient modeling of long-term temporal patterns [35,36].

2.3. Multi-Granularity Convolution

A single-scale TCN might not be sufficient to capture the multi-scale temporal correlations in motion sequences in human motion prediction tasks. To capture complicated short-term and long-term temporal connections, researchers have developed a method known as multi-granularity convolution, which can combines multi-scale information fusion [38]. By adjusting the convolutional size, CNN-based deep learning models can quickly gather feature information at various granularities, enabling more accurate decision making by combining and evaluating data from various scales. Recent achievements in the field of computer vision have fully exploited multi-granularity information fusion based on CNNs [39].

2.4. Attention Mechanisms

Additionally, understanding spatial relationships is essential when attempting to predict human motion. In order to overcome this constraint, attention mechanisms were incorporated into the model to extract the spatial correlations of joints. There is still untapped potential in the area of human motion prediction, despite the widespread use of attention mechanisms in natural language processing [40,41] and image processing [42]. Tang et al. [43] employed the attention module for information extraction along the temporal dimension, and Cai et al. [44] used it for global spatial dependency among joint trajectories. However, it is believed that the intrinsic three-dimensional coordinate information of the human body is crucial for spatial representation.

3. Approach

3.1. Problem Formulation

Our goal was to forecast future human posture sequences based on previous 3D human pose sequences. Three-dimensional joint positions were employed as the pose representation to prevent the ambiguity produced by the joint angle representation. A graphical representation of the human pose was created by analysing the properties of human joint positions over time. Let $x_{1:T} = [x_1, x_2, \ldots, x_T]$ represent the set of joint positions for T time steps, where $x_i \in \mathbb{R}^{J \times C}$, T specifies the number of input time steps; J the number of human pose joints; and $C = 3$ the feature dimension (x, y, z). Our goal was to anticipate the pose's future steps $x_{T+1:T+N} = [x_{T+1}, x_{T+2}, \ldots, x_{T+N}]$. We began by copying the latest pose $x_T N$ times to build a time series of length $T + N$, as described in the literature [25,45]. As a result, the goal became generating a time series of length $T + N$ from the input sequence $x_{1:T+N} = [x_1, x_2, \ldots, x_T; x_T, \ldots, x_T]$ to produce the output sequence $\hat{x}_{1:T+N}$, where x_i is commonly designated as the 3D coordinates of N body joints.

3.2. Overview

We employed a residual depth network consisting of DA-MgTCN modules to capture the global spatial correlation and multi-scale temporal dependence of human motion. Each DA-MgTCN module consisted of a two-branch attention structure module (DA) and a multi-granularity TCN module (MgTCN) connected in series to capture the inter-temporal dependence of historical motion sequences. The DA module was used to extract spatially significant information from joint-level and channel-level dimensions. A combination of multi-granularity convolution and TCN was used in the MgTCN module to increase the prediction quality and adapt to varied forms of human motion and multi-scale time. The complete model architecture was trained end-to-end, with global and local residual connections improving the deep neural network's performance. Each DA-MgTCN component is described in detail below.

$$Y_{DA-MgTCN} = MgTCN(DA(X)) \tag{1}$$

Figure 2 shows a detailed description of the module. The specifics of the DA and MgTCN modules are provided below.

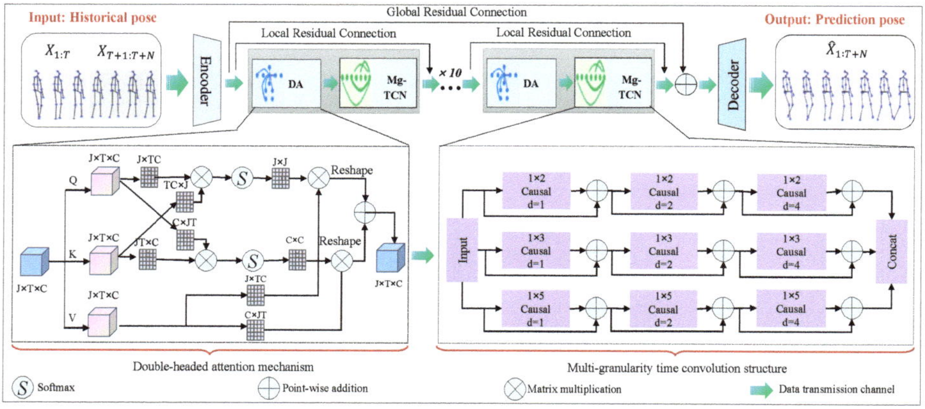

Figure 2. The whole architecture of our suggested solution for motion prediction, which employed an end-to-end framework. We encoded human poses $X_{1:T+N}$ and fed them into the DA-MgTCN, which was a series connection of DA and MgTCN modules. The DA module was used to extract spatially important information from dimensions at the joint and channel levels. The MgTCN was used to capture different scales of temporal dependencies. Finally, the decoder module recovered the time dimension length.

3.3. Dual Attention (DA)

The self-attentive mechanism is regarded as an efficient method for modeling remote dependencies. Tang et al. [43] and Cai et al. [44] used the attention module for information extraction along the temporal dimension and the modeling of global spatial dependencies, respectively. However, we observed that the 3D coordinate information from human joints is crucial for spatial representations.

As a result, we proposed a dual-attention module that took into account both joint-level attention and channel-level attention in order to extract joint-related and channel-related information for spatial correlation. The DA module is depicted in the lower left corner of Figure 2 and is described in detail below.

Given a human motion feature X, a linear transformation was first performed using the weight matrices W^q, W^k, and W^v to obtain the query Q, the key K, and the value V. These two branches shared the same embeddings: Q, K, and V. The embeddings are partially reorganized into $J \times CT$ (for Q, K, and V in the joint branch) and $C \times JT$ (for Q, K, and V in the channel branch) dimensions. The joint and channel attention were used to simultaneously mine the dependencies between joints in the space and channel dimensions. This was computed as follows:

$$\begin{cases} Q = W^q X, \\ K = W^k X, \\ V = W^v X \end{cases} \tag{2}$$

$$\begin{aligned} F_J &= Attention(Q_{(J)}, K_{(J)}, V_{(J)}) \\ &= softmax(\frac{Q_{(J)} K_{(J)}^T}{\sqrt{d_k}}) V_{(J)} \end{aligned} \tag{3}$$

$$\begin{aligned} F_C &= Attention(Q_{(C)}, K_{(C)}, V_{(C)}) \\ &= softmax(\frac{Q_{(C)} K_{(C)}^T}{\sqrt{d_k}}) V_{(C)} \end{aligned} \tag{4}$$

where $Q_{(J)}, K_{(J)}, V_{(J)}, Q_{(C)}, K_{(C)}$, and $V_{(C)}$ represent the deformations of the Q, K, and V matrices, respectively. W^q, W^k, and W^v are trainable weights, and d_k is the dimension of K. J and C denote joint-level and channel-level branches, respectively. F_J and F_C are the output features of the joint-level and channel-level networks. After obtaining the joint-level and channel-level features, we summed them one by one to obtain the spatially noticed feature representation $\hat{X}$ of the MgTCN, as shown in Equation (5):

$$\hat{X} = F_J \oplus F_C \tag{5}$$

After obtaining the spatially noticed feature representation $\hat{X}$ of the motion data, we could feed this representation into the subsequent layers of the network. This process helped to capture joint-level and channel-level contextual information, which is crucial for effective motion prediction modeling.

3.4. Multi-Granularity TCN (MgTCN)

To learn human motion temporal features efficiently, we extended the concept of temporal multi-granularity convolutional kernels to TCN networks and proposed MgTCNs for extracting temporal features at multiple scales for different motion styles. The MgTCN module is shown in the lower right corner of Figure 2 and consisted of multi-granularity causal convolution, dilated convolution, and residual blocks. There were three causal convolution channels in the MgTCN, each using kernels with granularity sizes of 2, 3, and 5 for feature extraction. Each channel consisted of three residual blocks connected in series. These units increased the perceptual field at a dilation rate of [1, 2, 4] and used ReLU as the activation function. In addition, a dropout unit was included in each residual block for regularization.

Causal convolution: The output at the tth timestamp for standard 1D convolution is calculated from the k elements around the previous layer with time step t, which is not reasonable for the human motion prediction task [31]. The goal of this research was to find the best function for generating human-like future poses based on previous motion capture sequences. As a result, the predicted pose at time step t could be derived only from all possible representations of previously observed frames and not from later poses. MgTCN's causal convolution ensured that only past data were used as the model input, preventing future information leakage. This was easily accomplished by shifting the standard convolution output by a few time steps, as shown in the equation below:

$$y[t] = \sum_{i=0}^{k} x[t-i] * w[i] \quad (6)$$

where $y[t]$ is the output, $x[t-i]$ are the inputs, $w[i]$ is the convolution weight at time step i, and k is the kernel size.

Dilated convolution: Causal convolution captures historical data inadequately. Increasing the network's depth or number of layers can help it capture historical data linearly. However, increasing the network depth exponentially increases the number of parameters, making network training more difficult. Oord et al. [35] suggested using dilation convolution to extend causal convolutional networks' receptive field to better capture historical information.

Dilated convolution is implemented by adding expansion parameters to the moving convolutional kernel. Compared to traditional deep convolutional networks, dilation convolution can obtain a larger receptive field without significantly increasing the number of parameters, thus capturing information over a longer time range. This approach can focus on both local details and motion trends over a longer time span when dealing with human motion prediction tasks.

Dilated causal convolution can be expressed by the following equation. For a filter $f = (0, \ldots, k-1)$ and $\mathbf{x} \in \mathbb{R}^T$, denoting the given 1D time-series input, the dilated convolution operation F on element s of the sequence is computed as:

$$F_{k,d}(\mathbf{x}) = \sum_{i=0}^{k-1} f(i) \cdot \mathbf{x}(s - d \cdot i) \quad (7)$$

where d is the dilation factor, k is the size of the filter, and the convolution kernel is restricted to slide only at the current position and to its left (i.e., past information).

The receptive field R for a three-layer convolution is calculated as:

$$R = 1 + (k-1) \times d_1 + (k-1) \times d_2 + (k-1) \times d_3 \quad (8)$$

where d_1, d_2, and d_3 are the dilation factors of the three-layer convolution, which are used to calculate the size of the receptive field.

Our TCN was calculated as:

$$\begin{cases} TCN_{k,d=1} = x + F_{k,d=1}(x), \\ TCN_{k,d=2} = TCN_{k,d=1} + F_{k,d=2}(TCN_{k,d=1}), (k = 2,3,5) \\ TCN_k = TCN_{k,d=4} = TCN_{k,d=2} + F_{k,d=4}(TCN_{k,d=2}) \end{cases} \quad (9)$$

Figure 3 shows an example with a three-layer causal expansion convolutional network (TCN). The TCN's elements in Figure 3 include a series of dilation causal convolutions with dilation factors $d = 1, 2, 4$ and a filter size $K = 3$.

Multi-granularity convolution: In order to handle complex, multi-action, multi-joint predictions of the human body, MgTCN required the use of convolutional kernel filters with different granularities to extract time-series features at different scales. This was necessary to meet the needs of short- and long-term predictions that require the capture of time-series features of different lengths. Three MgTCN time series were processed separately, which made it possible to combine multiple time granularities in the feature extraction process, which could better represent a large range of spatio-temporal features. Therefore, integrating time-series data with different time granularities to obtain better results is a challenge.

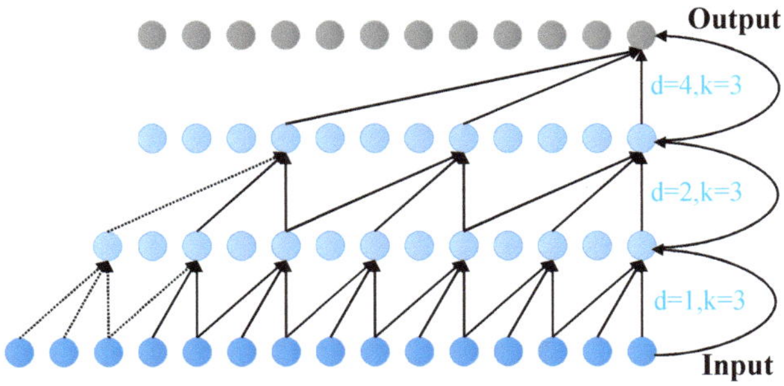

Figure 3. Architectural elements in a TCN (causal dilated convolutional network).

In order to handle the complex multi-action and multi-joint prediction of the human body, one must integrate time-series data with different time granularities to obtain better results. MgTCN used convolutional kernel filters with different granularities to extract time-series features at different scales. This satisfied the need for capturing short- and long-term forecasts of time-series features of different lengths. Three time series were treated separately in MgTCN, which allowed us to combine multiple temporal granularities in the feature extraction process to better represent certain large ranges of spatio-temporal features.

The MgTCN network output could be used to extract multi-temporal granularity features (short-term and long-term) using the aforementioned spatial and temporal feature extraction steps. We combined the data from these three TCN channels and used the equation below to make predictions in order to achieve the integration of the multi-granularity information.

$$Fusion = Cat(w_1 \times TCN_{k=2}, w_2 \times TCN_{k=3}, w_3 \times TCN_{k=5}) \quad (10)$$

$$MgTCN = g(Fusion) \quad (11)$$

where w_i is a learnable parameter to adjust the weights for different time periods, and $g(.)$ represents a mapping function that maps the fused features to the predicted values.

With this multi-granularity temporal convolution (MgTCN) method, we could both observe the general trend of human motion in the long-term and capture the outliers of short-term changes. This temporal correlation facilitated predictive power.

3.5. Global and Local Residual Connection

Residual connection skips a layer of the network and adds its output to the next layer's output. This eliminates gradient fading by propagating the gradient straight from the back layer to the front layer. This architecture simplifies neural network representation learning in deeper structures.

Figure 2 illustrates the use of global residual connections between the encoder and decoder modules and local residual connections in each DA-MgTCN module to enhance neural network training and deeper structural performance. This method assisted the network in capturing complex data patterns in human motion prediction.

3.6. Loss Function

To train our DA-MgTCN model, we employed an end-to-end training technique. The mean position per joint error (MPJPE) loss function between the anticipated motion sequence and the ground truth motion sequence was used to analyze the difference between the predicted outcomes and the true pose, which was defined as follows:

$$L = \frac{1}{N}\sum_{i=1}^{N}\sum_{j=1}^{T}||\hat{Y}_{i,j} - Y_{i,j}||_2^2 \quad (12)$$

where N is the number of human joints, T is the number of time steps in the future series, $\hat{Y}_{i,j} \in \mathbb{R}^C$ is the prediction of the ith joint at the jth time step, and $Y_{i,j}$ is the corresponding ground truth.

We optimized the loss function using the improved Adam method (AdamW [46]), which mitigated the overfitting problem by adding a weight decay term and could significantly improve the robustness of the model.

4. Experiments

In this section, we evaluate the performance of the proposed method using two large-scale human motion capture benchmark datasets: Human3.6M and CMU-Mocap.

4.1. Datasets

Human3.6M [47] is the largest existing human motion analysis database, consisting of 7 actors (S1, S5, S6, S7, S8, S9, and S11) performing 15 actions: walking, eating, smoking, discussing, directions, greeting, phoning, posing, purchases, sitting, sitting Down, taking photos, waiting, walking a dog, and walking together. Some actions are periodic, such as walking, while others are non-periodic, such as taking photos. Each pose includes 32 joints, represented in the form of an exponential map. By converting these into 3D coordinates, eliminating redundant joints, global rotation, and translation, the resulting skeleton retains 17 joints that provide sufficient human motion details. These joints include key ones that locate major body parts (e.g., shoulders, knees, and elbows). This strategy ensures that no crucial joints are overlooked. We downsampled the frame rate to 25 fps and used S5 and S11 for testing and validation, while the remaining five actors were used for training.

CMU-MoCap, available at http://mocap.cs.cmu.edu/, accessed on 13 June 2023, is a 3D human motion dataset released by Carnegie Mellon University that used 12 Vicon infrared MX-40 cameras to record the positions of 41 sensors attached to the human body, describing human motion. The dataset can be divided into six motion themes, including human interaction, interaction with environment, locomotion, physical activities and sports, situations and scenarios, and test motions.

These motion themes can be further subdivided into 23 sub-motion themes. The same data preprocessing method as in the literature [25] was adopted, simplifying each human body and reducing the motion rate to 25 frames per second. Furthermore, eight actions (basketball, basketball signals, directing traffic, jumping, running, soccer, walking, and washing the face) were selected from the dataset to evaluate the model's performance. No hyperparameters were adjusted in this dataset, and we only used the training and testing sets, applying a splitting method consistent with the common practice in the literature.

4.2. Implementation Details

All experiments in this paper were implemented using the PyTorch deep learning framework. The experimental environment was Ubuntu 20.04 with an NVIDIA A100 GPU. During the training process, the batch normalization size was set to 16, and the AdamW optimizer was used to optimize the model. The initial learning rate was set to 0.003, with decay by 5% every 5 epochs. The model was trained for 60 epochs, and each experiment was conducted three times. The average result was taken to ensure a more robust evaluation of the model's performance. The input motion prediction length was 25 frames (1000 ms), and the prediction generated 25 frames (1000 ms). The choice and configuration of the relevant hyperparameters are shown in Table 1.

Table 1. The choice and configuration of the relevant hyperparameters.

Hyperparameter/Config	Value
Optimizer	AdamW
Base learning rate	3×10^{-3}
Weight decay	10^{-2}
Optimizer momentum	$\beta_1 = 0.9, \beta_2 = 0.999$
Batch size	16
Warmup epochs	5
Epochs	60
Layer	10

4.3. Evaluation Metrics and Baselines

The same evaluation metrics as those used in existing algorithms [25,45] were employed for assessing model performance. The standard mean per joint position error (MPJPE) was used to measure the average Euclidean distance (in millimeters, mm) between the predicted joint 3D coordinates and the ground truth, as illustrated in Equation (12). In addition, to further illustrate the advantages of the method, we conducted a comparative analysis of our method with Res. sup. [17], convSeq2Seq [11], DMGNN [13], LTD [25], LPJP [44], Hisrep [48], MSR [49], and ST-DGCN [45].

4.4. Experimental Results and Analysis

Human3.6M: Based on the existing work, we divided the prediction results into short-term (80–400 ms) and long-term predictions (500–1000 ms). The experimental results are shown in Table 2, which demonstrates the joint position error and mean error for short-term (80 ms, 160 ms, 320 ms, 400 ms) and long-term (560 ms, 1000 ms) predictions for 15 kinds of movements. It was found that the existing methods usually showed high prediction accuracy when dealing with more periodic and regular movements, such as "walking" and "eating". However, when dealing with more random and irregular movements, such as "directions", "posing", and "purchases", the prediction accuracy decreased significantly. The algorithm proposed in this paper showed high prediction accuracy when dealing with highly complex, non-periodic, and irregular movements.

Table 2. Prediction of 3D joint positions in Human 3.6M for all actions. The best results are marked in bold.

Time (ms)	80	160	320	400	560	1000	80	160	320	400	560	1000
Action	Walking						Eating					
Res. sup. [17]	29.4	50.8	76.0	81.5	81.7	100.7	16.8	30.6	56.9	68.7	79.9	100.2
convSeq2Seq [11]	17.7	33.5	56.3	63.6	72.2	82.3	11.0	22.4	40.7	48.4	61.3	87.1
DMGNN [13]	17.3	30.7	54.6	65.2	73.4	95.8	11.0	21.4	36.2	43.9	58.1	86.7
LTD [25]	12.3	23.0	39.8	46.1	54.1	59.8	8.4	16.9	33.2	40.7	53.4	77.8
MSR [49]	12.2	22.7	38.6	45.2	52.7	63.0	8.4	17.1	33.0	40.4	52.5	77.1
Hisrep [48]	**10.0**	19.5	34.2	**39.8**	47.4	58.1	**6.4**	**14.0**	**28.7**	**36.2**	50.0	75.7
ST-DGCN [45]	10.2	19.8	34.5	40.3	48.1	56.4	7.0	15.1	30.6	38.1	51.1	76.0
Our model	10.1	**19.2**	**33.8**	40.2	**46.1**	**55.4**	7.0	14.3	30.2	38.5	**48.9**	**72.6**
Action	Smoking						Discussion					
Res. sup. [17]	23.0	42.6	70.1	82.7	94.8	137.4	32.9	61.2	90.9	96.2	121.3	161.7
convSeq2Seq [11]	11.6	22.8	41.3	48.9	60.0	81.7	17.1	34.5	64.8	77.6	98.1	129.3
DMGNN [13]	9.0	17.6	32.1	40.3	50.9	72.2	17.3	34.8	61.0	69.8	81.9	138.3
LTD [25]	7.9	16.2	31.9	38.9	50.7	72.6	12.5	27.4	58.5	71.7	91.6	121.5
MSR [49]	8.0	16.3	31.3	38.2	49.5	71.6	12.0	26.8	57.1	69.7	88.6	117.6
Hisrep [48]	7.0	14.9	29.9	36.4	47.6	69.5	10.2	**23.4**	**52.1**	65.4	86.6	119.8
ST-DGCN [45]	6.6	**14.1**	28.2	34.7	46.5	69.5	10.0	23.8	53.6	66.7	87.1	118.2

Table 2. *Cont.*

ours	**6.5**	14.6	**28.0**	**33.8**	**46.1**	**66.7**	**9.8**	24.2	54.5	**65.1**	**83.1**	**114.8**
Time (ms)	**80**	**160**	**320**	**400**	**560**	**1000**	**80**	**160**	**320**	**400**	**560**	**1000**
Action			**Directions**						**Greeting**			
Res. sup. [17]	35.4	57.3	76.3	87.7	110.1	152.5	34.5	63.4	124.6	142.5	156.1	166.5
convSeq2Seq [11]	13.5	29.0	57.6	69.7	86.6	115.8	22.0	45.0	82.0	96.0	116.9	147.3
DMGNN [13]	13.1	24.6	64.7	81.9	110.1	115.8	23.3	50.3	107.3	132.1	152.5	157.7
LTD [25]	9.0	19.9	43.4	53.7	71.0	101.8	18.7	38.7	77.7	93.4	115.4	148.8
MSR [49]	8.6	19.7	43.3	53.8	71.2	100.6	16.5	37.0	77.3	93.4	116.3	147.2
Hisrep [48]	7.4	18.4	44.5	56.5	73.9	106.5	**13.7**	**30.1**	**63.8**	**78.1**	**101.9**	138.8
ST-DGCN [45]	7.2	17.6	40.9	51.5	69.3	100.4	15.2	34.1	71.6	87.1	110.2	143.5
Our model	**6.9**	**17.0**	**40.7**	**49.0**	**68.0**	**98.5**	14.6	33.3	68.5	86.4	112.0	**135.9**
Action			**Phoning**						**Posing**			
Res. sup. [17]	38.0	69.3	115.0	126.7	141.2	131.5	36.1	69.1	130.5	157.1	194.7	240.2
convSeq2Seq [11]	13.5	26.6	49.9	59.9	77.1	114.0	16.9	36.7	75.7	92.9	122.5	187.4
DMGNN [13]	12.5	25.8	48.1	58.3	78.9	98.6	15.3	29.3	71.5	96.7	163.9	310.1
LTD [25]	10.2	21.0	42.5	52.3	69.2	103.1	13.7	29.9	66.6	84.1	114.5	173.0
MSR citedang2021msr	10.1	20.7	41.5	51.3	68.3	104.4	12.8	29.4	67.0	85.0	116.3	174.3
Hisrep [48]	8.6	18.3	39.0	49.2	67.4	105.0	**10.2**	**24.2**	**58.5**	75.8	107.6	178.2
ST-DGCN [45]	8.3	18.3	**38.7**	48.4	65.9	102.7	10.7	25.7	60.0	76.6	106.1	164.8
Our model	**8.3**	**18.1**	39.2	**47.9**	**64.5**	**95.6**	10.4	25.4	60.5	**74.8**	**103.2**	**162.2**
Action			**Purchases**						**Sitting**			
Res. sup. [17]	36.3	60.3	86.5	95.9	122.7	160.3	42.6	81.4	134.7	151.8	167.4	201.5
convSeq2Seq [11]	20.3	41.8	76.5	89.9	111.3	151.5	13.5	27.0	52.0	63.1	82.4	120.7
DMGNN [13]	21.4	38.7	75.7	92.7	118.6	153.8	11.9	25.1	44.6	50.2	60.1	104.9
LTD [25]	15.6	32.8	65.7	79.3	102.0	143.5	10.6	21.9	46.3	57.9	78.3	119.7
MSR [49]	14.8	32.4	66.1	79.6	101.6	139.2	10.5	22.0	46.3	57.8	78.2	120.0
Hisrep [48]	13.0	29.2	60.4	73.9	95.6	134.2	9.3	20.1	44.3	56.0	76.4	115.9
ST-DGCN [45]	**12.5**	**28.7**	60.1	73.3	95.3	133.3	8.8	19.2	42.4	53.8	74.4	116.1
Our model	12.6	29.1	**59.0**	**72.4**	**91.6**	**128.3**	**8.4**	**18.5**	**40.4**	**52.9**	**72.0**	**113.6**
Action			**Sitting Down**						**Taking Photo**			
Res. sup. [17]	47.3	86.0	145.8	168.9	205.3	277.6	26.1	47.6	81.4	94.7	117.0	143.2
convSeq2Seq [11]	20.7	40.6	70.4	82.7	106.5	150.3	12.7	26.0	52.1	63.6	84.4	128.1
DMGNN [13]	15.0	32.9	77.1	93.0	122.1	168.8	13.6	29.0	46.0	58.8	91.6	120.7
LTD [25]	16.1	31.1	61.5	75.5	100.0	150.2	9.9	20.9	45.0	56.6	77.4	119.8
MSR [49]	16.1	31.6	62.5	76.8	102.8	155.5	9.9	21.0	44.6	56.3	77.9	121.9
Hisrep [48]	14.9	30.7	59.1	72.0	97.0	**143.6**	8.3	18.4	40.7	51.5	72.1	**115.9**
ST-DGCN [45]	13.9	27.9	**57.4**	**71.5**	96.7	147.8	8.4	18.9	42.0	53.3	74.3	118.6
Our model	**13.8**	**27.0**	58.1	72.2	**95.7**	143.7	**8.2**	**18.1**	**40.6**	**51.2**	**70.9**	117.1
Action			**Waiting**						**Walking Dog**			
Res. sup. [17]	30.6	57.8	106.2	121.5	146.2	196.2	64.2	102.1	141.1	164.4	191.3	209.0
convSeq2Seq [11]	14.6	29.7	58.1	69.7	87.3	117.7	27.7	53.6	90.7	103.3	122.4	162.4
DMGNN [13]	12.2	24.2	59.6	77.5	106.0	136.7	47.1	93.3	160.1	171.2	194.0	182.3
LTD [25]	11.4	24.0	50.1	61.5	79.4	108.1	23.4	46.2	83.5	96.0	111.9	148.9
MSR [49]	10.7	23.1	48.3	59.2	76.3	106.3	20.7	42.9	80.4	93.3	111.9	148.2
Hisrep [48]	**8.7**	**19.2**	**43.4**	54.9	74.5	108.2	20.1	40.3	73.3	**86.3**	108.2	146.9
ST-DGCN [45]	8.9	20.1	43.6	54.3	72.2	103.4	18.8	39.3	73.7	86.4	**104.7**	139.8
Our model	9.2	19.9	43.6	**53.0**	**67.3**	**100.8**	**18.5**	**37.7**	**72.8**	87.6	105.8	**137.2**

Table 2. *Cont.*

Action	**Waiting Together**						**Average**					
Res. sup. [17]	26.8	50.1	80.2	92.2	107.6	131.1	34.7	62.0	101.1	115.5	97.6	130.5
convSeq2Seq [11]	15.3	30.4	53.1	61.2	72.0	87.4	16.6	33.3	61.4	72.7	90.7	124.2
DMGNN [13]	14.3	26.7	50.1	63.2	83.4	115.9	17.0	33.6	65.9	79.7	103.0	137.2
LTD [25]	10.5	21.0	38.5	45.2	55.0	65.6	12.7	26.1	52.3	63.5	81.6	114.3
MSR [49]	10.6	20.9	37.4	43.9	52.9	65.9	12.1	25.6	51.6	62.9	81.1	114.2
Hisrep [48]	8.9	**18.4**	35.1	41.9	52.7	64.9	10.4	22.6	47.1	58.3	77.3	112.1
ST-DGCN [45]	**8.7**	18.6	34.4	41.0	51.9	64.3	10.3	22.7	47.4	58.5	76.9	110.3
Our model	8.7	18.5	**33.5**	**40.5**	**50.8**	**61.4**	**10.2**	**22.3**	**46.9**	**57.7**	**75.1**	**106.9**

Our experimental results revealed that the proposed DA-MgTCN method outperformed most baseline methods in both short-term and long-term motion prediction. In particular, it can be observed from the experimental results that the proposed DA-MgTCN method outperformed most baseline methods in short-term motion prediction and improved more significantly in long-term prediction, with each MPJPE index reaching the optimum and obtaining excellent prediction results for both the 560 mm and 1000 mm MPJPE metrics. This success can be attributed to the ability of DA-MgTCN to fully capture spatial correlation and multi-granularity temporal features, which was a key factor in enhancing the model's prediction accuracy.

Qualitative comparison: We visualized the results of the aforementioned motion prediction to further assess the model's performance. Figure 4 illustrates the visualization results for actions including "walking", "discussion", "posing", and "sitting down". The first row in every subplot shows the ground truth pose sequences (in black), followed by the predicted poses (in blue), i.e., each row displays the prediction results of one model. From the visualization results, it was observed that the predictions generated by the DA-MgTCN method showed higher similarity to the actual sequences and exhibited lower distortion and better continuity between frames. This was due to the dual-branch spatial attention and multi-granularity temporal convolution modeling joint motion trajectories, which provided richer and smoother joint motion temporal context information. The model could sufficiently capture global spatial dependencies, allowing it to encode joint information with distant hidden dependencies. For example, in the "sitting down" motion visualization, the motion between the hands and feet was more coordinated and coherent. This demonstrated once again how well the suggested DA-MgTCN forecasted very complicated irregular movements and complex periodic motions.

CMU-MoCap: To further validate the generalization of the DA-MgTCN method, we compared its performance with existing algorithms on the CMU-MoCap dataset, including Res. sup. [17], convSeq2Seq [11], DMGNN [13], LTD [25], LPJP [44], MSR [49], and ST-DGCN [45]. The experimental results are shown in Table 3, presenting the mean per joint position error and corresponding average error for short-term and long-term predictions across eight actions. From the table, it can be observed that the DA-MgTCN method's short-term and long-term prediction accuracy was significantly higher than that of the other seven existing prediction algorithms, including Cai et al. [44] method, even when handling relatively complex non-periodic actions. The DA-MgTCN method improved the average prediction accuracy by about 1.5% in short-term prediction and 3% in long-term prediction, respectively, compared to the state-of-the-art ST-DGCN method. Thus, the comprehensive experimental results once again confirmed the effectiveness and generalization capabilities of the DA-MgTCN method.

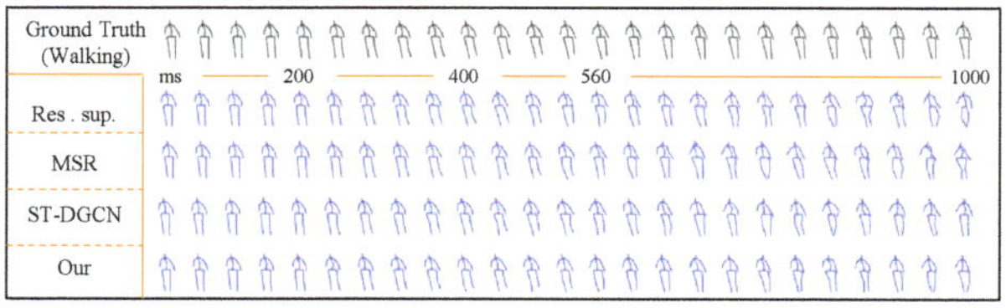

(**a**) Walking

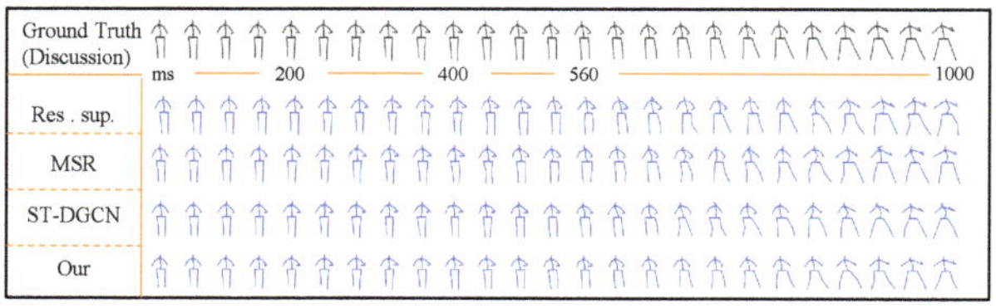

(**b**) Discussion

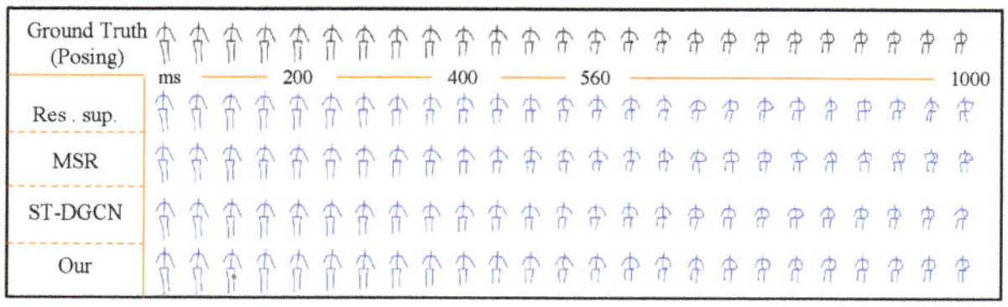

(**c**) Posing

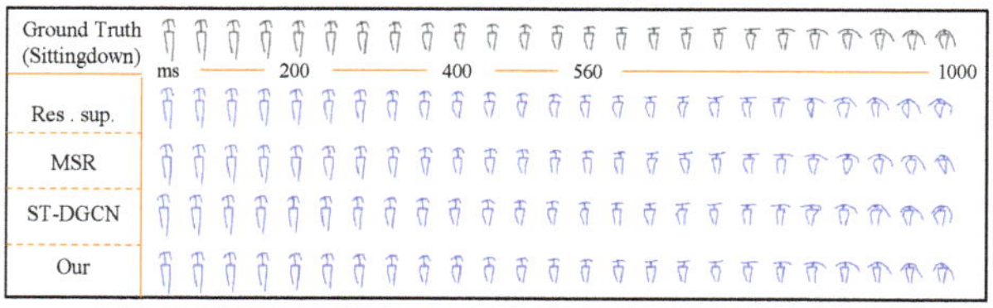

(**d**) Sitting down

Figure 4. Qualitative comparison.

Table 3. Short- and long-term prediction of 3D body poses on CMU-Mocap. All results are in millimeters. The best results are marked in bold.

Time (ms)	80	160	320	400	560	1000
Res. sup. [17]	24.0	43.0	74.5	87.2	105.5	136.3
convSeq2Seq [11]	12.5	22.2	40.7	49.7	—	84.6
DMGNN [13]	13.6	24.1	47.0	58.8	77.4	112.6
LTD [25]	9.3	17.1	33.0	40.9	55.8	86.2
LPJP [44]	9.8	17.6	35.7	45.1	-	93.2
MSR [49]	8.1	15.2	30.6	38.6	53.7	83.0
ST-DGCN [45]	7.6	14.3	29.0	36.6	50.9	80.1
Our model (DA-MgTCN)	**7.5**	**14.0**	**28.1**	**34.8**	**49.0**	**77.4**

4.5. Ablation Study

To deeply evaluate the contribution of each component in our model, we conducted a series of ablation experiments on the Human3.6M dataset. These experiments focused on the impact of the channel-attention (channel-att) and multi-grained (Mg) convolution modules on the model's performance. The results of the experiments are shown in Table 4.

Table 4. Influence of the channel-attention (channel-att) and multi-grained (Mg) convolution modules on the Human3.6M and CMU-MoCap datasets. On average, the two components of our model contributed to its accuracy. The best results are marked in bold.

		Human3.6M MPJPE (mm)						CMU-MoCap MPJPE (mm)					
Channel-att	**MgTCN**	**80**	**160**	**320**	**400**	**560**	**1000**	**80**	**160**	**320**	**400**	**560**	**1000**
√		10.4	22.9	48.1	59.0	75.9	108.7	7.7	14.5	29.0	36.1	51.9	81.7
	√	10.5	23.1	48.0	59.4	76.8	111.0	7.9	14.6	29.2	36.7	52.3	82.6
√	√	**10.2**	**22.4**	**46.9**	**57.7**	**74.8**	**106.7**	**7.5**	**14.0**	**28.1**	**34.8**	**49.0**	**79.4**

In terms of channel attention, the prediction accuracy significantly decreased when only joint attention was used without dual attention. The multi-granularity convolutional TCN module showed excellent performance in capturing long-term temporal dependence, thus improving the long-term prediction accuracy. Furthermore, when the channel-att or Mg module was removed, the error at 1000 ms increased by 1.9% and 4.0%, respectively, on the Human3.6M dataset, and by 2.9% and 4.0%, respectively, on the CMU-MoCap dataset. The best performance could be achieved by combining these two components. The multi-granularity model demonstrated better performance compared to the single-granularity model, especially for long prediction cycles. Additionally, the use of learnable weight parameters led to better prediction performance compared to fixed weights. This suggested that by designing a multi-granularity temporal structure, we could extract the temporal correlation between different time periods more effectively, thus improving the prediction performance.

Effects of the Number of DA-MgTCNs: Further, to validate the effect of multiple DA-MgTCNs in the model, we increased the number of DA-MgTCNs from 6 to 10 in step 2 and determined the prediction error and running time cost for both dataset predictions, as shown in Table 5. The experimental results showed that when 6 to 10 DA-MgTCNs were used, the predicted MPJPE decreased, while the time cost continued to increase. When 12 or 14 DA-MgTCNs were used, the prediction error remained stable at a lower level, but the time cost increased. Therefore, the use of 10 MgTCNs was chosen to achieve higher prediction accuracy and operational efficiency.

Table 5. The MPJPE of our model with different numbers of DA-MgTCNs for short-term and long-term prediction on Human3.6M and CMU-MoCap. The best results are marked in bold.

DA-	Human3.6M MPJPE (mm)						CMU-MoCap MPJPE (mm)					
MgTCNs	**80**	**160**	**320**	**400**	**560**	**1000**	**80**	**160**	**320**	**400**	**560**	**1000**
6	11.5	25.0	51.9	65.6	81.1	118.0	8.4	15.5	31.7	39.5	54.3	88.3
8	10.6	23.5	49.0	61.4	78.3	111.1	7.9	14.8	29.8	37.1	52.4	82.9
10	**10.2**	**22.3**	**46.9**	**57.7**	**74.8**	106.7	**7.5**	**14.0**	**28.1**	**34.8**	**49.0**	**77.4**
12	10.3	22.7	47.9	58.6	74.6	**106.5**	7.8	13.8	28.3	35.3	50.9	78.8
14	10.3	22.8	46.8	58.0	76.3	108.1	7.6	14.3	28.9	35.6	49.2	78.4

In summary, the experimental results in this paper revealed the importance of the DA-MgTCN method using the dual-attention and multi-granularity convolutional design in terms of performance improvement. Modeling joint motion trajectories using dual-attention, dual-branch spatial attention, and multi-granularity temporal convolution could provide richer and smoother temporal contextual information related to joint motion, which was conducive to adequately modeling spatial global dependencies and enabling the model to encode joint information with hidden dependencies at a distance, thus improving the overall performance of the model for both short-term and long-term motion prediction.

4.6. Limitations

In addition to the qualitative results presented in Figure 4, challenging cases encountered by the DA-MgTCN model were also investigated. Figure 5 illustrates an example of a predicted skeleton for the "walking a dog" action. It was evident that the last few frames did not perfectly align with the ground truth pose. This misalignment resulted from the high degree of uncertainty inherent in human motion, where a series of past poses can suggest various possible future outcomes. As a result, predicting long-term dependencies between joints and frames becomes more difficult. Furthermore, the experiments were constrained by more realistic data scenarios and experimental conditions, which may have posed challenges to our algorithm's validation. In the future, we will consider motion prediction in more intricate scenarios to investigate novel methods for multi-grain human motion prediction in multi-domain contexts. The aim is to enhance the adaptability and performance of the model.

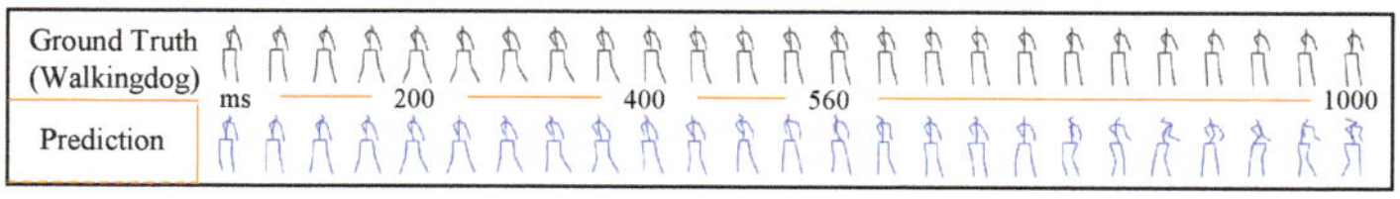

Figure 5. Example of the failure cases of our forecasting approach. The first line indicates the ground-truth 3D human motion. The second line, shown in blue, presents the predicted future motion.

5. Conclusions

In this paper, we proposed a novel human motion prediction method leveraging dual-channel attention and multi-granularity temporal convolutional networks (DA-MgTCNs) to accurately understand and analyze human motion. Our method combined a dual-attention mechanism and a multi-granularity temporal convolutional networks model to address the challenging problem of extracting inter-joint and intra-joint spatial features. Moreover, the multi-granularity temporal convolutional networks model facilitated the design of a TCN with different convolutional kernel granularities, enabling the learning of richer multi-scale temporal information and further enhancing the performance of the model. Extensive experiments were conducted on two large-scale datasets, Human3.6M and CMU-MoCap. The experimental results demonstrated that the proposed method significantly outperformed other approaches in both short-term and long-term prediction tasks, thus validating the effectiveness of the proposed algorithm. In future work, we aim to further optimize the network structure and parameter settings and extend the application of our model to spatio-temporal prediction tasks in real-world scenarios, such as robot perception and interaction.

Author Contributions: Conceptualization, B.H.; methodology, B.H.; software, B.H.; validation, B.H.; formal analysis, B.H.; investigation, B.H.; resources, B.H.; data curation, B.H.; writing—original draft preparation, B.H.; writing—review and editing, B.H.; visualization, B.H.; supervision, X.L.; project administration, X.L.; funding acquisition, X.L. All authors have read and agreed to the published version of the manuscript.

Funding: This work was supported in part by the National Natural Science Foundation of China under grants 62233003 and 62073072, the Key Projects of the Key R&D Program of Jiangsu Province under grants BE2020006 and BE2020006-1, and Shenzhen Natural Science Foundation under grants JCYJ20210324132202005 and JCYJ20220818101206014.

Institutional Review Board Statement: Not applicable.

Informed Consent Statement: Not applicable.

Data Availability Statement: The datasets generated and/or analyzed during the current study are publicly available. The Human3.6M dataset can be accessed through the reference in [47]. The CMU-MoCap dataset is publicly available and can be accessed online at http://mocap.cs.cmu.edu/ (accessed on 13 June 2023). The use of these datasets is governed by their respective usage policies.

Conflicts of Interest: The authors declare no conflict of interest.

Abbreviations

The following abbreviations are used in this manuscript:

DA	Dual attention
TCNs	Temporal convolutional networks
MgTCN	Multi-granularity TCN
GNNs	Graph neural networks

References

1. Chen, S.; Liu, B.; Feng, C.; Vallespi-Gonzalez, C.; Wellington, C. 3d point cloud processing and learning for autonomous driving: Impacting map creation, localization, and perception. *IEEE Signal Process. Mag.* **2020**, *38*, 68–86. [CrossRef]
2. Gui, L.Y.; Zhang, K.; Wang, Y.X.; Liang, X.; Moura, J.M.; Veloso, M. Teaching robots to predict human motion. In Proceedings of the 2018 IEEE/RSJ International Conference on Intelligent Robots and Systems (IROS), Madrid, Spain, 1–5 October 2018; pp. 562–567.
3. Koppula, H.S.; Saxena, A. Anticipating human activities using object affordances for reactive robotic response. *IEEE Trans. Pattern Anal. Mach. Intell.* **2015**, *38*, 14–29. [CrossRef] [PubMed]
4. Sheng, W.; Li, X. Multi-task learning for gait-based identity recognition and emotion recognition using attention enhanced temporal graph convolutional network. *Pattern Recognit.* **2021**, *114*, 107868. [CrossRef]
5. Kong, Y.; Wei, Z.; Huang, S. Automatic analysis of complex athlete techniques in broadcast taekwondo video. *Multimed. Tools Appl.* **2018**, *77*, 13643–13660. [CrossRef]
6. Dong, Y.; Li, X.; Dezert, J.; Zhou, R.; Zhu, C.; Wei, L.; Ge, S.S. Evidential reasoning with hesitant fuzzy belief structures for human activity recognition. *IEEE Trans. Fuzzy Syst.* **2021**, *29*, 3607–3619. [CrossRef]
7. Dong, Y.; Li, X.; Dezert, J.; Zhou, R.; Zhu, C.; Cao, L.; Khyam, M.O.; Ge, S.S. Multi-Source Weighted Domain Adaptation With Evidential Reasoning for Activity Recognition. *IEEE Trans. Ind. Inform.* **2022**, *19*, 5530–5542. [CrossRef]
8. Lehrmann, A.M.; Gehler, P.V.; Nowozin, S. Efficient nonlinear Markov models for human motion. In Proceedings of the Conference on Computer Vision and Pattern Recognition, Columbus, OH, USA, 23–28 June 2014; pp. 1314–1321.
9. Wang, J.M.; Fleet, D.J.; Hertzmann, A. Gaussian Process Dynamical Models for Human Motion. *IEEE Trans. Pattern Anal. Mach. Intell.* **2007**, *30*, 283–298. [CrossRef] [PubMed]
10. Taylor, G.W.; Hinton, G.E.; Roweis, S. Modeling human motion using binary latent variables. *Adv. Neural Inf. Process. Syst.* **2006**, *19*, 1345–1352.
11. Li, C.; Zhang, Z.; Lee, W.S.; Lee, G.H. Convolutional sequence to sequence model for human dynamics. In Proceedings of the IEEE Conference on Computer Vision and Pattern Recognition, Salt Lake City, UT, USA, 18–22 June 2018; pp. 2275–2284.
12. Li, M.; Chen, S.; Chen, X.; Zhang, Y.; Wang, Y.; Tian, Q. Symbiotic Graph Neural Networks for 3D Skeleton-Based Human Action Recognition and Motion Prediction. *IEEE Trans. Pattern Anal. Mach. Intell.* **2021**, *44*, 3316–3333. [CrossRef]
13. Li, M.; Chen, S.; Zhao, Y.; Zhang, Y.; Wang, Y.; Tian, Q. Dynamic multiscale graph neural networks for 3d skeleton based human motion prediction. In Proceedings of the IEEE/CVF Conference on Computer Vision and Pattern Recognition, Seattle, WA, USA, 13–19 June 2020; pp. 214–223.
14. Zhong, C.; Hu, L.; Zhang, Z.; Ye, Y.; Xia, S. Spatio-Temporal Gating-Adjacency GCN For Human Motion Prediction. In Proceedings of the Conference on Computer Vision and Pattern Recognition, New Orleans, LA, USA, 18–24 June 2022; pp. 6447–6456.
15. Fragkiadaki, K.; Levine, S.; Felsen, P.; Malik, J. Recurrent network models for human dynamics. In Proceedings of the 2015 IEEE International Conference on Computer Vision (ICCV), IEEE Computer Society, Santiago, Chile, 7–13 December 2015; pp. 4346–4354.
16. Jain, A.; Zamir, A.R.; Savarese, S.; Saxena, A. Structural-rnn: Deep learning on spatio-temporal graphs. In Proceedings of the 2016 IEEE Conference on Computer Vision and Pattern Recognition (CVPR), Las Vegas, NV, USA, 27–30 June 2016; pp. 5308–5317.
17. Martinez, J.; Black, M.J.; Romero, J. On human motion prediction using recurrent neural networks. In Proceedings of the IEEE Conference on Computer Vision and Pattern Recognition, Honolulu, HI, USA, 21–26 July 2017; pp. 2891–2900.
18. Liu, Z.; Wu, S.; Jin, S.; Liu, Q.; Lu, S.; Zimmermann, R.; Cheng, L. Towards natural and accurate future motion prediction of humans and animals. In Proceedings of the Conference on Computer Vision and Pattern Recognition, Long Beach, CA, USA, 15–20 June 2019; pp. 10004–10012.
19. Shu, X.; Zhang, L.; Qi, G.J.; Liu, W.; Tang, J. Spatiotemporal co-attention recurrent neural networks for human-skeleton motion prediction. *IEEE Trans. Pattern Anal. Mach. Intell.* **2021**, *44*, 3300–3315. [CrossRef]

20. Liu, Z.; Wu, S.; Jin, S.; Ji, S.; Liu, Q.; Lu, S.; Cheng, L. Investigating pose representations and motion contexts modeling for 3D motion prediction. *IEEE Trans. Pattern Anal. Mach. Intell.* **2022**, *45*, 681–697. [CrossRef]
21. Lebailly, T.; Kiciroglu, S.; Salzmann, M.; Fua, P.; Wang, W. Motion prediction using temporal inception module. In Proceedings of the Asian Conference on Computer Vision, Kyoto, Japan, 30 November–4 December 2020.
22. Cui, Q.; Sun, H.; Kong, Y.; Zhang, X.; Li, Y. Efficient human motion prediction using temporal convolutional generative adversarial network. *Inf. Sci.* **2021**, *545*, 427–447. [CrossRef]
23. Mao, W.; Liu, M.; Salzmann, M.; Li, H. Multi-level motion attention for human motion prediction. *Int. J. Comput. Vis.* **2021**, *129*, 2513–2535. [CrossRef]
24. Medjaouri, O.; Desai, K. Hr-stan: High-resolution spatio-temporal attention network for 3d human motion prediction. In Proceedings of the IEEE/CVF Conference on Computer Vision and Pattern Recognition, New Orleans, LA, USA, 18–24 June 2022; pp. 2540–2549.
25. Mao, W.; Liu, M.; Salzmann, M.; Li, H. Learning trajectory dependencies for human motion prediction. In Proceedings of the IEEE International Conference on Computer Vision, Seoul, Korea, 27 October–2 November 2019; pp. 4317–4326.
26. Shi, L.; Zhang, Y.; Cheng, J.; Lu, H. Decoupled spatial-temporal attention network for skeleton-based action recognition. *arXiv* **2020**, arXiv:2007.03263.
27. Aksan, E.; Kaufmann, M.; Cao, P.; Hilliges, O. A spatio-temporal transformer for 3d human motion prediction. In Proceedings of the 2021 International Conference on 3D Vision (3DV), London, UK, 1–3 December 2021, pp. 565–574.
28. Li, Y.; Wang, Z.; Yang, X.; Wang, M.; Poiana, S.I.; Chaudhry, E.; Zhang, J. Efficient convolutional hierarchical autoencoder for human motion prediction. *Vis. Comput.* **2019**, *35*, 1143–1156. [CrossRef]
29. Chiu, H.K.; Adeli, E.; Wang, B.; Huang, D.A.; Niebles, J.C. Action-agnostic human pose forecasting. In Proceedings of the 2019 IEEE Winter Conference on Applications of Computer Vision (WACV), Waikoloa Village, HI, USA, 7–11 January 2018; pp. 1423–1432.
30. Guo, X.; Choi, J. Human motion prediction via learning local structure representations and temporal dependencies. In Proceedings of the AAAI Conference on Artificial Intelligence, Honolulu, HI, USA, 27 January–1 February 2019; Volume 33, pp. 2580–2587.
31. Bai, S.; Kolter, J.Z.; Koltun, V. An empirical evaluation of generic convolutional and recurrent networks for sequence modeling. *arXiv* **2018**, arXiv:1803.01271.
32. Lea, C.; Flynn, M.D.; Vidal, R.; Reiter, A.; Hager, G.D. Temporal convolutional networks for action segmentation and detection. In Proceedings of the IEEE Conference on Computer Vision and Pattern Recognition, Honolulu, HI, USA, 21–26 July 2017; pp. 156–165.
33. Farha, Y.A.; Gall, J. Ms-tcn: Multi-stage temporal convolutional network for action segmentation. In Proceedings of the IEEE/CVF Conference on Computer Vision and Pattern Recognition, Long Beach, CA, USA, 15–20 June 2019; pp. 3575–3584.
34. Dauphin, Y.N.; Fan, A.; Auli, M.; Grangier, D. Language modeling with gated convolutional networks. In Proceedings of the International conference on machine learning. PMLR, Sydney, Australia, 6–11 August 2017; pp. 933–941.
35. van den Oord, A.; Dieleman, S.; Zen, H.; Simonyan, K.; Vinyals, O.; Graves, A.; Kalchbrenner, N.; Senior, A.; Kavukcuoglu, K. Wavenet: A generative model for raw audio. *arXiv* **2016**, arXiv:1609.03499.
36. Pavllo, D.; Feichtenhofer, C.; Grangier, D.; Auli, M. 3d human pose estimation in video with temporal convolutions and semi-supervised training. In Proceedings of the IEEE/CVF conference on Computer Vision and Pattern Recognition, Long Beach, CA, USA, 15–20 June 2019; pp. 7753–7762.
37. Yu, F.; Koltun, V. Multi-scale context aggregation by dilated convolutions. *arXiv* **2015**, arXiv:1511.07122.
38. Reis, M.S. Multiscale and multi-granularity process analytics: A review. *Processes* **2019**, *7*, 61. [CrossRef]
39. Yang, B.; Yang, J.; Ni, R.; Yang, C.; Liu, X. Multi-granularity scenarios understanding network for trajectory prediction. *Complex Intell. Syst.* **2023**, *9*, 851–864. [CrossRef]
40. Chorowski, J.K.; Bahdanau, D.; Serdyuk, D.; Cho, K.; Bengio, Y. Attention-based models for speech recognition. *arXiv* **2015**, arXiv:1506.07503.
41. Vaswani, A.; Shazeer, N.; Parmar, N.; Uszkoreit, J.; Jones, L.; Gomez, A.N.; Kaiser, Ł.; Polosukhin, I. Attention is all you need. *arXiv* **2017**, arXiv:1706.03762.
42. Xu, Y.; Yu, L.; Xu, H.; Zhang, H.; Nguyen, T. Vector sparse representation of color image using quaternion matrix analysis. *IEEE Trans. Image Process.* **2015**, *24*, 1315–1329. [CrossRef] [PubMed]
43. Tang, Y.; Ma, L.; Liu, W.; Zheng, W. Long-term human motion prediction by modeling motion context and enhancing motion dynamic. *arXiv* **2018**, arXiv:1805.02513.
44. Cai, Y.; Huang, L.; Wang, Y.; Cham, T.J.; Cai, J.; Yuan, J.; Liu, J.; Yang, X.; Zhu, Y.; Shen, X.; et al. Learning progressive joint propagation for human motion prediction. In Proceedings of the European Conference on Computer Vision, Glasgow, UK, 23–28 August 2020; pp. 226–242.
45. Ma, T.; Nie, Y.; Long, C.; Zhang, Q.; Li, G. Progressively Generating Better Initial Guesses Towards Next Stages for High-Quality Human Motion Prediction. In Proceedings of the Conference on Computer Vision and Pattern Recognition, New Orleans, LA, USA, 18–24 June 2022.
46. Loshchilov, I.; Hutter, F. Decoupled weight decay regularization. *arXiv* **2017**, arXiv:1711.05101.
47. Ionescu, C.; Papava, D.; Olaru, V.; Sminchisescu, C. Human3. 6m: Large scale datasets and predictive methods for 3d human sensing in natural environments. *IEEE Trans. Pattern Anal. Mach. Intell.* **2013**, *36*, 1325–1339. [CrossRef]

48. Mao, W.; Liu, M.; Salzmann, M. History Repeats Itself: Human Motion Prediction via Motion Attention. In Proceedings of the European Conference on Computer Vision, Glasgow, UK, 23–28 August 2020; pp. 474–489.
49. Dang, L.; Nie, Y.; Long, C.; Zhang, Q.; Li, G. MSR-GCN: Multi-Scale Residual Graph Convolution Networks for Human Motion Prediction. In Proceedings of the International Conference on Computer Vision, Montreal, QC, Canada, 10–17 October 2021; pp. 11467–11476.

sensors

MDPI

Article

Enhancement and Optimization of Underwater Images and Videos Mapping

Chengda Li [1], Xiang Dong [1,*], Yu Wang [2] and Shuo Wang [2]

[1] School of Electrical Engineering and Automation, Anhui University, Hefei 230601, China; z20201037@stu.ahu.edu.cn

[2] State Key Laboratory of Management and Control for Complex Systems, Institute of Automation, Chinese Academy of Sciences, Beijing 100190, China; yu.wang@ia.ac.cn (Y.W.); shuo.wang@ia.ac.cn (S.W.)

* Correspondence: xdong@ahu.edu.cn

Abstract: Underwater images tend to suffer from critical quality degradation, such as poor visibility, contrast reduction, and color deviation by virtue of the light absorption and scattering in water media. It is a challenging problem for these images to enhance visibility, improve contrast, and eliminate color cast. This paper proposes an effective and high-speed enhancement and restoration method based on the dark channel prior (DCP) for underwater images and video. Firstly, an improved background light (BL) estimation method is proposed to estimate BL accurately. Secondly, the R channel's transmission map (TM) based on the DCP is estimated sketchily, and a TM optimizer integrating the scene depth map and the adaptive saturation map (ASM) is designed to refine the afore-mentioned coarse TM. Later, the TMs of G–B channels are computed by their ratio to the attenuation coefficient of the red channel. Finally, an improved color correction algorithm is adopted to improve visibility and brightness. Several typical image-quality assessment indexes are employed to testify that the proposed method can restore underwater low-quality images more effectively than other advanced methods. An underwater video real-time measurement is also conducted on the flipper-propelled underwater vehicle-manipulator system to verify the effectiveness of the proposed method in the real scene.

Keywords: underwater image and video restoration; transmission map optimizer; saturation map; color correction

Citation: Li, C.; Dong, X.; Wang, Y.; Wang, S. Enhancement and Optimization of Underwater Images and Videos Mapping. *Sensors* **2023**, 23, 5708. https://doi.org/10.3390/s23125708

Academic Editor: Enrico Meli

Received: 4 June 2023
Revised: 17 June 2023
Accepted: 17 June 2023
Published: 19 June 2023

1. Introduction

The ocean is abundant in oil, combustible ice, fish stocks, and mineral resources, which has attracted the attention of more and more governments and researchers. Mainstream marine vehicles, such as underwater remotely operated vehicles (ROVs), autonomous underwater vehicles (AUVs), and commercial submarines, are equipped with deep-sea underwater imaging systems [1]. An underwater camera imaging system is a cost-effective resource that is used extensively because of its low cost, low technical requirements, easy transplantation, and convenient configuration compared with other techniques, such as multispectral stereoscopic imaging [2] and laser scanning imaging [3].

However, underwater images and video often display serious color cast, sharpness degradation, and contrast reduction affected by the complex underwater environment, absorption and scattering of light, suspended particles, and other factors [4]. Poor visibility brings inevitable problems in computer vision tasks, such as object recognition and object detection [5]. In a sense, how to restore underwater images and video is challenging and meaningful.

Existing underwater restoration methods are mostly based on the image formation model (IFM) [6–8], which describes the linear relationship among the direct component, the forward scattering component, and the backward scattering component. In this model, two crucial parameters, the background lights and the transmission map, were estimated to recover underwater haze-free images. He et al. [9] observed that in most local patches, the lowest pixel intensity of the R–G–B color channel based on land haze-free images is close to zero, and statistically summarized the dark channel prior (DCP), which was widely applied for underwater image restoration. Inspired and motivated by the imaging similarity between land-based images and underwater images, some typical image restoration methods based on the IFM and the DCP have been proposed. Chao et al. [10] directly applied the DCP to underwater images that can only handle the scattering problems effectively, but the problems of absorption were not solved. Considering the notable influence of red-light attenuation under most underwater circumstances on the dark channel, only the dark channel of the G–B color channels [11] was proposed to enhance underwater images, which improved the accuracy of the estimated transmission map. Galdran et al. [12] proposed a red channel compensation method as a variant of the DCP, which was described as the dark channel of the inverted R channel and G–B channel, and the pixel of the highest intensity in the R channel was used to estimate BL. Peng et al. [13] proposed a complex method based on image blurriness and light absorption (IBLA) to obtain more precise BL estimation and underwater scene-depth estimation. However, this method is quite time-consuming. Song et al. [14] proposed a new underwater dark channel prior (NUDCP), which included a BL estimation statistic model and the TM of the R–G–B channel estimation model combination of the compensation by the scene-depth map and the optimization by the saturation map to restore underwater images, but that parameter estimation was greatly affected by the environment. Liang et al. [15] proposed a backscattered light estimation that combined wavelength-dependent attenuation, image blurriness, and brightness priors based on a hierarchical searching technique.. Yu et al. [16] proposed a novel transmission map optimization model designed by Yin–Yang pair optimization to restore underwater images. This method can suppress background noise and enhance detailed texture features but simultaneously increases complexity and computing cost.

It is challenging to enhance and restore underwater images in distinct water types with different distortions. Available underwater image enhancement and restoration methods either cannot obtain the accurate BL and TM estimation or have low algorithm processing speed. To overcome these problems, we propose an effective and high-speed underwater image and video restoration method in this paper. The highlights of this paper are summarized as follows:

- An accurate and high-speed background light estimation method is proposed, which is not only suitable for distinct types of underwater images but also for low complexity. The proposed method is time-saving and adaptable for most underwater images.
- TM estimation with an improved optimizer is established. Integrating the compensation of the scene-depth map based on color attenuation prior (CAP) and the adaptive saturation map (ASM), an optimizer is designed to modify and refine the coarse TM. The proposed TM estimation method provides more accurate results and has lower complexity in different kinds of underwater images than other advanced models.
- An improved white balance (WB) algorithm is employed to improve the color cast and visibility for restored images. The gain factor is adaptively selected related to the restored underwater image intensity to avoid over- or under-correction.

The rest of this paper is organized as follows. The underwater image formation model and underwater image restoration based on the DCP are described in Section 2. In Section 3, the BL estimation and the design of the TM optimizer are depicted and explained in detail. The underwater image restoration method with color correction is presented. Section 4 reports the experimental results compared with other state-of-the-art methods. Section 5 discusses the results and conclusion.

2. Related Works

Underwater image restoration and degradation are mutually inverse operations when the light propagates in the water medium. Underwater image restoration based on the image formation model (IFM) is introduced briefly. The parameter estimations of global background lights and a transmission map are two vital parameters to restore underwater images based on the DCP, which can affect the results directly.

2.1. Underwater Image Formation Model

A simplified underwater image formation model is used to describe the haze image [17]:

$$I^C(x,y) = J^C(x,y)\, t^C(x,y) + B^C(1 - t^C(x,y)) \tag{1}$$

where $I^C(x,y)$ represents the image intensity captured by the camera from the scene at the position (x,y); J^C refers to the corresponding haze-free image; B^C denotes the background lights, which is a three-dimensional vector; $C \in \{R, G, B\}$ denotes one of the R–G–B color channels; and $t^C(x,y)$ denotes the transmission map, and it is an exponential attenuation function related to the spectral volume attenuation coefficient $\beta(x,y)$ and the scene depth $d(x,y)$, which can be expressed as follows:

$$t^C(x,y) = \mathrm{e}^{-\beta(x,y)\ d(x,y)} \tag{2}$$

2.2. Underwater Image Restoration Based on the DCP

Owing to the optical imaging similarities between land-based images and underwater images, the coarse transmission map (TM) is estimated based on the DCP under the statistical assumption $J_{\text{dark}}(x,y) = 0$.

First, we apply the minimum value operator on a local patch $\Omega(x,y)$ on both sides in Equation (1):

$$\min_{z\in\Omega(x,y)} I^C(z) = \min_{z\in\Omega(x,y)} J^C(z)t^C(z) + \min_{z\in\Omega(x,y)} B^C(1 - t^C(z)) \tag{3}$$

Generally, the background light B^C is assumed to be constant and positive, and both sides of Equation (3) are divided by B^C. Then, $t^C(z)$ is constant and continuous on a small patch, so we have

$$\frac{\min\limits_{z\in\Omega(x,y)} I^C(z)}{B^C} = \frac{\min\limits_{z\in\Omega(x,y)} J^C(z)}{B^C} t^C(z) + 1 - t^C(z) \tag{4}$$

Then, the minimum filter operator is applied among R–G–B color channels as follows:

$$\frac{\min\limits_{C}\min\limits_{z\in\Omega(x,y)} I^C(z)}{B^C} = \frac{\min\limits_{C}\min\limits_{z\in\Omega(x,y)} J^C(z)}{B^C} t^C(z) + 1 - \min_C t^C(z) \tag{5}$$

According to the dark channel prior, we get

$$\frac{\min\limits_{C}\min\limits_{z\in\Omega(x,y)} J^C(z)}{B^C} = 0 \tag{6}$$

Combining Equation (5) with Equation (6), the multiplicative term is deleted and the coarse transmission map of the R color channel $t^R(z)$ is expressed as follows:

$$t^R(z) = 1 - \frac{\min\limits_{C}\min\limits_{z\in\Omega(x,y)} I^C(z)}{B^C} \tag{7}$$

Finally, the restored underwater haze-free image $J^C(x,y)$ is calculated from

$$J^C(x,y) = \frac{I^C(x,y) - B^C}{\min(\max(t^C(x,y), 0.1), 0.9)} + B^C \tag{8}$$

The constants 0.1 and 0.9 are the lower and upper limits of the transmission map, respectively.

The coarse transmission maps of the R channel for typical underwater images are shown in Figure 1. The accurate R channel TMs based on the DCP are estimated for underwater images in Figure 1a–c, with uniform illumination and a gloomy foreground scene. On the contrary, in Figure 1d, the high-brightness target or white object near the camera is mistaken for the far scene, which causes the estimated transmission map of the region to be underestimated. In Figure 1e, because of some regions in which the red channel intensity is lower than the estimated background lights, such as the area behind the statue, the local TM is overestimated. Figure 1f shows the most inhomogeneous illumination scene in the shark back region and the poor TM estimation.

(a) (b) (c) (d) (e) (f)

Figure 1. Several classical underwater images and their TM of the R color channel based on the DCP. The upper layer is the original image, and the lower layer is their corresponding transmission map. (**a–f**) represent different underwater images.

On the one hand, the TM based on the DCP can be estimated precisely when the illumination of the scene from the light source is from far away, which is regarded as uniform illumination, just contrary to the artificial light. On the other hand, the estimated BL is also inseparable from the accurate TM estimation. However, it is extremely difficult to estimate the BL and the TM accurately in most real underwater images. Like the last three salient cases in Figure 1, the transmission map estimation is inaccurate, which affects the subsequent image restoration.

3. Problem Formulation

To solve the above problems, an improved background light estimation and transmission map optimizer are proposed to restore underwater images from the aspects of visibility, contrast, and color. The flow-process diagram of the whole method is shown in Figure 2. First, the underwater image color channels are judged whether to be seriously attenuated, and an appropriate method is chosen to estimate the BL. Secondly, a TM optimizer of the R color channel is designed to modify and refine the coarse TM based on the DCP. Then, the TMs of the remaining G and B channels can be calculated by the attenuation coefficient ratios of their channels to the red channel relatively. Using the estimated BL and TM, we can obtain a haze-free underwater image. Finally, an adaptive color correction method is applied to adjust and correct the brightness and color of the restored image.

3.1. Background Light Estimation

Recently, various methods for estimating background light have been developed, such as DCP_{RGB} or its variants [11], the quad-tree decomposition searching method [5,18], the BL candidate selection region [13], and BL estimation based on the statistic model [14], which worked well only for given underwater images, as it mostly had high computational complexity and was time consuming. Affected by the type of water, the degree of different color channel attenuation is different. The red channel is attenuated more severely than the G–B channel in these images captured in the open ocean. In the coastal oceans, the blue channel is more attenuated than the others.

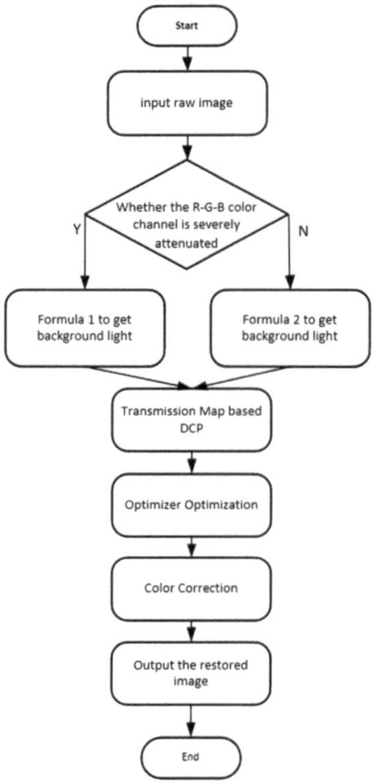

Figure 2. Flowchart of the proposed method.

Inspired by the paper in [15], we used Formula (9) to determine whether some color channel is a severe attenuation channel by calculating the ratios between the maximum channel pixel average and the channel pixel average of the input image and chose the relative method to compute BL. The BL estimation based on the statistical model can obtain an accurate BL of the raw image with no or little red component rapidly, which is not suitable for other channels with serious attenuation. Therefore, we extended it to other attenuation color channels, as shown in Equation (10). For underwater images with severe attenuation or darker intensity, it is more accurate to utilize an improved BL model based on statistic model to estimate the BL. For other situations, the DCP_{RGB} method is utilized. Thus, we took full advantage of the advantages of these two methods to obtain

$$BL = \begin{cases} BL1 \ , & \max(Avg(I^C))/\min(Avg(I^C)) \geq 2 \\ BL2 \ , & otherwise \end{cases} \tag{9}$$

Here,

$$BL1 = \begin{cases} 1.13 \times Avg^m + 1.11 \times Std^m - 25.6, & Avg(I^C) > \delta \\ \frac{140}{1+14.4 \times exp(-0.034 \times Med^n)}, & otherwise \end{cases} \tag{10}$$

$$BL2 = I^C(\text{argmax}\{\min_{\Omega(x,y)}(\min_C(I^C(x,y)))\}) \tag{11}$$

where δ is a threshold and set δ = 64, which was consulted in the literature [15]; m denotes the color channel, with an average intensity value greater than δ; and n denotes the opposite channel. *Avg*, *Std*, and *Med* denote the average value, the standard deviation, and the median value of the channel in the underwater image I, respectively. To avoid producing over- or under-estimation, the BL value was limited to between 5 and 250. Therefore, the ultimate estimated BL is:

$$B^C = \min(\max(B^{m,n}, 5), 250), C \in \{R, G, B\} \tag{12}$$

3.2. TM Optimizer Design

The design steps of the TM optimizer are depicted minutely in this part, which consists of two components: the correction of the scene-depth map and the optimization of the adaptive saturation map (ASM). For some special cases, our optimizer can eliminate and refine the unreasonable and error areas in the transmission map based on the DCP to obtain the accurate TM.

Firstly, the scene depth is estimated based on the color attenuation prior (CAP). Through the experiments on extensive foggy images, the scene depth can be described as a positively correlated function with the fog concentration and the difference value between image brightness and saturation [19]. The scene-depth estimation model is:

$$d(x,y) = \theta_0 + \theta_1\, v(x,y) + \theta_2\, s(x,y) + \tau(x,y) \tag{13}$$

where θ_0, θ_1, and θ_2 are coefficients; τ is a random parameter; $v(x,y)$ is the image brightness; and $s(x,y)$ is the image saturation. Those parameters can be calculated by supervised learning, and we directly followed the optimal solution in this paper [19]: θ_0 = 0.121779, θ_1 = 0.959710, θ_2 = −0.780245, τ = 0.041337.

The scene-depth maps of the underwater images might not be consistent with the reality in the corresponding region with the objects or targets with high brightness. Thus, the minimum filter is applied to settle those questions as follows:

$$d_f(x,y) = \min_{\Omega(x,y)} d(x,y) \tag{14}$$

The transmission map (TM) of R–G–B color channels is easily obtained as follows:

$$t^C(x,y) = e^{(-\beta^C d_f(x,y))},\ C \in \{R, G, B\} \tag{15}$$

Just the R channel is considered, and we set β^R = 1 and obtain:

$$t_d^R(x,y) = e^{-d_f(x,y)} \tag{16}$$

whereas the estimated TM in Equation (7) is related to the scene, intensity, and detail information of the raw image, which produces inaccurate estimation in specific cases easily. For example, some close background areas with a small intensity value of the R channel overestimate the TM value and are considered foreground [14]. The color attenuation prior (CAP) supposes that the near background area with a small intensity value has low brightness and saturation, and the TM value is correspondingly small, which is used to compensate for and refine the TM. The modified TM is expressed as:

$$t_m^R(x,y) = \min\{\mathrm{t}^R(x,y), t_d^R(x,y)\} \tag{17}$$

Secondly, when the underwater environment lacks light, we usually use artificial light (AL) to increase the view field of the external scene. Therefore, we ought to eliminate the effect of AL. However, a region with high brightness cannot indicate the emergence of artificial light. We found that the saturation value of a region illuminated by AL is low in HSV color space [12,14]. The image saturation map is expressed as follows:

$$S(x,y) = \begin{cases} 1, \max I^C(x,y) = 0 \\ \frac{\max I^C(x,y) - \min I^C(x,y)}{\max I^C(x,y)}, otherwise \end{cases} \tag{18}$$

Image saturation characterizes the vividness of color in an image, which can be used to reflect the effects of artificial lights. With an increase in white light, the channels lose saturation gradually, and the regions illuminated by artificial light force the brightness of pixels in the three channels to be closed, which results in the saturation value being very low. To some extent, the areas with low saturation in the image can be considered the areas illuminated by AL. Thus, we defined an adaptive saturation map (ASM) to describe this phenomenon as follows:

$$S_p(x,y) = 1 - \alpha\, S(x,y) \tag{19}$$

where α is the coefficient related to the mean value of the saturation map. We applied ASP to modify and optimize the TM to abate the intensity of the inhomogeneous illumination or artificial light region. Relevantly, the intensity in other areas has little impact.

$$t_f^R(x,y) = \max\{t_m^R(x,y), S_p(x,y)\} \tag{20}$$

Applying our optimizer to modify and refine TM, the inaccurate TMs of Figure 1d–f can be corrected and refined. Then, guided filtering [20] is applied to preserve the edges. The above inaccurate TMs are refined by our TM optimizer, as shown in Figure 3.

(a) (b) (c)

Figure 3. The above inaccurate TMs of the R channel are refined with our optimizer for Figure 1d–f. (**a**) The underestimated TM for the clay pot in the foreground region is accurately corrected; (**b**) the overestimated TM for the back of the statue in the background region is accurately corrected; (**c**) the TM in the shark back area with the non-uniform illumination is improved.

In Equation (2), the transmission map needs to be calculated on each color channel, and three channel transmission maps are not disrelated. Only one metric and two scale coefficients [21] need to be computed as follows:

$$\begin{cases} t_f^R(x,y) = e^{-\beta^R d(x,y)} \\ t_f^G(x,y) = e^{-\beta^G d(x,y)} = \left(e^{-\beta^R d(x,y)}\right)^{\frac{\beta^G}{\beta^R}} = \left(t_f^R(x,y)\right)^{\frac{\beta^G}{\beta^R}} \\ t_f^B(x,y) = e^{-\beta^B d(x,y)} = \left(e^{-\beta^R d(x,y)}\right)^{\frac{\beta^B}{\beta^R}} = \left(t_f^R(x,y)\right)^{\frac{\beta^B}{\beta^R}} \end{cases} \tag{21}$$

Since the TM is estimated on a small pitch $\Omega(x,y)$, the transmission map $t^C(x,y)$ easily demonstrates some color halos and block artifacts. Finally, the guided filter [20] is applied to solve these problems.

3.3. *Color Correction*

Although our method can effectively remove the fog in underwater images, some new problems also appear, such as the restored images having poor contrast, uneven color, and low brightness, which makes it hard to obtain valuable information. Inspired by the white balance algorithm [21], an improved color correction is proposed. Compared to the original method, its gain factor is selected automatically based on the brightness of the input image and can be expressed as:

$$\begin{cases} I_{out} = \frac{I_{in}}{V_p \times (m/m_{ref}) + \lambda_{as}} \\ m_{ref} = \sqrt{(m_R)^2 + (m_G)^2 + (m_B)^2} \end{cases} \quad (22)$$

where I_{in} and I_{out} denote the input restored image and the output corrected image; m_R, m_G, and m_B denote the mean value of the image I_{in} color channels; and V_p is the maximum intensity value of the image I_{in}. The parameter λ_{as} is the gain factor to adjust the color of the input image, whose range is 0 to 0.5. According to the color correction effect of the λ_{as} selection for the restored image, an auto-select gain factor mothed is proposed to obtain the desired λ_{as}. Our principle is as follows: When the maximum value of m_R, m_G, and m_B is greater than 0.45, the image I_{in} is brighter and λ_{as} is close to 0.5; in other words, if the maximum value is less than 0.45, the image is dark and λ_{as} is closed to 0, which can be expressed as:

$$\lambda_{as} = \begin{cases} 0.5 \times \tanh(m_2/m_1) \ , & m_2 > 0.45 \\ 0.5 \times \tanh(m_1/m_2) \ , & otherwise \end{cases} \quad (23)$$

where m_2 and m_1 are the maximum value and the minimum value of m_R, m_G, and m_B. The $tanh(\cdot)$ represents the hyperbolic tangent function.

Some typical underwater images restored by our method are shown in Figure 4 to explain the whole process of image restoration. Figure 4b,c represent the coarse TM based on DCP and the precise TM refined by the TM optimizer. Figure 4d shows the restored image without color correction. Figure 4e denotes the final enhanced image.

Figure 4. The whole processing of the proposed underwater image restoration method. (**a**) Raw image; (**b**) the TM of the R channel based on DCP; (**c**) refinement of the TM optimized by our TM optimizer; (**d**) the restored image; (**e**) the enhanced image with color correction.

4. Results and Evaluation

In this section, we analyze and evaluate the proposed underwater image restoration and enhancement method qualitatively and quantitatively compared with other state-of-the-art methods. All tested underwater images were from the underwater benchmark dataset [22] and all codes were from open-source codes, which were run on a Windows

11 PC with AMD Ryzen 9 5900HX with Radeon Graphics@3.30 GHZ, 6.00 GB, running Python 3.6.0.

4.1. Evaluation of Objectives and Approaches

The experiments were designed to verify the effectiveness and performance of our method from the following three perspectives:

(1) To prove the effectiveness of the TM optimizer;
(2) To prove the comprehensive performance of the proposed method;
(3) To test the real-time performance of underwater video.

In experiment (1), we implemented the evaluations by comparing them with other classical and representative underwater image restoration methods, which include the dark channel prior (DCP), the maximum intensity prior [23] (MIP), the underwater dark channel prior (UDCP), the image blurriness and light absorption (IBLA), the underwater light attenuation prior [24] (ULAP), and a new underwater dark channel prior (NUDCP).

To make the experimental results more convincing, four image-quality indexes were employed, which were as follows: two full-reference (FR) indexes (the peak signal-to-noise ratio (PSNR) and the structural similarity (SSIM)) and two non-reference (NR) indexes (the underwater image-quality measure [25] (UIQM) and the blind referenceless image spatial-quality evaluator [26] (BRISQUE)).

The PSRN index mainly measures the average ratio of signal maximum energy to signal noise energy and is widely used. A greater value indicates higher similarity. The SSIM index describes the structural similarity information between the enhanced image and the contrast reference image intuitively. A higher SSIM value represents better performance. Although the PSNR and SSIM indexes need contrast reference images, which are difficult to obtain, they can reflect the influence of the added noise of underwater restored images on the structure of the contrast reference image. The UIQM index takes colorfulness, sharpness, and contrast as the measurement components and combines the components linearly to assess image quality comprehensively. A higher value represents better visibility and quality. The BRISQUE index can measure the naturalness losses in underwater images. A lower value indicates a better result.

4.2. Performance of Transmission Map Optimizer for Single Underwater Image

First, the performance of our TM optimizer was verified with different kinds of underwater images, which included greenish and bluish underwater images. To express our performance objectively, we chose various kinds of restored underwater images compared with other existing methods shown in Figures 5 and 6, and their quantitative image-quality assessment indexes are recorded in Tables 1–4.

Figures 5 and 6 show the restored results obtained by various transmission map estimation methods for two greenish underwater images and two bluish underwater images, respectively. Underwater images are commonly greenish and bluish owing to the light absorption and scattering, which lose valuable details and cause color deviation. For those underwater images, our restored approach not only met the requirement of improving contrast but also removed the color cast.

In Figure 5(1,2), the UDCP method caused more serious color distortion and a darker background than the raw image. The DCP and MIP methods almost had no effect on improving contrast and color deviation. We noticed that the restored underwater images in Figure 5(1,2) by the IBLA, ULAP, and NUDCP methods seemed to have similar results in either improving the contrast or eliminating the degraded color, and the restored images were still low contrast and greenish, which was confirmed by the quantitative PSNR, SSIM, UIQM, and BRISQUE indexes from Tables 1 and 2. The contrast was improved evidently and the color distortion was removed in Figure 5(1f,2f), and the restored results were more realistic than others. Although the UIQM index wasn't the best, as shown in Table 2, the other indexes were sufficient to prove the effectiveness of the proposed TM optimizer.

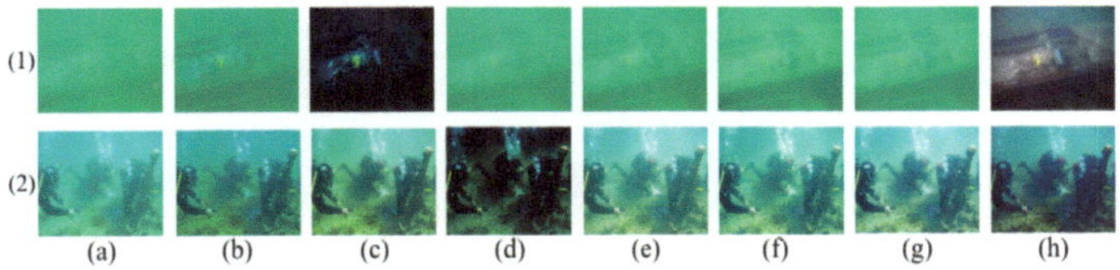

Figure 5. Image restoration results with different TM obtained for the greenish images. (**a**) Raw images; (**b**) DCP; (**c**) UDCP; (**d**) MIP; (**e**) ILBA; (**f**) ULAP; (**g**) NUDCP; (**h**) ours.

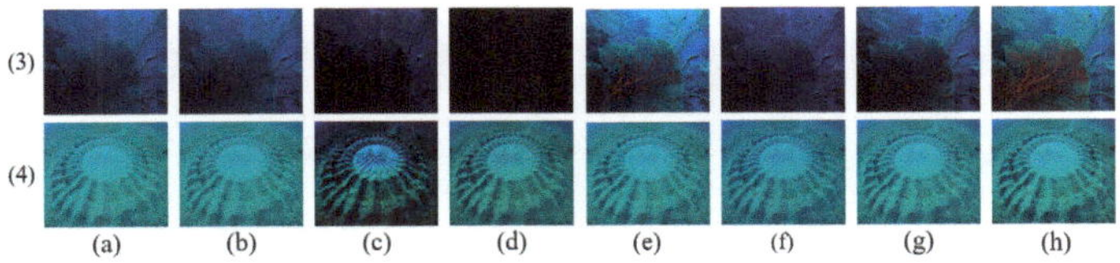

Figure 6. Image restoration results with different TM obtained for the bluish images. (**a**) Raw images; (**b**) DCP; (**c**) UDCP; (**d**) MIP; (**e**) ILBA; (**f**) ULAP; (**g**) NUDCP; (**h**) ours.

Table 1. Quantitative analysis of Figure 5(1).

Methods	Indexes			
	PSNR	**SSIM**	**UIQM**	**BRISQUE**
DCP	15.65	0.652	0.78	49.56
UDCP	17.66	0.251	0.35	44.56
MIP	18.53	0.734	1.21	44.27
IBLA	22.59	0.762	1.17	42.88
ULAP	23.96	0.759	1.27	41.18
NUDCP	25.54	0.763	1.29	42.57
Ours	**28.63**	**0.896**	**1.72**	**39.32**

The bold is to highlight the best indicator results.

Table 2. Quantitative analysis of Figure 5(2).

Methods	Indexes			
	PSNR	**SSIM**	**UIQM**	**BRISQUE**
DCP	18.32	0.676	1.31	18.73
UDCP	16.81	0.314	0.77	19.08
MIP	15.45	0.815	**1.68**	25.45
IBLA	23.57	0.786	1.59	15.37
ULAP	25.05	0.763	1.49	15.06
NUDCP	25.19	0.825	1.61	16.34
Ours	**27.99**	**0.852**	1.46	**14.65**

The bold is to highlight the best indicator results.

The image in Figure 6(3) is bluish and dark. From the restored results, we found that DCP, UDCP, MIP, and ULAP made the image less bright and a more serious bluish tone, and the SSIM index and UIQM index were low, which indicates that they produced little improvement. The IBLA and NUDCP methods improved brightness and contrast, but they were still insufficient compared with our method. The quantitative indicators of our method were still beaten slightly. For the slight bluish image in Figure 6(4), the DCP, IBLA, and ULAP methods still retained bluish illumination, which caused litter de-hazing and a restored effect. The MIP method not only did not improve the color distortion but also aggravated it. The quantitative evaluation shows that the proposed TM optimizer yielded the highest values of the PSNR, SSIM, UIQM, and BRISQUE indexes, which indicates that our method produced a more gratifying and effective performance than others.

Table 3. Quantitative analysis of Figure 6(3).

Methods	Indexes			
	PSNR	SSIM	UIQM	BRISQUE
DCP	16.15	0.354	0.73	61.81
UDCP	17.06	0.237	0.82	58.92
MIP	19.95	0.058	0.59	57.25
IBLA	23.09	0.501	1.19	53.35
ULAP	24.05	0.297	0.87	55.69
NUDCP	27.02	0.401	1.07	54.32
Ours	**28.24**	**0.651**	**1.31**	**52.81**

The bold is to highlight the best indicator results.

Table 4. Quantitative analysis of Figure 6(4).

Methods	Indexes			
	PSNR	SSIM	UIQM	BRISQUE
DCP	18.28	0.575	1.09	45.62
UDCP	16.71	0.548	1.09	42.38
MIP	20.01	0.609	1.14	42.83
IBLA	24.43	0.628	1.19	39.53
ULAP	26.24	0.611	1.14	39.58
NUDCP	28.58	0.625	1.22	39.95
Ours	**28.99**	**0.736**	**1.41**	**38.62**

The bold is to highlight the best indicator results.

4.3. Comprehensive Performance for Single Underwater Image

Our proposed algorithm aims to enhance the contrast, eliminate color deviation, and improve the visibility of the underwater image. Thus, the proposed color correction method was carried out on the restored images as post-processing to restore underwater images overall.

To evaluate the performance of the proposed method comprehensively, we compared with the DCP, UDCP, MIP, IBLA, ULAP, and NUDCP. These methods add histogram equalization (HE) to correct the color of restored and enhanced images. Since the NUDCP method designed its color correction algorithm, we continued to employ it and meanwhile named it NU_CC. For convenience, the proposed method without color correction was named O_WCC, and the overall restoration was called "ours".

The quantitative evaluation and data indexes are recorded in Table 5, showing the average values of 100 raw images, which were picked randomly from the underwater benchmark datasets. Figures 7 and 8 show two kinds of typical underwater image restoration results. From Table 5, we discovered that after they were handled by the HE algorithm, the DCP, UDCP, MIP, and IBLA methods produced distinct and valuable improvements in image quality; however, sometimes it was just the opposite.

The overall restoration results for the underwater images restored by different methods can be seen in Figures 7 and 8. To increase the comparison, two opposite results were selected to illustrate the necessity of color correction. The DCP, UDCP, MIP, and IBLA methods often failed to eliminate the greenish and bluish color in Figures 7b–e and 8b–e. Color correction can effectively improve brightness, remove the color cast, and enhance contrast. However, after being enhanced by HE, as shown in Figures 7h–l and 8h–l, the restored underwater images demonstrated high brightness and their colors were overcorrected and oversaturated, which masked significant information in the images. Although the color correction method can improve the UIQM index value of the restored images, as shown in Table 5, the unnatural images did not satisfy our requirements. Our method could effectively overcome the image degradation to obtain the haze-free underwater images shown in Figures 7g and 8g. The final image was enhanced and restored by our improved white balance method, shown in Figures 7n and 8n, considering the uniqueness and correlation of each color channel and image brightness, and were neither oversaturated nor over-enhanced, which is very consistent with human vision.

Table 5. Quantitative analysis of underwater image enhancement and restoration based on different methods.

Methods	Indexes			
	PSNR	SSIM	UIQM	BRISQUE
DCP	18.86	0.37	0.99	45.33
UDCP	17.79	0.39	0.82	43.91
MIP	20.81	0.52	1.02	38.89
IBLA	22.59	0.65	1.38	34.77
NUDCP	25.39	0.71	1.46	29.14
O_WCC	28.54	0.77	1.61	36.67
DCP + HE	19.08	0.41	1.58	38.96
UDCP + HE	19.43	0.49	1.63	38.91
MIP + HE	21.98	0.62	1.74	35.05
IBLA + HE	22.66	0.71	1.79	32.11
ULAP + HE	24.66	0.68	**1.81**	30.69
NU_CC	26.04	0.75	1.68	28.63
Ours	**28.75**	**0.85**	1.75	**27.67**

The bold is to highlight the best indicator results.

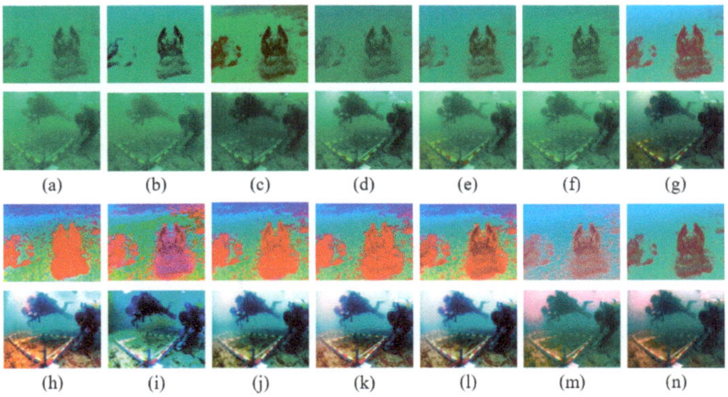

Figure 7. Comparative results for greenish images. (**a**) Raw images; (**b**) DCP; (**c**) UDCP; (**d**) MIP; (**e**) IBLA; (**f**) NUDCP; (**g**) O_WCC; (**h**) DCP + HE; (**i**) UDCP + HE; (**j**) MIP + HE; (**k**) IBLA + HE; (**l**) ULAP + HE; (**m**) NU_CC; (**n**) ours.

Lastly, we selected some different underwater images from 60 challenging underwater images, including greenish underwater images, bluish underwater images, turbid underwater images, dark underwater images, and low-visibility underwater images. The ultimate results of restoration by different enhancement and restoration methods are shown in Figure 9 to show the overall performance of our proposed method.

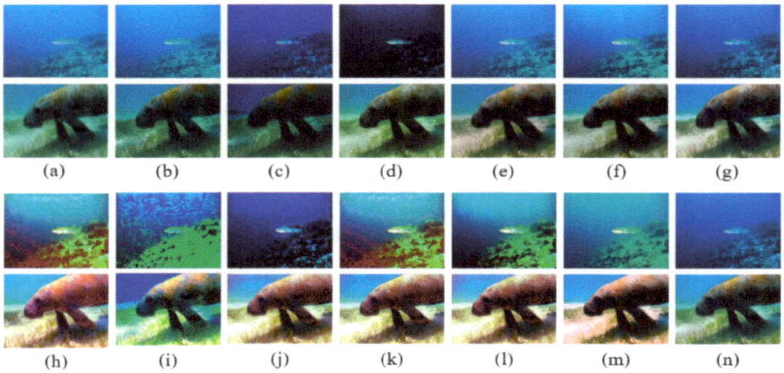

Figure 8. Comparative results for bluish images. (**a**) Raw images; (**b**) DCP; (**c**) UDCP; (**d**) MIP; (**e**) IBLA; (**f**) NUDCP; (**g**) O_WCC; (**h**) DCP + HE; (**i**) UDCP + HE; (**j**) MIP + HE; (**k**) IBLA + HE; (**l**) ULAP + HE; (**m**) NU_CC; (**n**) ours.

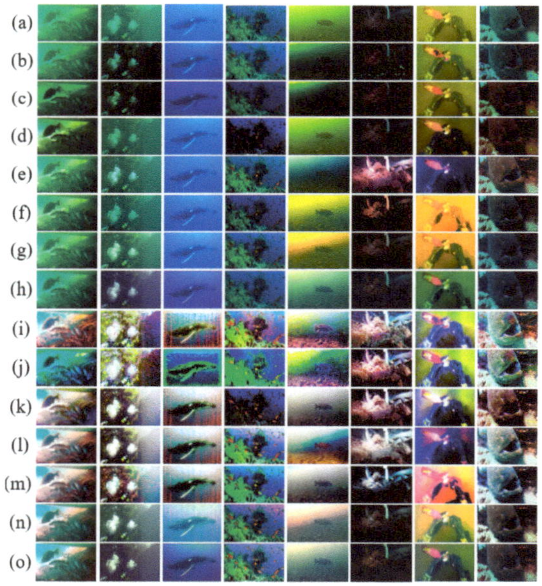

Figure 9. Comparative results for challenge underwater images. (**a**) Raw images; (**b**) DCP; (**c**) UDCP; (**d**) MIP; (**e**)IBLA; (**f**) ULAP; (**g**) NUDCP; (**h**) O_WCC; (**i**) DCP + HE; (**j**) UDCP + HE; (**k**) MIP + HE; (**l**) IBLA + HE; (**m**) ULAP + HE; (**n**) NU_CC; (**o**) ours.

4.4. Enhancement for Underwater Video

Our method is intended to carry out underwater or undersea image and video restoration on underwater mobile systems, such as AOVs and UAVs [27,28]. In general, the hardware resource and computational ability of those devices show pretty poor performance, so the component of real time is challenging and significant. We tested our proposed underwater image restoration method in real time on a flipper-propelled underwater vehicle-manipulator system [29], and the size of the underwater R–G–B image obtained by the system webcam was 312 pixels × 554 pixels. In the marine environment, a two-minute underwater video with 3000 frames was enhanced and restored by different methods, which included the DCP, UDCP, MIP, ULAB, NUDCP, and ours. The real-time measurement results are listed in Table 6.

Table 6. The processing speed of enhancement and restoration for underwater videos.

	Test Time (s)	TPF (ms)	FPS
DCP	975	982	1.0
UDCP	2169	723	1.4
MIP	4758	1586	0.6
ULAP	2016	672	1.5
NUDCP	1155	385	2.6
Ours	**288**	**96**	**10.4**

TPF: time per frame, *FPS*: frame per second. The bold is to highlight the best indicator results.

The DCP, UDCP, MIP, and ULAP were only applied to enhance and restore a single image, and their mean processing speed for underwater video was much slower and could not be used for real-time enhancement and restoration. The NUDCP method could improve the speed of underwater video enhancement, but our processing time was 90 ms–100 ms per underwater image and the average restoration speed was more than 10 FPS, which is nearly four times as fast as the NUDCP because some functions were optimized and the parameters could be solved quickly. Our method can improve the processing speed of underwater video restoration greatly and can be applied to engineering tasks in real time.

5. Conclusions

This paper proposes an efficient underwater image and video restoration method that contains a fast and practical background light estimation method, an accurate transmission map estimation based on a TM optimizer, and an adaptive color correction algorithm. The designed method can be applied to a variety of environments and can meet real-time requirements. The quality of the restored underwater image and video depends on the accuracy of the estimated BL and the TM. Meanwhile, the BL and the TM are also closely connected, and the accuracy of the BL influences the TM estimation. Thus, an accurate BL estimation method is particularly significant. To verify the performance and the real time of our proposed method for underwater image and video both qualitatively and quantitatively, we designed several corresponding experiments to explain and demonstrate.

The proposed method performed better at enhancing underwater image and video. Because the background light cannot be estimated completely and accurately under special circumstances, such as a scene with generous artificial light and lots of sediment, our method may obtain the wrong transmission map, which makes the restored images have a color imbalance. For most attenuated underwater images, our method can obtain a better result. This will be instrumental in improving the visual perception ability of underwater vehicles.

We will continue to study how to accurately estimate the BL and optimize the TM for complex underwater environments in the future. Meanwhile, future research will pay attention to the real-time processing speed of underwater images and video to meet the physical demands of GPU programming to provide a visual basis for complex underwater operations.

Author Contributions: Conceptualization, X.D. and C.L.; methodology, X.D., C.L., Y.W. and S.W.; software, C.L.; validation, X.D. and C.L.; formal analysis, X.D. and C.L.; investigation, C.L.; resources, X.D.; data curation, X.D. and Y.W.; writing—original draft preparation, C.L.; writing—review and editing, X.D. and C.L.; visualization, C.L.; supervision, X.D.; project administration, X.D. and C.L.; funding acquisition, X.D. All authors have read and agreed to the published version of the manuscript.

Funding: This research was partially supported by the National Key Research and Development Project Monitoring and Prevention of Major Natural Disasters Special Program under grant No. 2020YFC1512202, partly supported by the National Natural Science Foundation of China under grant No. 62273342, and partly supported by the Key Research and Development Project of Anhui Province under grant No. 2022a05020035.

Institutional Review Board Statement: Not applicable.

Informed Consent Statement: Not applicable.

Data Availability Statement: Not applicable.

Acknowledgments: Special thanks to the Institute of Automation, Chinese Academy of Sciences, for providing the experimental site.

Conflicts of Interest: The authors declare no conflict of interest.

Abbreviations

The following abbreviations are used in this manuscript:

DCP	dark channel prior
BL	background lights
TM	transmission map
ASM	adaptive saturation map
IFM	image formation model
ROV	remotely operated vehicles
AUV	autonomous underwater vehicles
CAP	color attenuation prior
AL	artificial light
WB	white balance
MIP	maximum intensity prior
UDCP	underwater dark channel prior
IBLA	image blurriness and light absorption
ULAP	underwater light attenuation prior
NUDCP	new underwater dark channel prior
PSNR	peak signal-to-noise ratio
SSIM	structural similarity
UIQM	underwater image-quality measure
BRISQUE	blind referenceless image spatial-quality evaluator
HE	histogram equalization
NU_CC	NUDCP with color correction
O_WCC	our method without color correction

References

1. Bae, I.; Hong, J. Survey on the Developments of Unmanned Marine Vehicles: Intelligence and Cooperation. *Sensors* **2023**, *23*, 4643. [CrossRef]
2. Kazemzadeh, F.; Haider, S.A.; Scharfenberger, C.; Wong, A.; Clausi, D.A. Multispectral stereoscopic imaging device: Simultaneous multiview imaging from the visible to the near-infrared. *IEEE Trans. Instrum. Meas.* **2014**, *63*, 1871–1873. [CrossRef]
3. Han, M.; Lyu, Z.; Qiu, T.; Xu, M. A review on intelligence dehazing and color restoration for underwater images. *IEEE Trans. Syst. Man Cybern. Syst.* **2018**, *50*, 1820–1832. [CrossRef]
4. Shahani, K.; Song, H.; Mehdi, S.R.; Sharma, A.; Tunio, G.; Qureshi, J.; Khaskheli, N. Design and testing of an underwater microscope with variable objective lens for the study of benthic communities. *J. Mar. Sci. Appl.* **2021**, *20*, 170–178. [CrossRef]
5. Pan, H.; Lan, J.; Wang, H.; Li, Y.; Zhang, M.; Ma, M.; Zhang, D.; Zhao, X. UWV-Yolox: A Deep Learning Model for Underwater Video Object Detection. *Sensors* **2023**, *23*, 4859. [CrossRef]
6. Zhu, D. Underwater Image Enhancement Based on the Improved Algorithm of Dark Channel. *Mathematics* **2023**, *11*, 1382. [CrossRef]
7. Park, E.; Sim, J.Y. Underwater image restoration using geodesic color distance and complete image formation model. *IEEE Access* **2020**, *8*, 157918–157930. [CrossRef]
8. Wu, Z.; Ji, Y.; Song, L.; Sun, J. Underwater Image Enhancement Based on Color Correction and Detail Enhancement. *J. Mar. Sci. Eng.* **2022**, *10*, 1513. [CrossRef]

9. He, K.; Sun, J.; Tang, X. Single image haze removal using dark channel prior. *IEEE Trans. Pattern Anal. Mach. Intell.* **2010**, *33*, 2341–2353. [PubMed]
10. Chao, L.; Wang, M. April. Removal of water scattering. In Proceedings of the 2010 2nd International Conference on Computer Engineering and Technology, Chengdu, China, 16–19 April 2010; IEEE: New York, NY, USA, 2010; Volume 2, pp. V2–V35.
11. Drews, P.; Nascimento, E.; Moraes, F.; Botelho, S.; Campos, M. Transmission estimation in underwater single images. In Proceedings of the IEEE International Conference on Computer Vision Workshops, Sydney, Australia, 2–8 December 2013; pp. 825–830.
12. Galdran, A.; Pardo, D.; Picón, A.; Alvarez-Gila, A. Automatic red-channel underwater image restoration. *J. Vis. Commun. Image Represent.* **2015**, *26*, 132–145. [CrossRef]
13. Peng, Y.T.; Cosman, P.C. Underwater image restoration based on image blurriness and light absorption. *IEEE Trans. Image Process.* **2017**, *26*, 1579–1594. [CrossRef]
14. Song, W.; Wang, Y.; Huang, D.; Liotta, A.; Perra, C. Enhancement of underwater images with statistical model of background light and optimization of transmission map. *IEEE Trans. Broadcast.* **2020**, *66*, 153–169. [CrossRef]
15. Liang, Z.; Ding, X.; Wang, Y.; Yan, X.; Fu, X. Gudcp: Generalization of underwater dark channel prior for underwater image restoration. *IEEE Trans. Circuits Syst. Video Technol.* **2021**, *32*, 4879–4884. [CrossRef]
16. Yu, K.; Cheng, Y.; Li, L.; Zhang, K.; Liu, Y.; Liu, Y. Underwater Image Restoration via DCP and Yin–Yang Pair Optimization. *J. Mar. Sci. Eng.* **2022**, *10*, 360. [CrossRef]
17. Guo, Q.; Xue, L.; Tang, R.; Guo, L. Underwater image enhancement based on the dark channel prior and attenuation compensation. *J. Ocean Univ. China* **2021**, *16*, 757–765. [CrossRef]
18. Yu, H.; Li, X.; Lou, Q.; Lei, C.; Liu, Z. Underwater image enhancement based on DCP and depth transmission map. *Multimed. Tools Appl.* **2020**, *79*, 20373–20390. [CrossRef]
19. Zhu, Q.; Mai, J.; Shao, L. A fast single image haze removal algorithm using color attenuation prior. *IEEE Trans. Image Process.* **2015**, *24*, 3522–3533.
20. Yang, M.; Sowmya, A.; Wei, Z.; Zheng, B. Offshore underwater image restoration using reflection-decomposition-based transmission map estimation. *IEEE J. Ocean. Eng.* **2019**, *45*, 521–533. [CrossRef]
21. Li, X.; Lei, C.; Yu, H.; Feng, Y. Underwater image restoration by color compensation and color-line model. *Signal Process. Image Commun.* **2022**, *101*, 116569. [CrossRef]
22. Li, C.; Guo, C.; Ren, W.; Cong, R.; Hou, J.K.; Wong, S.; Tao, D. An underwater image enhancement benchmark dataset and beyond. *IEEE Trans. Image Process.* **2019**, *29*, 4376–4389. [CrossRef]
23. Carlevaris-Bianco, N.; Mohan, A.; Eustice, R.M. Initial results in underwater single image dehazing. In Proceedings of the Oceans 2010 Mts/IEEE Seattle, Washington, DC, USA, 20–23 September 2010; IEEE: New York, NY, USA, 2010; pp. 1–8.
24. Song, W.; Wang, Y.; Huang, D.; Tjondronegoro, D. A rapid scene depth estimation model based on underwater light attenuation prior for underwater image restoration. In Proceedings of the Pacific Rim Conference on Multimedia, Hefei, China, 21–22 September 2018; Springer: Cham, Switzerland, 2018; pp. 678–688.
25. Anwar, S.; Li, C. Diving deeper into underwater image enhancement: A survey. *Signal Process. Image Commun.* **2020**, *89*, 115978. [CrossRef]
26. Mittal, A.; Moorthy, A.K.; Bovik, A.C. No-reference image quality assessment in the spatial domain. *IEEE Trans. Image Process.* **2012**, *21*, 4695–4708. [CrossRef] [PubMed]
27. Monterroso Muñoz, A.; Moron-Fernández, M.-J.; Cascado-Caballero, D.; Diaz-del-Rio, F.; Real, P. Autonomous Underwater Vehicles: Identifying Critical Issues and Future Perspectives in Image Acquisition. *Sensors* **2023**, *23*, 4986. [CrossRef]
28. Hu, K.; Wang, T.; Shen, C.; Weng, C.; Zhou, F.; Xia, M.; Weng, L. Overview of Underwater 3D Reconstruction Technology Based on Optical Images. *J. Mar. Sci. Eng.* **2023**, *11*, 949. [CrossRef]
29. Wang, R.; Wang, S.; Wang, Y.; Cai, M.; Tan, M. Vision-based autonomous hovering for the biomimetic underwater robot—RobCutt-II. *IEEE Trans. Ind. Electron.* **2018**, *66*, 8578–8588. [CrossRef]

sensors

MDPI

Article

The Objective Dementia Severity Scale Based on MRI with Contrastive Learning: A Whole Brain Neuroimaging Perspective

Yike Zhang [1,2], Wenliang Fan [3], Xi Chen [1,2], Wei Li [1,2,*] and on behalf of the for Alzheimer's Disease Neuroimaging Initiative [†]

1 School of Artificial Intelligence and Automation, Huazhong University of Science and Technology, Wuhan 430074, China
2 Image Processing and Intelligent Control Key Laboratory of the Education Ministry of China, Wuhan 430074, China
3 Department of Radiology, Union Hospital, Tongji Medical College, Huazhong University of Science and Technology, Wuhan 430074, China
* Correspondence: liwei0828@hust.edu.cn
† Data used in the preparation of this article were obtained from the Alzheimer's Disease Neuroimaging Initiative (ADNI) database (adni.loni.usc.edu). As such, the investigators within the ADNI contributed to the design and implementation of ADNI and provided data but did not participate in the analysis or writing of this report. A complete listing of ADNI investigators can be found at http://adni.loni.usc.edu/wp-content/uploads/how_to_apply/ADNI_Acknowledgement_List.pdf.

Citation: Zhang, Y.; Fan, W.; Chen, X.; Li, W.; on behalf of the for Alzheimer's Disease Neuroimaging Initiative. The Objective Dementia Severity Scale Based on MRI with Contrastive Learning: A Whole Brain Neuroimaging Perspective. *Sensors* **2023**, *23*, 6871. https://doi.org/10.3390/s23156871

Academic Editor: Yuemin Zhu

Received: 29 June 2023
Revised: 25 July 2023
Accepted: 29 July 2023
Published: 2 August 2023

Abstract: In the clinical treatment of Alzheimer's disease, one of the most important tasks is evaluating its severity for diagnosis and therapy. However, traditional testing methods are deficient, such as their susceptibility to subjective factors, incomplete evaluation, low accuracy, or insufficient granularity, resulting in unreliable evaluation scores. To address these issues, we propose an objective dementia severity scale based on MRI (ODSS-MRI) using contrastive learning to automatically evaluate the neurological function of patients. The approach utilizes a deep learning framework and a contrastive learning strategy to mine relevant information from structural magnetic resonance images to obtain the patient's neurological function level score. Given that the model is driven by the patient's whole brain imaging data, but without any possible biased manual intervention or instruction from the physician or patient, it provides a comprehensive and objective evaluation of the patient's neurological function. We conducted experiments on the Alzheimer's disease Neuroimaging Initiative (ADNI) dataset, and the results showed that the proposed ODSS-MRI was correlated with the stages of AD 88.55% better than all existing methods. This demonstrates its efficacy to describe the neurological function changes of patients during AD progression. It also outperformed traditional psychiatric rating scales in discriminating different stages of AD, which is indicative of its superiority for neurological function evaluation.

Keywords: medical robots; disease evaluation; Alzheimer's disease; contrastive learning

1. Introduction

Alzheimer's disease (AD) is a progressive neurodegenerative disease with an insidious onset. It is the most common cause of dementia and one of the most expensive and deadliest diseases of the century [1]. Clinical manifestations include memory impairment, aphasia, apraxia, agnosia, impairment of visuospatial skills, executive dysfunction, personality changes, and behavioral changes. The cause of AD is still unclear; however, its progression is usually accompanied by a decline in neurological function. Therefore, in clinical diagnosis, physicians attempt to evaluate subtle changes in the progression of the disease stage by evaluating the level of neurological function and recommending an appropriate treatment plan. The Psychiatric Rating Scale is one of the auxiliary evaluation methods commonly used by physicians. There are many kinds of psychiatric rating scales for Alzheimer's disease; however, these scales are not sufficiently accurate and comprehensive, which leads to the low credibility of the traditional scales. Moreover, they lack objectivity and are not a reliable basis for clinical diagnosis and treatment. Structural magnetic resonance imaging (sMRI) is another auxiliary evaluation method that allows physicians to diagnose based on the degree of brain atrophy in MRI images. However, this approach is highly dependent on the physician's subjective judgment and does not provide a detailed quantitative evaluation.

Deep learning is a powerful method that has been applied to medical images and automated accomplishing targeted tasks in an end-to-end way. There have been many published studies in recent years that have successfully applied this approach to medical image analysis tasks and generally achieved significant improvements [2–4]. Most previous studies focused on the prediction of disease categories, classifying patients' images into discrete categories. However, for many types of diseases, the stages may exhibit a continuous variation in severity that can change over time [5–7]. These different stages of a disease are often classified into ordinal levels (e.g., normal, mild, moderate, and severe), but variations within these ordinal levels are usually underappreciated, and deep learning methods have been applied to evaluate disease stages in limited cases. This may be owing to the absence of the gold standard and representative samples for determining the AD stage.

In the field of deep learning, evaluating the differences between images can be expressed as a metric learning problem. The contrastive learning strategy is commonly employed for metric learning, and it is often used to evaluate the similarity between images. This strategy trains a network using contrastive loss functions and considers the metric between the network output vectors as the evaluation criterion for the image similarity metric. Therefore, we utilized a contrastive learning strategy to evaluate the sMRI images of patients with AD to complete the disease stage discrimination.

In this paper, an evaluation model is proposed based on contrastive learning to evaluate the neurological function of patients with AD. The evaluation model takes the sMRI images of the patient's brain as the network input, uses a twin residual neural network as the network backbone, and trains the network with a supervised contrastive loss function. The supervised contrastive learning loss function effectively utilized the medical image information of patients with AD, as well as their label information. In the evaluation phase, we abstract the neurological level difference into the vector distance calculated from the twin neural network to obtain the neurological function score of patients with AD.

The main contributions of this paper are as follows:

(1) We propose an objective dementia severity scale based on MRI (ODSS-MRI) using a contrastive learning framework to evaluate neurological function during AD progression, which is driven by the whole brain image of the patient. This effectively ensures that the evaluation score will not be affected by any possible biased subjective factors during the testing of the physician or patient.
(2) The score obtained from the proposed ODSS-MRI is highly correlated with the stages of AD, indicating that ODSS-MRI can accurately describe the neurological function changes during AD progression.
(3) The ODSS-MRI scores perform much better on discriminating different stages of AD progression compared to traditional psychiatric rating scales, indicating its advantage in the neurological function evaluation of AD.

2. Related Work

In this section, we briefly introduce the current methods for the evaluation of the neurological function in AD and contrastive learning methods related to the evaluation proposed in this paper.

2.1. Evaluation of Neurological Function in Alzheimer's Disease

Due to the continuous growth of the aging population in China, the number of patients with AD is increasing, and an accurate and refined disease stage evaluation of AD has become a critical problem in clinical treatment. Clinically, AD begins with lesions on the brain. However, patients often do not notice any associated changes initially, and only after years of accumulation do patients exhibit obvious symptoms. Moreover, doctors cannot simply use medical imaging and clinical diagnostic data to diagnose the patient's real situation. Usually, physicians evaluate the neurological function of the patients to diagnose the stage of AD. Currently, one of the main methods for this purpose is neuropsychological testing, which allows physicians to measure the neurological function of patients by observing their performance in tasks involving multiple brain functions, such as movement, language, attention, memory, and thinking, and quantitatively evaluating their neurological function to obtain a score within a certain range [8–13]. The common traditional neuropsychological tests include the Mini-Mental State Examination (MMSE) [8], Clinical Dementia Rating (CDR) [9], Functional Activities Questionnaire (FAQ) [10], Alzheimer's Disease Assessment Scale-Cognitive (ADAS-COG) [11], and Rey Auditory Verbal Learning Test (RAVLT) [12]. The MMSE primarily screens for cognitive impairment by quizzing both orientation and memory, whereas the CDR screens for mild cognitive impairment and classifies clinical dementia by quantifying life skills. Although neuropsychological tests can visually reflect different aspects of patients' brain function, each scale suffers from the problem of incomplete evaluation of neurological function due to different emphases on neurological function evaluation, which in turn affects the diagnosis of the severity of the disease. At the same time, the traditional psychiatric rating scales utilize a question-and-answer format with cooperation between physicians and patients, which may lead to a lack of patient cooperation during the evaluation stage, and the score is also affected by the biased subjectivity of the physician.

With the rapid development of medical imaging technology in recent years, neuroimaging has gradually become one of the main methods to evaluate the neurological function of patients, with structural MRI being the most frequently used [14,15]. Structural MRI is a mature medical imaging technique, which has become another important method for physicians to evaluate the AD stage of patients because of its safety, non-invasiveness, reproducibility, and sensitivity to morphological changes in the brain. During AD, the patient's brain atrophy gradually increases [16]. Based on this characteristic, physicians usually assess the patient's condition based on the degree of brain atrophy and the volume of gray and white matter in the sMRI images; however, due to the individual differences of patients, it is difficult for physicians to accurately diagnose the disease progression of patients.

2.2. Alzheimer's Diagnosis Based on Deep Learning

With the great popularity of deep learning, many researchers have tried to apply this approach to study the pathogenesis of AD, diagnose the stages of AD, and so on. For example, Suk et al. [17] combined MRI, PET, MMSE, and other data to train a classifier and reported promising experimental results in their paper; Ortiz et al. [18] achieved more than 90% accuracy on AD and Normal Control (NC) classification problems by training multiple classifiers followed by voting; Lian et al. [5] proposed a hierarchical fully convolution network to automatically identify local regions with the discriminatory ability in whole-brain sMRI, and on this basis, multi-scale feature representation was integrated to construct a hierarchical classification model for AD diagnosis.

So far, many studies have achieved good results on the classification problem of AD, but it is not enough to just solve the classification problem. The discrete label obtained by prediction lacks information on disease-stage changes. Few studies have applied deep learning methods to the problem of neurological function evaluation in AD; however, some studies use contrastive learning strategies to evaluate the severity of other diseases. For example, Li et al. [19] applied the contrastive learning strategy to continuous disease severity evaluation and disease change detection in retinopathy and osteoarthritis, and the output of the trained twin neural network achieved a correlation of more than 0.85 with the expert sort order on the 100-image test dataset ranked by experts; Li et al. [20] applied the contrastive learning strategy to the detection of psoriasis severity and located skin lesions in the image. The severity scores obtained by the trained network framework achieved better results than the severity index method commonly used in psoriasis. Contrastive learning has the potential to solve the problem of neurological function evaluation in AD. In general, deep learning has achieved better results than traditional methods in diagnosing the stage of AD; however, no deep learning method is available for the problem of neurological function evaluation in AD.

2.3. Contrastive Learning

Contrastive learning is a type of metric learning, which focuses on learning the common features between similar instances and distinguishing the differences between dissimilar instances [21–27]. Compared to traditional deep learning, contrastive learning does not need to focus on the details of instances but only needs to learn to distinguish data at the abstract semantic level of the feature space. Therefore, its optimization is simple, and the model can be more robust. The goal of contrastive learning is to train an encoder that can encode similar data of the same class while making the encoding results of different classes of data as different as possible. This approach can be divided into self-supervised contrastive learning and supervised contrastive learning depending on whether artificial annotations are used in the training stage.

Self-supervised contrastive learning is often used for network pre-training by constructing agent tasks to enable models to learn robust features of large-scale data without using human annotations [28–31]. Self-supervised contrastive learning constructs positive and negative sample pairs by data transformation so that the model learns to aggregate similar samples and mutually reject dissimilar samples in the feature space, thus aggregating similar samples together in the feature space. In recent years, many self-supervised contrastive learning algorithms have been proposed [21–24], and the models pre-trained using contrastive learning methods perform well on many downstream tasks, such as image classification, image segmentation, and face recognition.

Another approach in the contrastive learning domain is supervised contrastive learning. It was proposed in recent years by Khosla P et al. [25]. Supervised contrastive learning added artificial annotation information to the loss function based on self-supervised contrastive learning. Generally, in self-supervised contrast learning, an anchor sample has only one positive sample and all the other samples are regarded as negative samples. In the case of supervised contrastive learning, each anchor sample can have multiple positive and negative samples using labels. This incorporates the intra-class relationship of the data into the loss function, and in this way, the encoder can consider the intra-class similarity between the data during the training stage, rather than compressing similar data into one point of the feature space as in the self-supervised contrastive learning.

3. Materials and Methods

3.1. Materials

3.1.1. Participants

The sMRI data used in this experiment were obtained from the ADNI program (adni.loni.usc.edu, accessed on 24 June 2022), which was established in 2003 and led by Principal Investigator Michael W. Weiner, MD. ADNI is a longitudinal multicenter study aimed at developing clinical, imaging, genetic, and biochemical biomarkers for the early detection and follow-up of AD. This program provides researchers with a variety of neuroimaging data, such as magnetic resonance imaging, functional magnetic resonance imaging, magnetic resonance diffusion tensor imagery, and positron emission scanning images, which is currently the most widely used public database of AD. We obtained 2997 sMRI images from 633 subjects in the ADNI dataset, including 149 AD, 263 mild cognitive impairment (MCI), and 221 NC subjects. Table 1 shows the demographic information of the subjects included in our dataset, including the category, number of subjects, sex (male/female), age (average), and the number of sMRI images.

Table 1. Subject information statistics.

Stage of Disease	AD	MCI	NC
Number	149	263	221
Sex (male/female)	79/70	156/107	115/106
Age	76.3	75.8	77.4
Number of Images	542	1425	1030

3.1.2. Data Pre-Processing

Data pre-processing was performed to ensure that the model was capable of effectively learning pattern information in the dataset. In the data-preprocessing stage, we obtained the image from the ADNI website after gradient nonlinear correction, non-uniform correction, and histogram sharpening. To quickly process a large amount of sMRI data, we used the Matlab-based CAT12 toolbox for batch processing of the data. First, we performed an anterior commissure-posterior commissure (AC-PC) correction of the image data and obtained standard brain atlas images using the AC-PC line as the baseline. Subsequently, we de-skulled and removed invalid areas of the images, preserving only the brain locations, after which we aligned the images with the Montreal Neurological Institute (MNI) standard template. Finally, we modulated the images to ensure spatial consistency and preserve variability.

After the pre-processing stage, we can obtain whole brain images with a size of $113 \times 137 \times 113$. Subsequently, we sliced the 3D image. Considering that the middle part of the image contains richer brain information, we sliced the middle layer (69/137) of the 3D whole brain image along the coronal plane. Finally, we obtained the 2D slice dataset for this study and divided the dataset into the training dataset (80%) and the test dataset (20%) based on the subjects and added the image from the first patient participation test to the test dataset as historical data to the training dataset.

3.2. *Methods*

In this section, we propose an objective dementia severity scale based on MRI using a contrastive learning strategy for AD neurological evaluation problems. The proposed evaluation model is driven by MRI and uses a contrastive learning framework to evaluate the neurological function of patients with AD. The model comprehensively considers the neurological functional information of the patient's whole brain and does not require any additional knowledge of neurology and pathology, effectively ensuring that the obtained functional scores are not influenced by biased subjective factors of the physician and the patient. The model structure is shown in Figure 1. The model utilizes a two-channel structure, in which the test sample is to be evaluated. The MRI images of the healthy person were randomly selected as the baseline sample. They are utilized as input of two channels, and then the samples are entered into a Residual Networks (ResNet) model with shared weights to obtain the high-dimensional feature vectors of the sample to be evaluated and the baseline healthy person sample. We then map the high-dimensional feature vectors into a low-dimensional feature space using a Projection Head. Finally, the cosine similarity is used to calculate the similarity between the low-dimensional feature vectors of the sample to be evaluated and the baseline healthy subjects. The similarity is expanded by a hundredfold to obtain the final neurological function evaluation score.

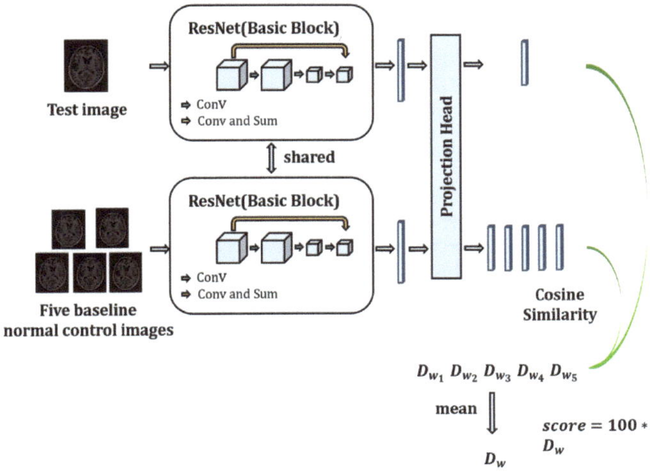

Figure 1. Schematic of the AD neurological function evaluation model based on contrastive learning. (∗ means multiplication).

1. Main structure of the model

The two-channel structure of the ODSS-MRI model is a weight-sharing residual neural network, as shown in Figure 2. First, the network simultaneously mines potentially valid pattern information from the data in both input channels and performs weight sharing during optimization to ensure that the samples in the two channels are mapped into the same feature space. Second, the model adapts the residual neural network as the feature encoder, which utilizes a skip connection, as shown in Figure 2, so that gradient diffusion does not occur when the depth of the network increases. At the same time, the network parameters are easier to optimize, and the gradient information propagates more easily during the back-propagation stage.

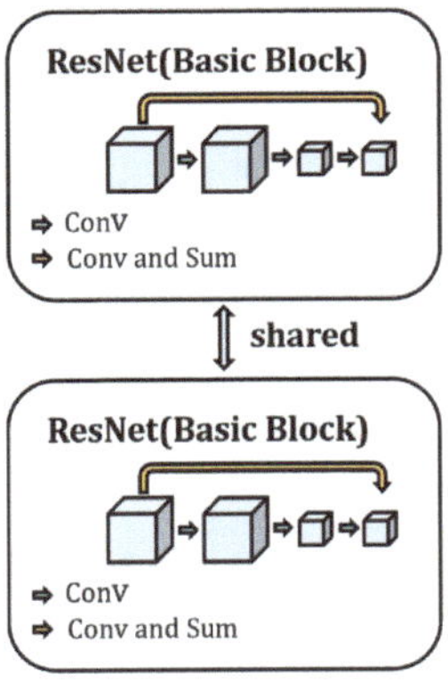

Figure 2. Schematic of the residual neural network with weight sharing.

2. Model Training

The model is trained using the supervised contrastive strategy. As shown in Figure 3, there are two channels, and each batch of images is copied once and fed into the two channels. Then the ResNet model is used as a feature encoder to extract features from the input images. ResNet uses the network structure of skip connection to significantly increase the depth of the network, which can extract semantically informative features from images. The fully connected layer is removed from ResNet and only the 1024-dimensional feature vector output obtained from the average pooling layer is used. A Projection Head is added after the encoder, which maps the 1024-dimensional feature vector space to a 128-dimensional feature vector space, and then we normalize the 128-dimensional feature vectors to keep them on a unit hypersphere. The essence of the projection head is a multi-layer perceptron consisting of linear fully connected layers and the ReLU function to add nonlinearity, and the number of fully connected layers of the projection head is set to one in the model. The study [24] (SimCLR) has shown that the addition of a learnable nonlinear transformation between the feature representation of the image and the contrastive loss can significantly improve the quality of the learned representation feature vector of the images. Finally, the supervised contrastive loss is used to train the model to fully utilize the image and label information of the data.

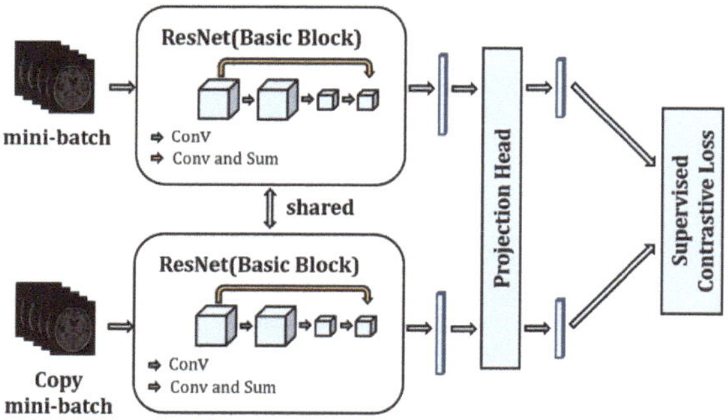

Figure 3. Schematic of the model training stage.

3. Supervised contrastive learning loss function

In the contrastive learning domain, the purpose of the loss function is to allow the model learning to cluster similar samples in the feature space and mutually exclude dissimilar samples so that similar samples are aggregated together in the feature space. Most contrastive loss functions do not use the label information, but supervised contrastive learning adds label information to the loss function based on contrastive learning, as shown in Equations (1) and (2), where N represents the amount of image data in a mini-batch, y represents the label of the image, z represents the feature vector, and τ represents the temperature parameter.

$$L_{SupCon} = \sum_{i=1}^{2N} L_{supcon}^{i} \tag{1}$$

$$L_{supcon}^{i} = -\frac{1}{2N_{y_i} - 1} \sum_{j=1}^{2N} \mathrm{I}_{i \neq j} * \mathrm{I}_{y_i \neq y_j} * log \frac{\exp\left(z_i * \frac{z_j}{\tau}\right)}{\sum_{k=1}^{2N} \mathrm{I}_{i \neq k} * \exp\left(z_i * \frac{z_k}{\tau}\right)} \tag{2}$$

The supervised contrastive loss function is a generalization of the contrastive loss function. Supervised contrastive loss treats data with the same label as positive pairs, and those with different labels as negative samples. This expands the number of positive pairs while maintaining a sufficient number of negative pairs. The loss function calculates the similarity for all positive pairs and then performs a weighted average, which contrasts with a large number of negative pairs. In addition, by increasing the number of positive samples, the network can better characterize intra-class similarities.

4. Neurological function scoring

After training the model, a distance metric function needs to be chosen to calculate the distance between the two output vectors, in order to transform the output feature vectors into neurological function scores. There are many distance metric functions, such as Euclidean distance and Mahalanobis Distance. However, the more similar the images in this framework, the higher the similarity of the two vectors corresponding to the output, and the smaller the distance between the images. Therefore, this framework utilizes the cosine distance to measure the distance between the two vectors. The neurological function score is calculated using Equation (3):

$$score = 100 * \frac{z_i * z_j}{\max(|z_i|, |z_j|)} \tag{3}$$

Figure 4 shows the specific process of neurological function scoring. We randomly select five baseline healthy individuals from the training dataset as the baseline for comparison with the test dataset. This approach assumes that healthy individuals exhibit a relatively consistent level of neurological function. However, given that subtle differences may exist between healthy individuals' nervous systems, they are eliminated by averaging the five scores obtained.

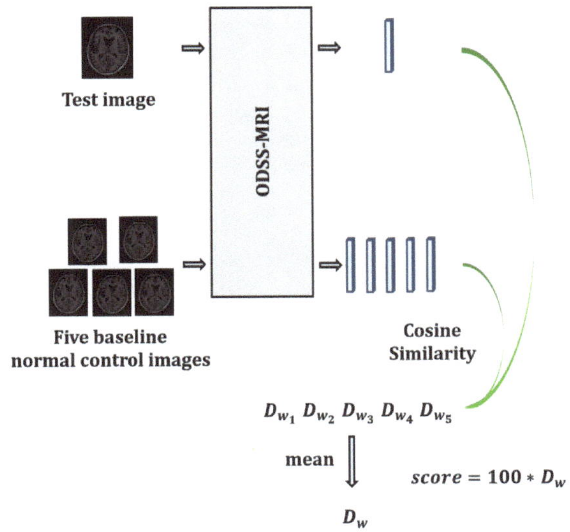

Figure 4. Neurological function scoring. (∗ means multiplication).

4. Results

This section compares the ODSS-MRI with the traditional Psychiatric Rating Scales and analyzes the results.

4.1. Correlation Analysis

In the first part of this experiment, we compared the correlation between ODSS-MRI, traditional psychiatric rating scales, and the APOE allele gene with disease stage, and the results are shown in Table 2. The results showed that the scores obtained using the proposed ODSS-MRI have a Pearson correlation of 88.55% and a Spearman correlation of 87.12% with the stage of AD, which were 17.74% and 1.35% higher than CDR with the best performance in the traditional mental rating scale, respectively. This indicated that the ODSS-MRI can better describe the neurological function changes during AD progression compared to traditional psychiatric rating scales. The correlation between the NPI-Q, the GDS scale, and the APOE allele with the disease stage was low, ranging from 20% to 40%, which is far lower than the ODSS-MRI, and cannot be used as a reliable basis for identifying the disease stage of the subject.

Table 2. Correlation analysis between different indicators and disease stages.

	Pearson Correlation (%)	Spearman Correlation (%)
APOE4 A1	27.74	28.078
APOE4 A2	33.78	33.89
NPI-Q	40.04	50.59
GDS	23.25	31.96
FAQ	65.73	74.60
MMSE	57.956	68.51
CDR	70.81	85.77
ODSS-MRI	88.55 ± 0.86	87.12 ± 1.3

The second part of the experiment examined the correlation between ODSS-MRI and the MMSE/CDR, as shown in Figure 5. The abscissa of the two scatterplots is the ODSS-MRI and the ordinate is the MMSE and CDR scale scores, respectively. On this basis, we used the ODSS-MRI to fit the MMSE and CDR scale scores. The experimental results showed that the ODSS-MRI model score was negatively correlated and positively correlated with the MMSE and CDR scores, respectively, with correlation values of 55.56% and 71.54%, respectively. The MMSE and CDR scale scores are the most commonly used psychiatric score scales with high diagnostic accuracy in clinical diagnosis of AD. The ODSS-MRI model scores achieved a high correlation with both, which showed that the neurological function score obtained by ODSS-MRI can reflect the patient's neurological function level to a certain extent.

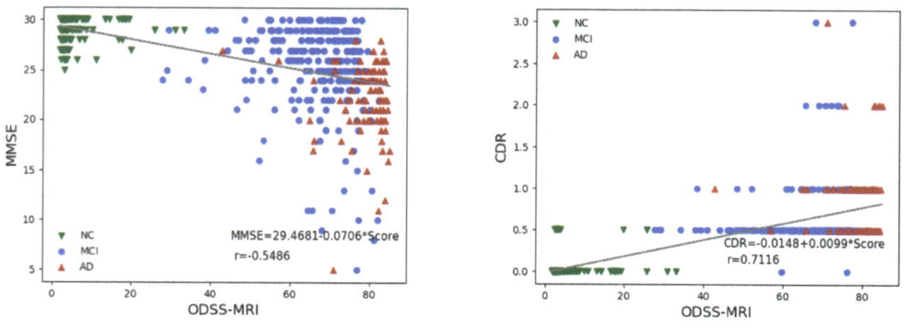

(a) (b)

Figure 5. ODSS-MRI score and the MMSE (**a**) and CDR (**b**) scale score scatter plot. (∗ means multiplication).

4.2. Differential Analysis

To evaluate whether the ODSS-MRI exhibited significant differences and the degree of difference between the NC, MCI, and AD groups, Student's *t*-tests were performed for NC/MCI, MCI/AD, and NC/AD scores, and the *p*-values were shown in Table 3. The results showed that the APOE4 allele, NPI-Q, and GDS scores were not significantly different between NC and MCI groups, which means that they could not distinguish NC from MCI well. The *p*-values of the remaining clinical test indicators were less than 0.0001. The *p*-values of ODSS-MRI between NC/MCI, MCI/AD, and NC/AD groups were much smaller than those of other psychiatric scale scores or indicators, which indicates that the neurological function scores obtained by the ODSS-MRI model exhibited more variability between the groups compared with other indicators and was more effective in discriminating different stages of AD progression. In particular, the ODSS-MRI showed much more variability between MCI/AD and NC/AD groups than other scores, indicating that the ODSS-MRI model scores were significantly different between AD and the other groups.

Table 3. The *p*-value of differential analysis using different indicators.

	NC vs. MCI	MCI vs. AD	NC vs. AD
APOE4 A1	2.06×10^{-2}	4.13×10^{-8}	7.43×10^{-11}
APOE4 A2	3.83×10^{-4}	7.73×10^{-10}	2.65×10^{-14}
NPI-Q	3.57×10^{-3}	2.31×10^{-15}	1.94×10^{-11}
GDS	7.38×10^{-1}	7.92×10^{-8}	2.67×10^{-5}
FAQ	2.14×10^{-11}	1.69×10^{-35}	2.22×10^{-24}
MMSE	1.09×10^{-11}	2.07×10^{-31}	1.07×10^{-33}
CDR	2.73×10^{-7}	6.08×10^{-84}	3.53×10^{-34}
ODSS-MRI	2.52×10^{-18}	3.40×10^{-272}	9.03×10^{-141}

4.3. Sensitivity, Specificity, and Accuracy

In this experiment, the FAQ, MMSE, CDR, and ODSS-MRI were used as classification indicators to calculate the sensitivity, specificity, and accuracy in the two-classification and three-classification tasks, respectively, and the results are shown in Tables 4 and 5. In the two-classification task of NC/MCI, all four scale scores had high specificity, but the sensitivity of both the FAQ and MMSE was only 20–40%. This indicated that both of them found it difficult to accurately identify MCI patients mixed in NC. In the MCI/AD two-classification task, the specificity of both FAQ and MMSE was low, and the sensitivity of CDR was considerably low (only 1%). This indicated that the CDR scores of AD were mixed with MCI. Although ODSS-MRI did not yield the best results in the sensitivity and specificity experiments, both values were high, and this method yield the highest accuracy (89.42%), approximately 16% higher than the next highest index. In the NC/AD task, all scales performed well because of the simplicity of the classification task. Overall, ODSS-MRI, while performing the best in several tasks, was close to the best-performing indicators in others. Except for the NC/MCI task with 1.79% lower than the CDR scale, the other two tasks achieved the highest accuracy, especially in the accuracy experiment.

In the three-classification task of NC/MCI/AD, the sensitivity of the FAQ and MMSE for MCI was not high and the accuracy was low compared with the CDR and ODSS-MRI, indicating that the FAQ and MMSE are not appropriate for identifying the stages of AD, especially for MCI. The sensitivity of CDR for AD and the specificity of MCI were lower than the other approaches, which indicated that the CDR scale score was significantly weak for the recognition of AD. This is consistent with the characteristics of CDR in two-classification tasks. Compared with other psychiatric rating scales, ODSS-MRI exhibited good sensitivity, specificity, and accuracy in the recognition of NC, MCI, and AD, and there were no obvious defects in the recognition of each stage of AD. The accuracy was slightly lower than CDR in the recognition of NC, and the recognition effect of the other two was the best, especially for MCI recognition accuracy (89.57%), which was much higher than that of the other psychiatric rating scales.

Table 4. Sensitivity, specificity, and accuracy of evaluation indicators in the two-classification task.

Tasks	Scores	Sensitivity(%)	Specificity(%)	Accuracy(%)
NC/MCI	FAQ	32.09	99.35	68.06
	MMSE	29.63	95.43	63.21
	CDR	99.25	94.87	97.41
	ODSS-MRI	100.00	89.85	95.62
MCI/AD	FAQ	88.33	33.33	50.79
	MMSE	76.09	40.88	55.02
	CDR	1.03	99.25	73.15
	ODSS-MRI	75.73	94.55	89.42
NC/AD	FAQ	91.38	100.00	97.63
	MMSE	93.33	100.00	98.10
	CDR	100.00	100.00	100.00
	ODSS-MRI	100.00	100.00	100.00

Table 5. Sensitivity, specificity, and accuracy of evaluation indicators in the three-classification task.

Tasks	Scores	Sensitivity (%)	Specificity (%)	Accuracy (%)
FAQ	AD	81.54	77.01	77.68
	MCI	19.55	96.35	57.86
	NC	99.35	66.32	77.90
MMSE	AD	72.16	82.66	80.85
	MCI	20.74	89.46	56.56
	NC	95.43	62.40	73.94
CDR	AD	1.03	99.57	82.56
	MCI	98.52	63.70	80.43
	NC	94.87	99.46	97.86
ODSS-MRI	AD	75.73	96.82	93.04
	MCI	94.55	85.00	89.57
	NC	89.85	100.00	96.52

4.4. Scatter and Box-Line Plots

In this experiment, to show the distribution of neurological function scores in Alzheimer's disease patients at each stage, the MMSE, CDR, and ODSS-MRI scores of patients were plotted as scatter plots and box-line plots, as shown in Figure 6. The experimental results showed that the MCI score was distributed in all intervals from the scatter plot. Although a significant difference was observed between the mean value of the MCI group and the mean of the NC and AD groups from the box plot, the main body of the distribution overlapped more with the distribution of the NC and AD groups. This indicated that the MMSE scores were almost indistinguishable for MCI patients. Unlike the MMSE scale scores, the CDR scale scores were able to distinguish NC well, but could not distinguish MCI from AD, while the CDR scores were not sufficiently fine-grained and only divided into seven levels. The ODSS-MRI scores exhibited good differentiation between NC, MCI, and AD compared with MMSE and CDR scores, and the degree of response to subtle changes in neurological level was much higher than that of CDR. Based on the box plot, the mean difference between the three groups was evident, and the data distribution of the subjects overlapped less and was evenly distributed. In summary, for the MMSE and CDR scores, no clear boundaries were observed between groups in each stage of AD, and the CDR scores had fewer levels, which cannot effectively reflect the changes in AD progression. The ODSS-MRI effectively described the subtle changes in neurological function during the AD progression and exhibited better intra-class distribution dispersion. This indicated that the ODSS-MRI was better suitable to describe the AD progression.

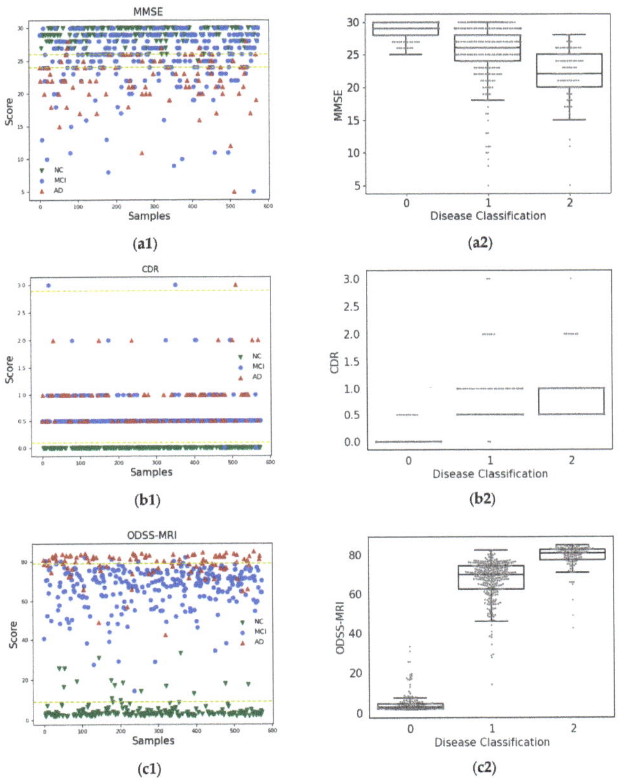

Figure 6. Scatter plots for MMSE (**a1**), CDR (**b1**), and ODSS-MRI (**c1**) and Box-line plots for MMSE (**a2**), CDR (**b2**), ODSS-MRI (**c2**).

4.5. Statistical Information Table

To further evaluate the interpretation of AD progression by the ODSS-MRI, we counted statistical information such as variance, interquartile difference, and range of the ODSS-MRI model score after normalization and compared them with the normalized MMSE and CDR scores, and the statistical results are shown in Table 6. The results show that the variance, standard deviation, and quartile deviation of ODSS-MRI are larger than those of MMSE and CDR, and the variation ratio is also higher, which indicates that the dispersion degree of ODSS-MRI scores is higher than that of the other two. The scores are not stacked on a score value, which indicated that ODSS-MRI is better suitable to respond to subtle changes in AD progression.

4.6. Pareto Analysis

To further illustrate the distribution of the ODSS-MRI, MMSE, and CDR scores, we plotted the Pareto analysis of the three groups, and the results are shown in Figure 7. The figure shows that ODSS-MRI is distributed within each interval and the cumulative curve is relatively smooth. Conversely, in MMSE and CDR, 90% of the score data are concentrated in two intervals and the cumulative curve is considerably steep. This indicates that the MMSE and CDR scores are not sufficiently granular, with severe data accumulation and insufficient refinement for the evaluation of the neurological function of the patient. The experiment showed that the ODSS-MRI scores were more granular compared to the traditional psychiatric rating scales.

Table 6. Comparison of dispersion of ODSS-MRI, MMSE, and CDR.

	ODSS-MRI	MMSE	CDR
Variance	0.151101	0.027904	0.022673
Standard deviation	0.388717	0.167044	0.150577
Quartile deviation	0.849023	0.2	0.166667
Variation ratio	0.998261	0.801418	0.510676

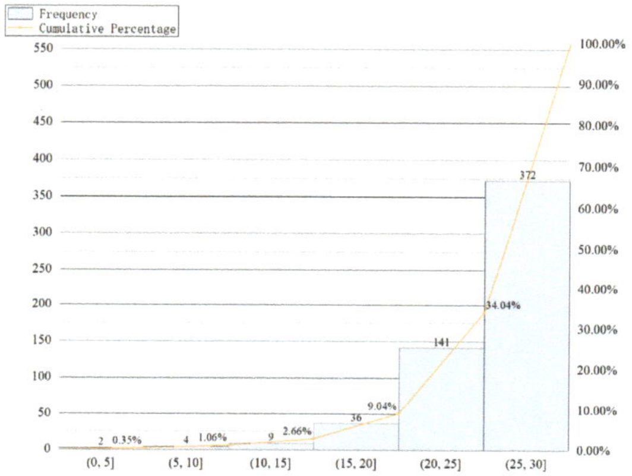

(a)

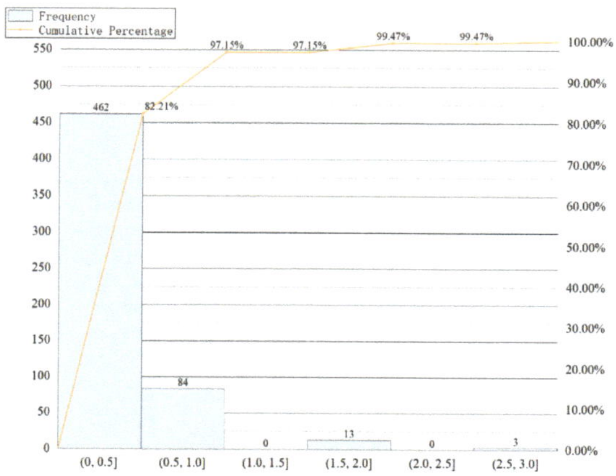

(b)

Figure 7. *Cont.*

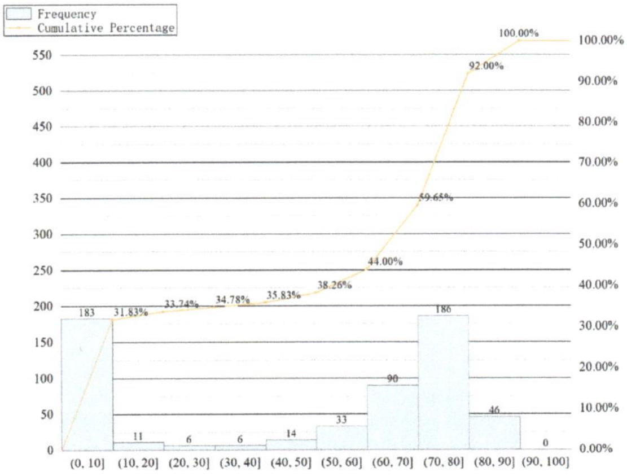

(c)

Figure 7. Pareto analysis chart of MMSE (**a**), CDR (**b**), and ODSS-MRI (**c**).

5. Discussion and Conclusions

In this paper, we proposed a neurological function evaluation model, based on contrastive learning for the evaluation of the progression of AD. Compared with traditional psychiatric rating scales, the data-driven ODSS-MRI ensures objectivity and consistency because it is completely based on medical imaging data and is not affected by biased subjectivity from physicians with different experiences and patients in different statuses during disease severity testing. The neurological function score obtained by the model has a Pearson correlation coefficient of 88.55% and a Spearman correlation coefficient of 87.12% with the stages of AD, which are much higher than those of traditional psychiatric rating scales. This indicates that the ODSS-MRI is superior at describing AD progression. Moreover, the ODSS-MRI exhibits high accuracy in discriminating different progressive stages of AD compared to existing methods, which demonstrated its utility in terms of the neurological function evaluation of AD patients. Above all, this study developed a new method for the neurological function evaluation in AD, which proposed an objective dementia severity scale only based on the whole brain imaging data of patients with contrastive learning and improved the reliability and consistency of existing psychiatric rating scales. This method can potentially be applied to other disease severity evaluation tasks.

Author Contributions: Conceptualization, W.L.; data curation, W.F.; formal analysis, Y.Z.; funding acquisition, W.L.; investigation, Y.Z.; methodology, Y.Z.; project administration, X.C.; resources, W.L.; software, Y.Z.; supervision, W.L.; validation, X.C.; writing—original draft, Y.Z.; writing—review and editing, W.L.; and Alzheimer's Disease Neuroimaging Initiative provided the data. All authors have read and agreed to the published version of the manuscript.

Funding: This work was supported by the National Natural Science Foundation of China (61473131, 60905024).

Informed Consent Statement: Informed consent was obtained from all subjects involved in the study.

Data Availability Statement: Data used in the preparation of this article were obtained from the Alzheimer's Disease Neuroimaging Initiative (ADNI) database (adni.loni.usc.edu, accessed on 24 June 2022). As such, the investigators within the ADNI contributed to the design and implementation of ADNI and provided data but did not participate in the analysis or writing of this report. A complete listing of ADNI investigators can be found at http://adni.loni.usc.edu/wp-content/uploads/how_to_apply/ADNI_Acknowledgement_List.pdf, accessed on 24 June 2022.

Acknowledgments: Data collection and sharing for this project was funded by the Alzheimer's Disease Neuroimaging Initiative (ADNI) (National Institutes of Health Grant U01 AG024904) and DOD ADNI (Department of Defense award number W81XWH-12-2-0012). ADNI is funded by the National Institute on Aging, the National Institute of Biomedical Imaging and Bioengineering, and through generous contributions from the following: AbbVie, Alzheimer's Association; Alzheimer's Drug Discovery Foundation; Araclon Biotech; BioClinica, Inc.; Biogen; Bristol-Myers Squibb Company; CereSpir Inc.; Cogstate; Eisai Inc.; Elan Pharmaceuticals, Inc.; Eli Lilly and Company; EuroImmun; F. Hoffmann-La Roche Ltd. and its affiliated company Genentech, Inc.; Fujirebio; GE Healthcare; IXICO Ltd.; Janssen Alzheimer Immunotherapy Research & Development, LLC.; Johnson & Johnson Pharmaceutical Research & Development LLC.; Lumosity; Lundbeck; Merck & Co., Inc.; Meso Scale Diagnostics, LLC.; NeuroRx Research; Neurotrack Technologies; Novartis Pharmaceuticals Corporation; Pfizer Inc.; Piramal Imaging; Servier; Takeda Pharmaceutical Company; and Transition Therapeutics. The Canadian Institutes of Health Research provided funds to support ADNI clinical sites in Canada. Private-sector contributions are facilitated by the Foundation for the National Institutes of Health (www.fnih.org). The grantee organization is the Northern California Institute for Research and Education, and the study is coordinated by the Alzheimer's Therapeutic Research Institute at the University of Southern California. ADNI data are disseminated by the Laboratory for Neuro Imaging at the University of Southern California.

Conflicts of Interest: The authors declare no competing interest.

References

1. Thies, W.; Bleiler, L. 2013 Alzheimer's disease facts and figures. *Alzheimers Dement.* **2013**, *9*, 208–245.
2. Zhou, Z.; Sodha, V.; Rahman Siddiquee, M.M.; Feng, R.; Tajbakhsh, N.; Gotway, M.B.; Liang, J. Models genesis: Generic autodidactic models for 3d medical image analysis. In Proceedings of the Medical Image Computing and Computer Assisted Intervention–MICCAI 2019, 22nd International Conference, Shenzhen, China, 13–17 October 2019.
3. Saini, M.; Susan, S. VGGIN-Net: Deep Transfer Network for Imbalanced Breast Cancer Dataset. *IEEE/ACM Trans. Comput. Biol. Bioinform.* **2023**, *20*, 752–762. [CrossRef]
4. Khader, F.; Mueller-Franzes, G.; Arasteh, S.T.; Han, T.; Haarburger, C.; Schulze-Hagen, M.; Schad, P.; Engelhardt, S.; Baebler, B.; Foersch, S. Medical Diffusion--Denoising Diffusion Probabilistic Models for 3D Medical Image Generation. *arXiv* **2022**, arXiv:2211.03364.
5. Lian, C.; Liu, M.; Zhang, J.; Shen, D. Hierarchical fully convolutional network for joint atrophy localization and Alzheimer's disease diagnosis using structural MRI. *IEEE Trans. Pattern Anal. Mach. Intell.* **2018**, *42*, 880–893. [CrossRef] [PubMed]
6. Shao, W.; Peng, Y.; Zu, C.; Wang, M.; Zhang, D.; Alzheimer's Disease Neuroimaging Initiative. Hypergraph based multi-task feature selection for multimodal classification of Alzheimer's disease. *Comput. Med. Imaging Graph.* **2020**, *80*, 101663. [CrossRef] [PubMed]
7. Saratxaga, C.L.; Moya, I.; Picón, A.; Acosta, M.; Moreno-Fernandez-De-Leceta, A.; Garrote, E.; Bereciartua-Perez, A. MRI deep learning-based solution for Alzheimer's disease prediction. *J. Pers. Med.* **2021**, *11*, 902. [CrossRef]
8. Folstein, M.F.; Folstein, S.E.; Mchugh, P.R. "Mini-Mental State": A Practical Method for Grading the Cognitive State of Patients for the Clinician. *J. Psychiatr. Res.* **1975**, *12*, 189–198. [CrossRef] [PubMed]
9. Berg, L. Clinical Dementia Rating (CDR). *Psychopharmacol. Bull.* **1988**, *24*, 637.
10. Pfeffer, R.I.; Kurosaki, T.T.; Harrah, C.H.; Chance, J.M.; Filos, S. Measurement of Functional Activities in Older Adults in the Community. *J. Gerontol.* **1982**, *37*, 323–329. [CrossRef]
11. Rosen, W.G.; Mohs, R.C.; Davis, K.L. A New Rating Scale for Alzheimer's Disease. *Am. J. Psychiatry* **1984**, *141*, 1356–1364.
12. Schmidt, M. *Rey Auditory Verbal Learning Test: A Handbook*; Western Psychological Services: Los Angeles, CA, USA, 1996.
13. Prasher, V.; Farooq, A.; Holder, R. The Adaptive Behaviour Dementia Questionnaire (ABDQ): Screening questionnaire for dementia in Alzheimer's disease in adults with Down syndrome. *Res. Dev. Disabil.* **2004**, *25*, 385–397. [CrossRef]
14. Dickerson, B.C. Advances in quantitative magnetic resonance imaging-based biomarkers for Alzheimer disease. *Alzheimers Res. Ther.* **2010**, *2*, 21. [CrossRef]
15. Zhou, Y.; Tan, C.; Wen, D.; Sun, H.; Han, W.; Xu, Y. The biomarkers for identifying preclinical Alzheimer's disease via structural and functional magnetic resonance imaging. *Front. Aging Neurosci.* **2016**, *8*, 92. [CrossRef] [PubMed]
16. Qing, Z.; Chen, F.; Lu, J.; Lv, P.; Li, W.; Liang, X.; Wang, M.; Wang, Z.; Zhang, X.; Zhang, B.; et al. Causal structural covariance network revealing atrophy progression in Alzheimer's disease continuum. *Hum. Brain Mapp.* **2021**, *42*, 3950–3962. [CrossRef] [PubMed]
17. Heung-Il, S. Deep Learning-Based Feature Representation for AD/MCI Classification. *Med. Image Comput. Comput. Assist. Interv. MICCAI* **2013**, *16*, 583–590.
18. Andres, O. Ensembles of Deep Learning Architectures for the Early Diagnosis of the Alzheimer's Disease. *Int. J. Neural Syst.* **2016**, *26*, 1650025.

19. Li, M.D.; Chang, K.; Bearce, B.; Chang, C.Y.; Huang, A.J.; Campbell, J.P.; Brown, J.M.; Singh, P.; Hoebel, K.V.; Erdogmus, D.; et al. Siamese neural networks for continuous disease severity evaluation and change detection in medical imaging. *NPJ Digit. Med.* **2020**, *3*, 48. [CrossRef]
20. Li, Y.; Wu, Z.; Zhao, S.; Wu, X.; Kuang, Y.; Yan, Y.; Ge, S.; Wang, K.; Fan, W.; Chen, X.; et al. PSENet: Psoriasis severity evaluation network. *Proc. AAAI Conf. Artif. Intell.* **2020**, *34*, 800–807. [CrossRef]
21. Wu, Z.; Xiong, Y.; Yu, S.X.; Lin, D. Unsupervised feature learning via non-parametric instance discrimination. In Proceedings of the IEEE Conference on Computer Vision and Pattern Recognition, Salt Lake City, UT, USA, 18–23 June 2018; pp. 3733–3742.
22. He, K.; Fan, H.; Wu, Y.; Xie, S.; Girshick, R. Momentum contrast for unsupervised visual representation learning. In Proceedings of the IEEE/CVF Conference on Computer Vision and Pattern Recognition, Seattle, WA, USA, 13–19 June 2020; pp. 9729–9738.
23. Chen, T.; Kornblith, S.; Norouzi, M.; Hinton, G. A simple framework for contrastive learning of visual representations. International conference on machine learning. *PMLR* **2020**, *119*, 1597–1607.
24. Chen, X.; He, K. Exploring simple siamese representation learning. In Proceedings of the IEEE/CVF Conference on Computer Vision and Pattern Recognition, Nashville, TN, USA, 19–25 June 2021; pp. 15750–15758.
25. Khosla, P.; Teterwak, P.; Wang, C.; Sarna, A.; Tian, Y.; Isola, P.; Maschinot, A.; Liu, C. Supervised contrastive learning. *Adv. Neural Inf. Process. Syst.* **2020**, *33*, 18661–18673.
26. Chen, T.; Kornblith, S.; Swersky, K.; Norouzi, M.; Hinton, G. Big self-supervised models are strong semi-supervised learners. *Adv. Neural Inf. Process. Syst.* **2020**, *33*, 22243–22255.
27. Wang, X.; Zhang, R.; Shen, C.; Kong, T.; Li, L. Dense contrastive learning for self-supervised visual pre-training. In Proceedings of the IEEE/CVF Conference on Computer Vision and Pattern Recognition, Nashville, TN, USA, 19–25 June 2021; pp. 3024–3033.
28. Han, T.; Xie, W.; Zisserman, A. Self-supervised co-training for video representation learning. *Adv. Neural Inf. Process. Syst.* **2020**, *33*, 5679–5690.
29. Grill, J.B.; Strub, F.; Altché, F.; Tallec, C.; Richemond, P.; Buchatskaya, E.; Doersch, C.; Pires, B.A.; Guo, Z.; Azar, M.; et al. Bootstrap your own latent-a new approach to self-supervised learning. *Adv. Neural Inf. Process. Syst.* **2020**, *33*, 21271–21284.
30. Xie, S.; Gu, J.; Guo, D.; Qi, C.R.; Guibas, L.; Litany, O. Pointcontrast: Unsupervised pre-training for 3d point cloud understanding. In Proceedings of the Computer Vision–ECCV 2020, 16th European Conference, Glasgow, UK, 23–28 August 2020; Proceedings, Part III 16. Springer International Publishing: Berlin/Heidelberg, Germany, 2020; pp. 574–591.
31. Tian, Y.; Krishnan, D.; Isola, P. Contrastive multiview coding. In Proceedings of the Computer Vision–ECCV 2020, 16th European Conference, Glasgow, UK, 23–28 August 2020; Proceedings, Part XI 16. Springer International Publishing: Berlin/Heidelberg, Germany, 2020; pp. 776–794.

sensors

MDPI

Article

Path Following and Collision Avoidance of a Ribbon-Fin Propelled Underwater Biomimetic Vehicle-Manipulator System

Yanbing He [1], Xiang Dong [1,*], Yu Wang [2] and Shuo Wang [2]

[1] School of Electrical Engineering and Automation, Anhui University, Hefei 230601, China; z20201027@stu.ahu.edu.cn
[2] State Key Laboratory of Management and Control for Complex Systems, Institute of Automation, Chinese Academy of Sciences, Beijing 100190, China; yu.wang@ia.ac.cn (Y.W.); shuo.wang@ia.ac.cn (S.W.)
* Correspondence: xdong@ahu.edu.cn

Abstract: This paper addresses the problem of path following and dynamic obstacle avoidance for an underwater biomimetic vehicle-manipulator system (UBVMS). Firstly, the general kinematic and dynamic models of underwater vehicles are presented; then, a nonlinear model predictive control (NMPC) scheme is employed to track a reference path and collision avoidance simultaneously. Moreover, to minimize the tracking error and for a higher degree of robustness, improved extended state observers are used to estimate model uncertainties and disturbances to be fed into the NMPC prediction model. On top of this, the proposed method also considers the uncertainty of the state estimator, while combining a priori estimation of the Kalman filter to reasonably predict the position of dynamic obstacles during short periods. Finally, simulations and experimental results are carried out to assess the validity of the proposed method in case of disturbances and constraints.

Keywords: underwater biomimetic vehicle-manipulator system; nonlinear model predictive control; extended state observers (ESOs); collision avoidance; path following

Citation: He, Y.; Dong, X.; Wang, Y.; Wang, S. Path Following and Collision Avoidance of a Ribbon-Fin Propelled Underwater Biomimetic Vehicle-Manipulator System. *Sensors* **2023**, *23*, 7061. https://doi.org/10.3390/s23167061

Academic Editor: Enrico Meli

Received: 29 June 2023
Revised: 6 August 2023
Accepted: 7 August 2023
Published: 9 August 2023

1. Introduction

Over the past few decades, humans have explored and exploited a great number of marine resources; autonomous underwater vehicles (AUVs) have been playing an essential role in underwater operations for marine water quality monitoring as well as oil and gas exploration and other resource surveys [1,2]. With the increasing popularity of underwater operations in the marine environment and the more complex and diverse working environment, the human requirements for underwater robots have further increased, and biomimetic underwater vehicles (BUVs) have attracted a lot of attention from researchers. Through natural selection and continuous self-evolution, fish have evolved near-perfect body mechanisms and control modes of underwater motion. Compared with traditional propeller-equipped underwater robots, BUVs have more maneuverability and can be used to achieve fine attitude adjustment [3,4], especially for low-speed cruising in some particularly restricted areas. In addition, because of the undulating fin movements, they cause less noise and less pollution to the ocean [5].

For the past twenty years, research on the design and motion control of BUVs has received rapid development. Heriot-Watt University's Sfakiotakis used the parallel bellows actuator as an early undulatory-fin device [6]. Thereafter, Delft University of Technology designed a biomimetic autonomous underwater vehicle with an undulating fin [7]. Niu et al. successfully generated a motion modality by investigating the use of the central pattern generator (CPG) approach for motion learning of an Anguilliform robot fish.

They applied it to a BUV [8]. Wang et al. proposed a backstepping (BP) technique for a bio-inspired robot with undulatory fins to achieve closed-loop depth and course control [9]. Most researchers, however, are more focused on undulatory fin control or some closed-loop path tracking, while BUVs in the actual operation process will be subject to external forces caused by surges, and other interferences and obstacle-avoidance problems are not well-solved.

In recent years, the research commonly has addressed the unknown disturbances of underwater vehicles by designing observers to estimate the unknown disturbances and then compensating for the estimated disturbances by using controllers, which include high-gain observers and an extended state observer (ESO) [10]. In [11], the researchers proposed a motion control system composed of a MIMO PID controller and a high-gain observer, which considers the presence of unmodeled plant dynamics of ROV, measurement errors, noise, and environmental disturbances. In [12], the team proposed a neural extended state observer (ESO) to address the system parameters' uncertainty and environmental disturbances. In [13], they proposed a linear extended state observer (LESO) applied in the case of sideslip caused by wind, current, and other disturbances in the environment. In addition to the uncertain dynamics and disturbances, another noteworthy problem is the physical constraints for ribbon-fin-propelled underwater vehicles. In recent years, the model predictive control (MPC) algorithm has become an effective optimization method to handle constrained systems [14]. A novel robust MPC method was introduced in [15] to solve the roll constraint for following a straight path. In [16], the team proposed a path-following collision-avoidance MPC method that approximates obstacle shapes as convex polygons and combines the arising norm equality with the dual collision-avoidance inequality constraints.

Most of the above literature for model uncertainty and unknown disturbances in the case of underwater robots' or surface vessels' path tracking problems has put forward valid solutions, but few can take into account the obstacle-avoidance problem in the control algorithm, especially in BUVs. In the natural ocean environment, once there is a disturbance, and once deviated from the predetermined trajectory, the risk of collision will greatly increase; underwater robots must safely avoid a collision, and maintain tracking on the predetermined path, which is a tough challenge for the underdriven underwater robot with only surge force and yaw moment. In addition, the control input of an underwater biomimetic vehicle-manipulator system (UBVMS) that relies on fluctuating fins for propulsion is different from that of conventional AUVs and surface vessels, which makes the model uncertainty and the estimation of unknown external disturbances more complex.

In this paper, we propose a concept of integrating an underwater biomimetic vehicle manipulator system (UBVMS) path tracking problem with the multi-object/obstacle-avoidance problem based on the framework of MPC. From a practical perspective, the actual physical platform limitations are incorporated into the constraints of the optimization problem by using the complete nonlinear dynamics constraints of the UBVMS. In experimental tests, since the scan period of the sonar commonly used for underwater obstacle detection is much longer than the sampling period of the NMPC, the speed of obstacle motion is predicted a priori using the Kalman filter. In the case of predicted collisions, not only does it improve the safety of the UBVMS but it also deviates from the defined path as little as possible. To further improve the robustness of obstacle avoidance, the collision cost term in the objective function is shaped using the uncertainty of the state estimation observer. For the consideration of environmental disturbances and the uncertainty of the UBVMS, the total external disturbances, excluding the effects of surge force and yaw moment, are computed using the introduction of extended state observers (ESOs), and the total disturbance term is finally incorporated into the prediction term in NMPC. When the NMPC control cycle ends and the UBVMS reaches a new position, the following optimal input is recalculated based on the current state, and the desired trajectory. Thus, the NMPC algorithm with disturbance compensation, and with obstacle-avoidance constraints incorporated into the control, can be practically applied to the UBVMS even in realistic harsh natural marine environments. Finally, we validate the effectiveness of the algorithm through simulation and experiments with a physical prototype.

2. UBVMS Model

2.1. Design of the UBVMS

Bio-inspired long-fin propulsion is introduced into a UVMS (underwater vehicle manipulator system). Compared with the traditional propeller propulsion, it is smaller and can achieve smooth control of the body. The bionic fin propulsors provide excellent stability for underwater operations, and the propeller can provide sufficient power to ensure better maneuverability in the actual water with disturbances. As shown in Figure 1, the UBVMS has two symmetrical long-bionic-fin propulsors, four vertically oriented small propellers, and two flippers at the end of the tail to help improve maneuverability. The UBVMS head is equipped with a depth camera, altimeter, sonar, and many other sensors to help acquire more information about the underwater environment while conducting underwater operations. More details are described in [17]. The physical parameters of UBVMS are in Table 1.

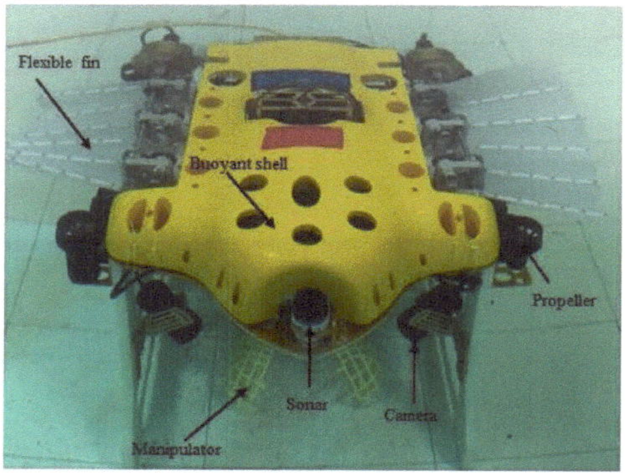

Figure 1. UBVMS.

Table 1. Physical parameters of UBVMS.

Parameter	Value	Parameter	Value
Mass	55 kg	Buoyancy	54.3 kg
Manipulator mass	2.5 kg	Fin length	0.58 m
Length	0.95 m	Maximum frequency	20 Hz
Width	0.7 m	Maximum linear velocity	0.43 m
Height	0.6 m	Maximum steering speed	0.89 rad/s

2.2. General Kinematics and Dynamics of Underwater Vehicle

The UBVMS has good stability due to its large metacentric height, so it is reasonable to neglect the motion in pitch and roll [9]. With bionic fin propulsors only, the maximum speed of the UBVMS is approximately 0.4 m/s, and the nonlinear damping is negligible at low speeds in the laboratory pool. Based on that, the 3-DOF kinematic and dynamic model can be presented as follows [18]:

$$\dot{x} = \begin{bmatrix} J(\varphi)v \\ M^{-1}(-C(v)v - Dv + \tau + \tau_d) \end{bmatrix} = f(x,u) \quad (1)$$

where $x = [\eta^T \; v^T]^T \in \mathbb{R}^6$ is a state vector, $\eta = [x \; y \; \varphi]^T \in \mathbb{R}^3$, $x(m)$ and $y(m)$ are the positions, $\varphi(rad)$ is the heading angle with respect to the earth-fixed frame, and $v = [u \; v \; r]^T \in \mathbb{R}^3$ represents the velocities in surge (m/s), sway (m/s), and yaw (rad/s), decomposed in the body-fixed frame. $\tau = [f_u\ 0\ f_r]^T$, $f_u(N)$, and $f_r(N \cdot m)$ are the surge forces and yaw moment produced by the bionic fin propulsors. $\tau_d = [\tau_{du}\ \tau_{dv}\ \tau_{dr}]^T$, $\tau_{du}(N)$, $\tau_{dv}(N)$, and $\tau_{dr}(N \cdot m)$ describe the external forces and moment acting on surge, sway, and yaw, respectively. Based on the previous reasonable assumptions, the system inertial mass matrix (invertible) M and damping matrix D are a diagonal matrix and Coriolis and centrifugal matrix $C(v)$, which are given by:

$$M \triangleq \begin{bmatrix} m_{11} & 0 & 0 \\ 0 & m_{22} & 0 \\ 0 & 0 & m_{33} \end{bmatrix}, D \triangleq \begin{bmatrix} d_{11} & 0 & 0 \\ 0 & d_{22} & 0 \\ 0 & 0 & d_{33} \end{bmatrix}, \quad C \triangleq \begin{bmatrix} 0 & 0 & -m_{22}v \\ 0 & 0 & m_{11}u \\ m_{22}v & -m_{11}v & 0 \end{bmatrix} \tag{2}$$

where $J(\varphi) \in SO(3)$ is the rotation matrix from body-fixed velocities ν to earth-fixed frame velocities $\dot{\eta}$, and has the following structure:

$$J(\varphi) = \begin{bmatrix} \cos(\varphi) & -\sin(\varphi) & 0 \\ \sin(\varphi) & \cos(\varphi) & 0 \\ 0 & 0 & 1 \end{bmatrix} \tag{3}$$

The assumptions in this paper are as follows:

1. The UBVMS states x, y, and φ can be measured.
2. The total unknowns τ_{du}, τ_{dv}, and τ_{dr} are bounded, namely $|\tau_{du}| \leq \tau_{dumax}$, $|\tau_{dr}| \leq \tau_{drmax}$, $|\tau_{dv}| \leq \tau_{dvmax}$ and $|\dot{\tau}_{du}| \leq |\dot{\tau}_{dumax}|$, $|\dot{\tau}_{dr}| \leq |\dot{\tau}_{drmax}|$, $|\dot{\tau}_{dv}| \leq |\dot{\tau}_{dvmax}|$.

3. Problem Formulation

3.1. Nonlinear Model Predictive Control

NMPC is a numerical optimization-based control method to achieve the desired optimal performance by systematically ensuring that different constraints' control is satisfied. The system cost function is constructed, which refers to the error between the state and the desired system state for the next sampling periods, combined with the physical constraints of the UBVMS; then, the cost function is minimized to obtain the optimal control sequence [19,20]. The first element of the control vector sequence is then transformed and fed into the propeller, repeating optimal control at the next sample.

Based on the previously defined system state vectors x and control vectors u, we can obtain the following optimization problem and solve it online:

$$\begin{aligned} &\min_{x(t),u(t)} J(x,u) = \textstyle\int_{t=0}^{T} \{J_x(x(t), x_{ref}(t)) + J_u(\Delta u(t)) + J_c(x(t))\}dt \\ &\text{s.t. } \dot{x} = f(x,u), (\text{t}) \in U,\ x(0) = x(t_0) \end{aligned} \tag{4}$$

where f represents system continuous nonlinear equations, J_x is a cost function of the error with the desired state x_{ref} of the system, J_u and J_c control input and collision cost penalties, respectively, and U is a set of feasible control vector inputs. Since the NMPC is to be used with a practical system, the thrust of the bionic fin propulsor is limited, and such hard constraint must be considered at every control cycle. The input control vector constraints are as follows:

$$\text{U} = \left\{ \text{u} \in \mathbb{R}^3 \middle| \begin{bmatrix} f_{u\text{min}} \\ f_{v\text{min}} \\ f_{r\text{min}} \end{bmatrix} \leq u \leq \begin{bmatrix} f_{u\text{max}} \\ f_{v\text{max}} \\ f_{r\text{max}} \end{bmatrix} \right\} \tag{5}$$

It is worth noting here that the UBVMS is an underdriven underwater vehicle and, in fact, $f_{v\text{min}} = f_{v\text{max}} = 0$. Now, for a detailed description of the cost function part in

Equation (4), the first term of the cost function $J_x(x(t),\ x_{ref}(t))$ represents the state cost, which penalizes deviating from the desired state x_{ref}; the details are as follows:

$$J_x(x(t), x_{ref}(t)) = \left|\left|x(t) - x_{ref}(t)\right|\right|^2_{Q_x} \tag{6}$$

where $Q_x \in \mathbb{R}^{6\times 6}$ represents positive definite weight matrices. The second term of the cost function $J_u(\Delta u(t))$ is to ensure the smoothness of the system control vector throughout the control process u, related to the previous control action; the details are as follows:

$$J_u(\Delta u(t)) = \ ||\ u(t+T) - u(t)\ ||^2_{Q_u} \tag{7}$$

The last term of the cost function is to expect the UBVMS to have a sufficiently safe distance from dynamic obstacles when working underwater, and we introduce a collision cost term similar to the logistic function, which ensures that the cost function is continuously differentiable. Formally,

$$J_c(x(t)) = \sum_{i=1}^{n} \frac{Q_{c,i}}{1 + \exp k_i(d_i(t) - r_{th,i}(t))} \tag{8}$$

where $Q_{c,i}$ is a design weight, $d_i(t)$ is the Euclidean distance from the $i - th$ dynamic obstacle, and $k_i > 0$ is a positive constant that denotes the smoothness of the obstacle avoidance trajectory. As the value of k_i increases, the cost function becomes very sharp, which can make the optimization problem more challenging to solve, so the value of k_i can be chosen to be adjusted according to different environmental situations [21]. The global reference path is provided by the global planner; we use a cubic spline to interpolate, calculate the reference yaw and divide the whole reference trajectory into segments of the same size; a sequence of path segments connecting path points M forms a global reference path $\updownarrow$. The UBVMS chooses the closest point as the tracking target, as shown in Figure 2.

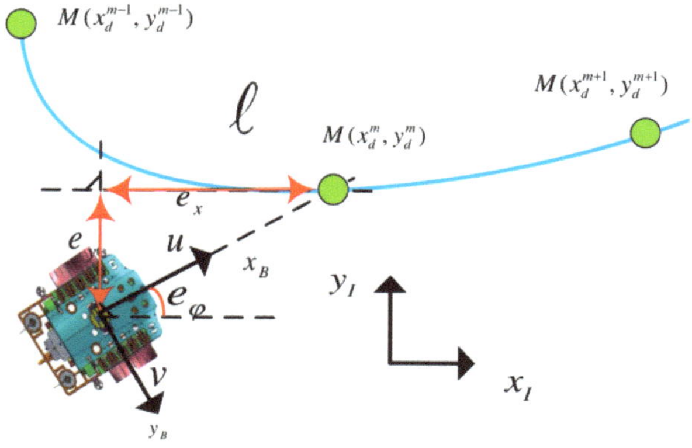

Figure 2. The global guidance principle.

3.2. *Uncertainty Consideration and Disturbance Estimation*

For the design of a UBVMS control system, system uncertainty is a fundamental issue that cannot be avoided, coming from the internal (structure or parameters) elements of the system on the one hand and the external (environmental disturbances) elements on the other [22]. NMPC is a model-based controller, and it is essential to improve the model's accuracy as much as possible, so before applying it to practical objects, we need to consider the impact on the control quality due to the model's internal uncertainty and external environmental disturbances. In addition, NMPC is also based on the model to make future state predictions. Then, in the optimization based on the prediction, if the disturbance is not compensated for in time, the control effect will become terrible, and even the system will become unstable. We add disturbance terms to the prediction model of NMPC, such as τ_{du}, τ_{dr} and τ_{dv} shown in Equation (1). The disturbance observer will compute the disturbance in each control cycle. When operating in a complex marine environment, the operational performance of marine vehicles is susceptible to external environmental disturbances. The disturbance observers are widely applied in various marine applications and have a practical significance [23]. In [22], the estimation ability of ESO for uncertainty is discussed and analyzed, and the researchers show that ESO can deal with a large number of disturbances. ESO is the core of ADRC proposed by [24], and then Gao proposed an improved linear ADRC control algorithm, LADRC, based on the classical ADRC idea [25]. In [26], LESO is introduced to estimate the uncertain parameters and external disturbances, and the surge velocity is calculated.

In our work, ESO is used to estimate the velocity and external disturbances in the motion coordinate system, and the nonlinear function in ESO is modified by introducing a hyperbolic tangent function to replace the original nonlinear function, which is a bounded singular function with more adjustable parameters than the nonlinear function and can be used to flexibly adjust the limits of the error magnitude as follows:

$$\beta \tanh(\lambda z) = \frac{e^{\lambda z} - e^{-\lambda z}}{e^{\lambda z} + e^{-\lambda z}} \tag{9}$$

λ and β are positive parameters that can be used to adjust the limit range of the amplitude and the rate of change in the amplitude, respectively. According to the UB-VMS dynamic model (see Equation (1)), the disturbance moment observer of the external environment in the yaw direction can be:

$$\begin{cases} \dot{\hat{\varphi}} = \hat{r} - l_0 \tanh(\lambda_1(\hat{\varphi} - \varphi)) \\ \dot{\hat{r}} = \hat{\tau}_{dr} - l_1 \tanh(\lambda_1(\hat{\varphi} - \varphi)) + \frac{(m_{11} - m_{22})}{m_{33}} uv - \frac{d_{33}}{m_{33}} r + b_0 \frac{f_d}{m_{33}} \\ \dot{\hat{\tau}}_{dr} = -l_2 \tanh(\lambda_1(\hat{\varphi} - \varphi)) \end{cases} \tag{10}$$

where l_0, l_1, and l_2 are positive parameters, which could adjust the feedback range of estimation errors. Increasing their values within a certain range can improve the estimation accuracy, but it also brings oscillations as a result [27]. $\hat{r}$, $\hat{\tau}_{dr}$, and $\hat{\varphi}$ represent the respective estimated values, and b_0 is the compensation factor.

$$\begin{cases} \dot{\hat{\zeta}} = \hat{u} - l_3 \tanh(\lambda_2(\hat{\zeta} - \zeta)) \\ \dot{\hat{u}} = \hat{\tau}_{du} - l_4 \tanh(\lambda_2(\hat{\zeta} - \zeta)) + \frac{m_{22}}{m_{11}} vr - \frac{d_{11}}{m_{11}} u + b_1 \frac{f_u}{m_{11}} \\ \dot{\hat{\tau}}_{du} = -l_5 \tanh(\lambda_2(\hat{\zeta} - \zeta)) \end{cases} \tag{11}$$

$$\begin{cases} \dot{\hat{\chi}} = \hat{v} - l_6 \tanh(\lambda_3(\hat{\chi} - \chi)) \\ \dot{\hat{v}} = \hat{d}_v - l_7 \tanh(\lambda_3(\hat{\chi} - \chi)) - \frac{m_{11}}{m_{22}} ur - \frac{d_{22}}{m_{22}} v \\ \dot{\hat{\tau}}_{dv} = -l_8 \tanh(\lambda_3(\hat{\chi} - \chi)) \end{cases} \tag{12}$$

$\zeta = u \cdot t$ and $\chi = v \cdot t$ denote the displacement of the surge velocity and sway velocity in the body-fixed frame, respectively, where $l_3, \ldots, l_8$ are positive parameters, $\hat{u}$, $\hat{\tau}_{du}$, $\hat{\zeta}$, $\hat{v}$, $\hat{\tau}_{dv}$, and $\hat{\chi}$ represent the respective estimated values, and b_1 is the compensation factor, u^* is the optimal control input for the solution. Notably, the estimates for the disturbance terms are added to the prediction module of the controller NMPC and updated by ESOs at each sampling cycle. There is a more intuitive representation in Figure 3 and Algorithm 1.

Algorithm 1: NMPC for UBVMS Algorithm

Initialize system parameters;
Repeat;
    Measure and compute estimated velocities, along with system state $x(t_0)$ disturbance term updates from ESOs;
    Compute future predicted states $\hat{x}(t+nT_s), n = 1, \ldots, N$ and predicted obstacle states (17);
        Solve the OCP (4) and obtain the desired optimal solution sequence $u^*(t+nT_s), n = 1, \ldots, N-1$;
    Allocate the desired optimal solution u_0^* and convert to bionic fin propulsor;
    If sonar scan completed, **then**
        Kalman filter update: compute $r_{th,k}(t)$ and update P_t^-, P_t;
    Else $P_t^- = P_t^-$, $P_t = P_t$;
    End if
    Compute the desired reference path and save system data;
    Update: $nmpciter = nmpciter + 1$;
Until stop.

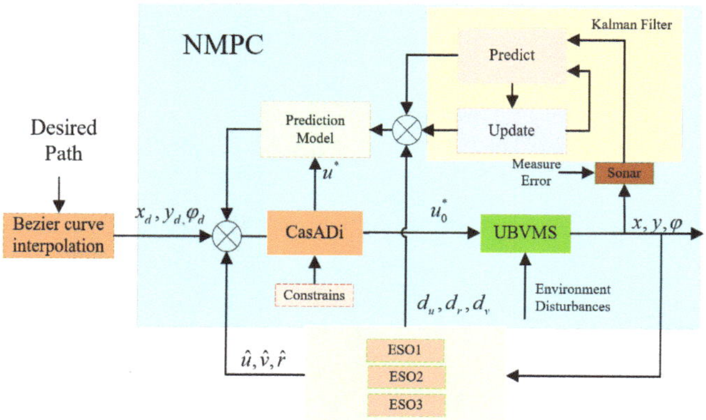

Figure 3. Diagram of the proposed control scheme.

3.3. *Robust Collision Avoidance*

In the previous section, we discussed the uncertainty in the UBVMS; in order to further improve the robustness of control in underwater environments, the uncertainty of state estimation of dynamic obstacles in water is considered, inspired by the literature [28]. With non-negligible uncertainty in predicting obstacle states due to the inaccuracy of the prediction model and the interference of external environmental measurement noise, we should be more conservative and allow the system to approach the obstacle by increasing $r_{th,i}(t)$ along the prediction horizon. In the natural marine underwater environment, dynamic obstacles are hardly ever in actual uniform motion, so to simulate the trajectory of dynamic obstacles closer to the real situation, we chose the uniform motion mode with Gaussian noise interference as follows:

$$p_i^{obs}(t) = p_i^{obs}(t_0) + (v_i(t_0) + w_0)(t - t_0) \tag{13}$$

where w_0 is a random number $w_0(t) \sim N(0, Q_0(t))$, and $v_i(t_0)$ and $p_i^{obs}(t_0)$ are the velocity and position at the current moment, respectively. To obtain an accurate estimation of the obstacle state and compute the uncertainty in the state estimator of objects in the process, for the proposed obstacle motion model (12), we choose to perform uncertainty propagation based on a Kalman filter (KF) [28]. The KF can be formulated as:

Predict:

$$\begin{cases} \hat{x}_t^- = F\hat{x}_{t-1} \\ P_t^- = FP_{t-1}F^T + Q_t \end{cases} \quad (14)$$

Update:

$$\begin{cases} K_t = P_t^- H^T \left(HP_{t-1}H^T + R_t\right)^{-1} \\ \hat{x}_t = \hat{x}_t^- + K_t(Z_t - H\hat{x}_t^-) \\ P_t = (I - K_t H)\, P_t^- \end{cases} \quad (15)$$

where $x_t = [x^{obs}, y^{obs}, v^{obs}]^T$ is a state vector, P_t^- and P_t represent the a priori (predicted) and a posteriori state vectors, respectively, and $Z_t = [x^o, y^o, v^o]^T$ is the measurement vector acquired by employing sonar at moment t. F is the state transition Jacobian matrix, and H is the observation matrix. Q_t is the covariance matrix of the motion obstacle system noise, and R_t is the covariance matrix of the observation noise, and we must note that all of these are positive parameters. The reasonable selection of Q_t and R_t is crucial for the development of the Kalman filter [29]. We express the uncertainty of the $i - th$ dynamic obstacle at moment t by solving the equation $P_t^- = FP_{t-1}F^T + Q_t$ with maximum eigenvalue $\delta_k(t)$. $r_{th,i}(t)$ can be calculated using the following equation:

$$r_{th,i}(t) = r_{th} + 10\delta_i(t) \quad (16)$$

where r_{th} represents constant parameters, and the use of an enclosing sphere to approximate the uncertainty ellipsoid makes the bound more conservative [28]. Compared to the travel space of a mobile robot on land, the moveable space of a UBVMS is ample, and it is wiser to convert the complex elliptical distance calculation to a circular distance calculation in terms of computational complexity. In addition, since usually the sonar scanning time to ascertain the obstacle position is relatively long, the time to obtain the updated obstacle position is much longer than the NMPC sampling period. Since we cannot determine the latest state of the obstacle movement in every control cycle, to carry out a safer motion control, reasonable use of the Kalman filter a priori information for obstacle position prediction is critical. The obstacle trajectories are predicted as follows:

$$p_{obs\ i}^{pre}(t_0 + n \cdot T_s) = p_{obs\ i}(t_0) + n \cdot T_s \cdot v^{o-}(t_0)\ ,\ \mathrm{n} \in (1, \ldots, N) \quad (17)$$

where $P_{obs\ i}(t_0)$ denotes the obstacle position at moment t_0, T_s is the NMPC sampling period, and $v(t_0)^{o-}$ denotes the $i - th$ obstacle prior velocity obtained by Kalman filtering at moment t_0. In the NMPC prediction module, along with the number of prediction steps n from 1 to N, the obstacle prediction position is also updated until the Kalman filter is updated for the prior velocity. The proposed NMPC algorithm is briefly described in Algorithm 1.

The control structure of the proposed system is schematically shown in Figure 3. CasADi is an open-source software framework for numerical optimization that is well suited to nonlinear optimization problems, especially those constrained by differential equations [30,31]. In the simulation and experiment part of this paper, CasADi solver plugin ipopt is selected to solve the optimal control problem.

4. Simulations and Experimental Results

4.1. Simulation Results

The physical parameters for the UBVMS are chosen as [9]: $m_{11} = 57.5$, $m_{22} = 61.3$, $m_{33} = 1.15$, $d_{11} = 53$, $d_{22} = 58$, and $d_{33} = 3.1$. The initial value of the system state vector is 0. The NMPC-related control parameters are as follows: $\mathrm{T_s} = 0.2$ s, $N_p = 25$, $N_c = 1$, and weighting matrices $Q_x = diag\{55.0, 55.0, 5.0, 1.0, 800, 5.0\}$ and $Q_u = \mathrm{diag}\{10.0, 25.0\}$. The collision penalty weights and prediction penalty weights are $Q_c = 20$, $Q_{cp} = 10$. The desired surge velocity in the body-fixed frame is 0.2 m/s, and $l_1 = 3.6$, $l_2 = 4.32$, $l_3 = 1.728$, $l_4 = 2.4$, $l_5 = 1.92$, $l_6 = 1.516$, $l_7 = 2.4$, $l_8 = 1.92$, and $l_9 = 0.528$. The external disturbance is $\tau_{du} = 0.5\sin(0.75 \times 0.1t)$, $\tau_{dv} = 0.4\sin(0.75 \times 0.1t)$, $\tau_{dr} = 0.3\sin(0.75 \times 0.12t)$.

Dynamic obstacles are initially positioned as follows: $(x_{t_0}^{obs}, y_{t_0}^{obs}) = (6, 3)$. The radius of the UBVMS in the simulation is 0.3 m, and the radius of dynamic obstacle expansion is 0.5 m. The initial velocity of the obstacle is -0.15 m/s, as well as adding to the Gaussian distribution $N(0,\ 0.02)$. In the multi-obstacle simulation, the initial positions of the three dynamic obstacles are as follows: (5.4, 4.0), (4.15, 1.0), (5.0, 6.0). To better verify the effectiveness of the proposed method, the three dynamic obstacles make a linear motion with initial velocity -0.15 m/s and an additional Gaussian distribution $N(0,\ 0.02)$ at the approach of the UBVMS.

Figure 4 shows the more critical moments of the BUMVS in the path tracking process. The blue curve is the desired path, the small red circle is the UBVMS, the large one is the dynamic obstacle, and the marker symbol above the red circle indicates the yaw angle at this moment. The yellow circle indicates the posterior position based on the Kalman filter. In Figure 4a, it can be seen that because there is an obstacle in front of the right of the UBVMS at this moment, it chooses to deviate from the intended trajectory to the left to avoid a collision; notably, in Figure 4b, when the obstacle trajectory is predicted by combining a priori messages in the KF, the controller chooses to drive to the right with lower probability of collision. Figure 4c,d show that the UBVMS returns to the original desired path soon after avoiding this collision.

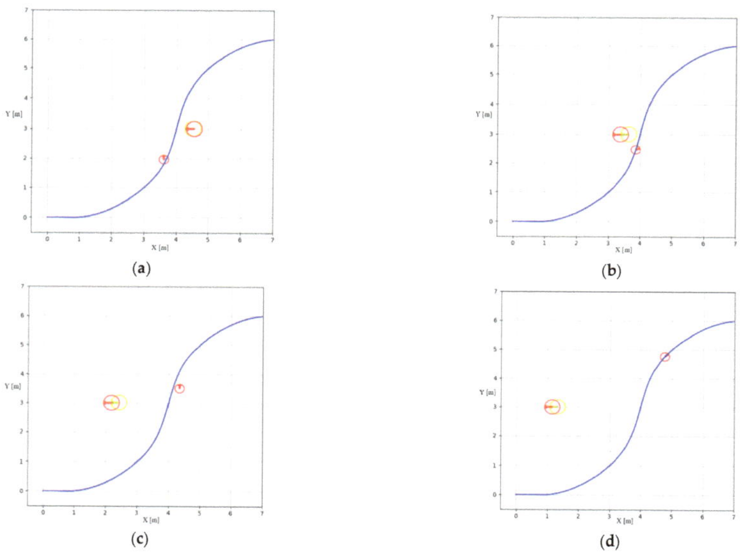

Figure 4. Process snapshots. (**a**) t = 35.02 s. (**b**) t = 37.20 s. (**c**) t = 46.00 s. (**d**) t = 52.6 s.

In Figure 5, note that there is always obvious error in path tracking without the ESO disturbance observer, while the NMPC with disturbance observation compensation can achieve error-free tracking. In addition, it can be seen that when the UBVMS has access to the predicted future trajectory of the obstacle, it chooses the opposite direction of motion to the obstacle (positive x-axis direction) to avoid collisions. The surge, sway, and yaw velocities are presented in Figure 6a–c, respectively. Here, it is interesting to note that the surge velocity quickly reaches the desired speed of 0.2 m/s after avoiding the obstacle, and sway velocity is also maintained at approximately 0 m/s. Figure 6d–f show the ability of ESOs to estimate time-varying disturbances in the external environment; although the input control vector forces and moments cause some oscillations during sharp changes, the disturbances can still be approximately calculated in a short time. Figure 7a and 7b illustrate the surge thrust and yaw moment computed by the CasADi solver, respectively. When constraining the variation in the control vector Δu in the objective function, the control vector is smoother, and the effect is obvious for the yaw moment.

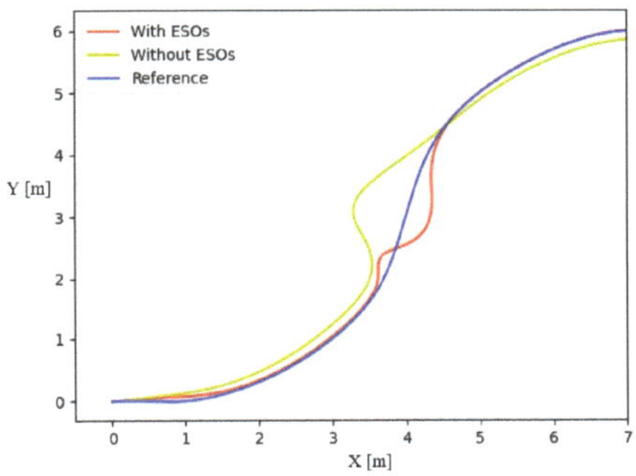

Figure 5. The curve path tracking.

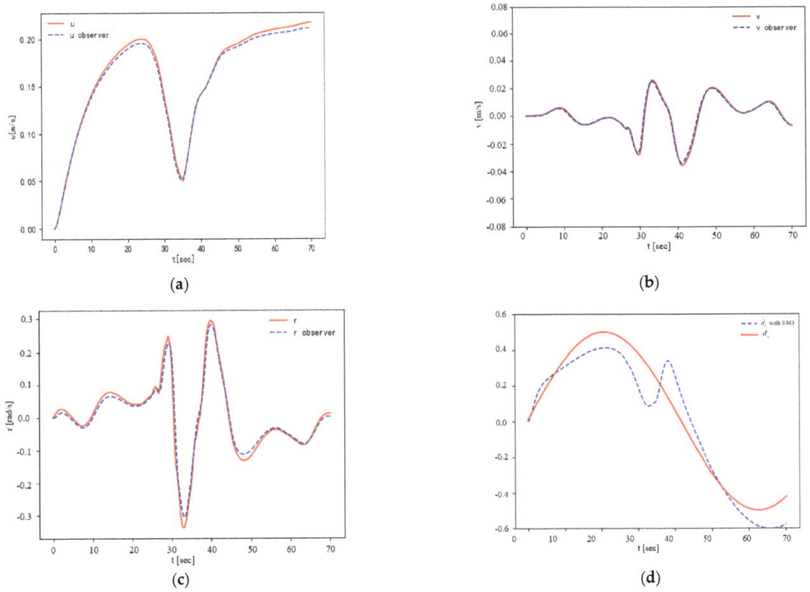

Figure 6. *Cont.*

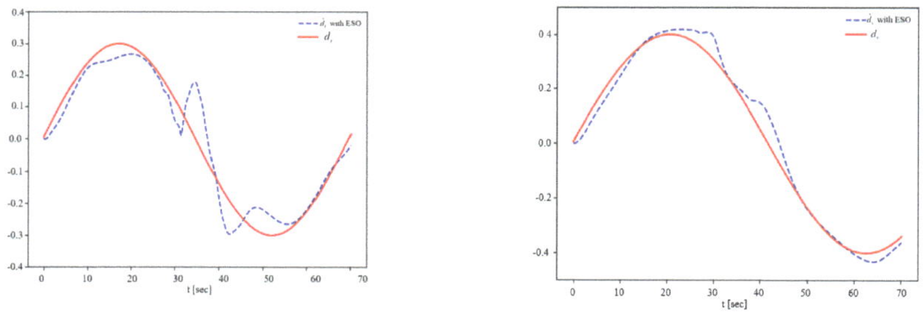

(e) (f)

Figure 6. Estimation of velocity and disturbance. (**a**) Surge velocity. (**b**) Sway velocity. (**c**) Yaw velocity. (**d**) $\hat{d}_u(t)$. (**e**) $\hat{d}_r(t)$. (**f**) $\hat{d}_v(t)$.

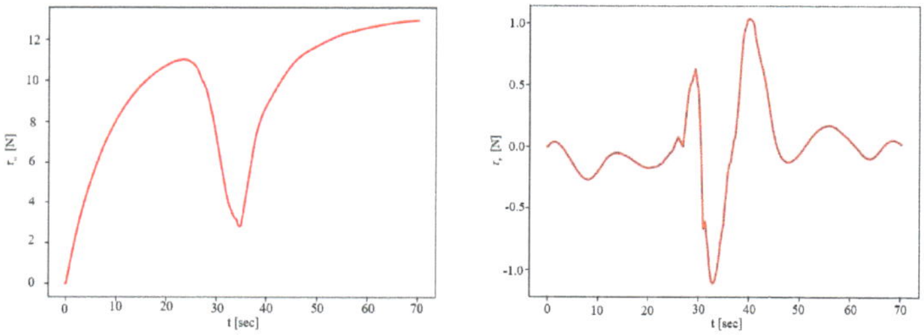

(**a**) (**b**)

Figure 7. Surge force and yaw moment. (**a**) Surge force. (**b**) Yaw moment.

Figure 8 shows the results of the Kalman filter estimating and predicting the velocity of an obstacle with a non-uniform motion. In the same scenario as Figure 4 above, we add multiple dynamic obstacles near the predetermined desired trajectory of the UBVMS and, for each added obstacle, we carry out ten consecutive simulation experiments. The results of the scenario experiments are shown in Table 2.

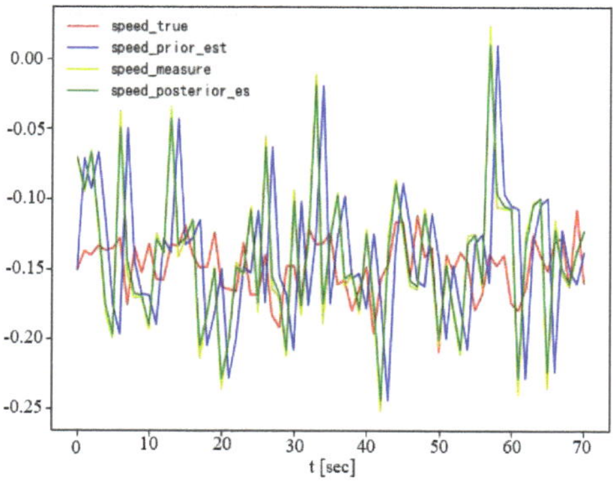

Figure 8. Obstacle velocity produced by KF estimator.

Table 2. Multi-obstacle simulation results.

Number of Obstacles	Without Priori (Predicted)			With Priori (Predicted)		
	Clearance Mean (m)	Traveled Distance Mean (m)	%Failures (Collisions)	Clearance Mean (m)	Traveled Distance Mean (m)	%Failures (Collisions)
1	0.0487	11.6114	60%	**0.3100**	**11.4018**	**100%**
2	0.0172	**10.7272**	20%	**0.3245**	10.8660	**90%**
3	0.0574	**10.8160**	20%	**0.3295**	11.3869	**60%**

The simulation results show that when a priori prediction is added to NMPC, the situation is much better than without it. First, we have fewer collisions, although we still fail four times with three dynamic obstacles added; this is probably more related to the maneuverability of the UBVMS itself, which has better stability but does not move very fast. Second, it maintains a sufficient safety distance, and in the case of only one obstacle, it not only has a 100% success rate but also has a shorter driving distance.

4.2. Experimental Results

To further verify the performance of the proposed algorithm, we conducted experiments on the UBVMS in an indoor pool. The position and yaw angle of the UBVMS (x, y, φ) were obtained by the global camera located above the pool, the data were transmitted via wired fiber optic cable, the velocity was calculated using an ESO differential tracker, and the sonar acquired the position of the obstacle.

Figure 9 shows the process of the UBVMS performing linear tracking and avoiding moving obstacles. It can be observed that the proposed control system can enable the UBVMS to smoothly perform path tracking as well as prevent a collision. When the sonar detects the obstacle at a close distance, the surge of collision cost of the objective function drives the UBVMS to deviate from the desired trajectory to avoid the collision, and then it returns to the desired path largely without overshoot and achieves zero error before reaching the target end point. The starting position of the UBVMS is the origin in geodesic coordinates, the initial velocity is 0, and the tracking path is a straight line of length 2 m.

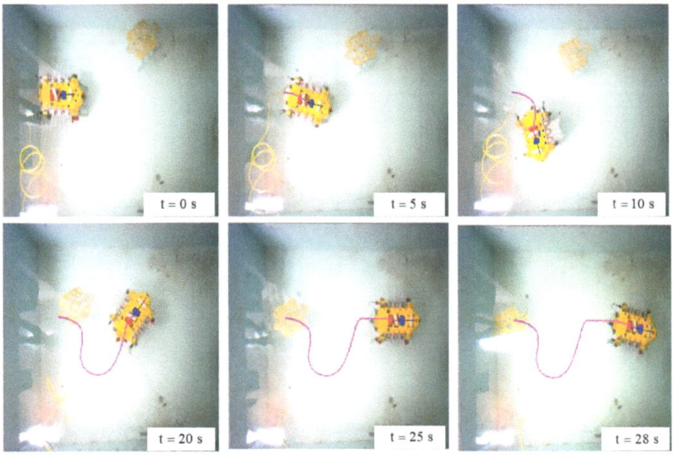

Figure 9. Screenshot sequence of the complete experiment.

It is noteworthy that due to the relatively small size of the indoor swimming pool, when the UBVMS moves at the maximum frequency (20 Hz), the additional interference caused by the reflected wave in the swimming pool cannot be ignored. Therefore, when reaching the end of the desired path, the UBVMS still has a slight trajectory offset due to inertia and disturbance. The calculated $\tau = [f_u \; 0 \; f_r]^T$ is multiplied by the respective scale factor to complete the mapping from force and torque to bilateral fin frequency. Figure 10b shows the time evolution of the control frequency of bilateral fins.

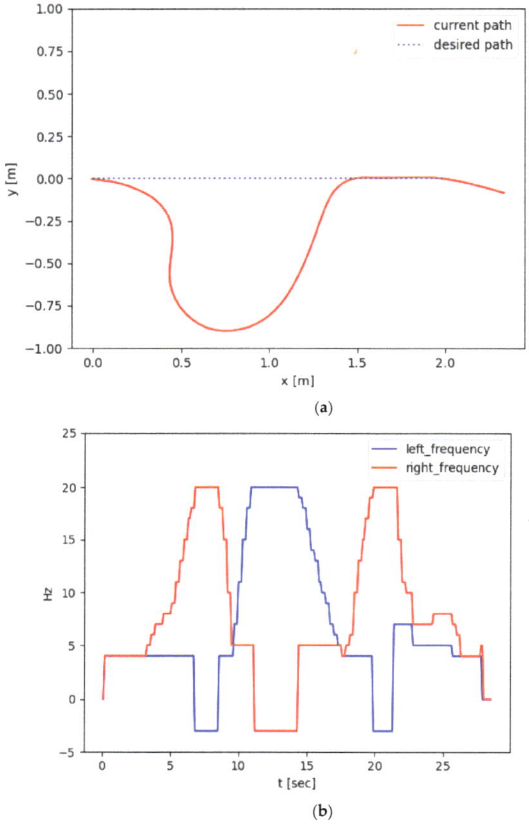

Figure 10. Experimental results. (**a**) Desired path and current path. (**b**) Control frequency of bilateral fins.

5. Conclusions

In this article, we have investigated the path following and dynamic obstacle avoidance of the UBVMS propelled by undulatory ribbon-fins with model uncertainties, external environmental disturbances, and input constraints. To achieve minimized path tracking error and robust obstacle avoidance, a general kinematic/dynamic model of the underwater vehicle was developed; external forces were modeled as unmeasured disturbances and were estimated using three ESOs fed into the NMPC prediction model. Moreover, we considered the uncertainty of the state estimator, combining the a priori estimation of the Kalman filter to predict the position of dynamic obstacles. The simulation results show that the proposed method can track the curved trajectory and achieve multi-obstacle avoidance successfully in the event of time-varying disturbances. Our experiments showed that this path-following and collision-avoidance approach for the UBVMS is valid and maintains system stability.

Author Contributions: Conceptualization, X.D. and Y.H.; methodology, X.D., Y.H., Y.W. and S.W.; software, Y.H.; validation, X.D. and Y.H.; formal analysis, X.D. and Y.H.; investigation, Y.H.; resources, X.D.; data curation, X.D. and Y.W.; writing—original draft preparation, Y.H.; writing—review and editing, X.D. and Y.H.; visualization, Y.H.; supervision, X.D., Y.W. and S.W.; project administration, X.D. and Y.H.; funding acquisition, X.D., Y.W. and S.W. All authors have read and agreed to the published version of the manuscript.

Funding: This research was partially supported by the National Key Research and Development Project "Monitoring and Prevention of Major Natural Disasters Special Program" under grant no. 2020YFC1512202; partly supported by the National Natural Science Foundation of China under grant no. 62273342; and partly supported by the Key Research and Development Project of Anhui Province under grant no. 2022a05020035.

Institutional Review Board Statement: Not applicable.

Informed Consent Statement: Not applicable.

Data Availability Statement: Not applicable.

Acknowledgments: Special thanks to the Institute of Automation, Chinese Academy of Sciences, for providing the experimental site.

Conflicts of Interest: The authors declare no conflict of interest.

Abbreviations

The following abbreviations are used in this manuscript:

UBVMS	underwater biomimetic vehicle-manipulator system
AUV	autonomous underwater vehicles
BUV	biomimetic underwater vehicles
CPG	central pattern generator
NMPC	nonlinear model predictive control
ESO	Extended state observers
KF	Kalman Filter

References

1. Wu, Z.; Liu, J.; Yu, J.; Fang, H. Development of a novel robotic dolphin and its application to water quality monitoring. *IEEE/ASME Trans. Mechatron.* **2017**, *22*, 2130–2140. [CrossRef]
2. Richmond, K.; Flesher, C.; Lindzey, L.; Tanner, N.; Stone, W.C. Sunfish®: A human-portable exploration auv for complex 3d environments. In Proceedings of the OCEANS 2018 MTS/IEEE Charleston, Charleston, SC, USA, 22–25 October 2018; pp. 1–9.
3. Lauder, G.V.; Anderson, E.J.; Tangorra, J.; Madden, P.G. Fish biorobotics: Kinematics and hydrodynamics of self-propulsion. *J. Exp. Biol.* **2007**, *210*, 2767–2780. [CrossRef] [PubMed]
4. Neveln, I.D.; Bai, Y.; Snyder, J.B.; Solberg, J.R.; Curet, O.M.; Lynch, K.M.; MacIver, M.A. Biomimetic and bio-inspired robotics in electric fish research. *J. Exp. Biol.* **2013**, *216*, 2501–2514. [CrossRef]
5. Zhou, C.; Low, K.H. Design and locomotion control of a biomimetic underwater vehicle with fin propulsion. *IEEE/ASME Trans. Mechatron.* **2011**, *17*, 25–35. [CrossRef]
6. Sfakiotakis, M.; Laue, D.M.; Davies, B.C. An experimental undulating-fin device using the parallel bellows actuator. In Proceedings of the 2001 ICRA, IEEE International Conference on Robotics and Automation (Cat. No. 01CH37164), Seoul, Republic of Korea, 21–26 May 2001; Volume 3, pp. 2356–2362.
7. Vercruyssen, T.G.A. Phase Resolved PIV Analysis of an Undulating Fin: Experimental Investigation of the Galatea Propulsion Mechanism. Master's Thesis, Delft University of Technology, Delft, The Netherlands, August 2010.
8. Niu, X.; Xu, J.; Ren, Q.; Wang, Q. Locomotion learning for an anguilliform robotic fish using central pattern generator approach. *IEEE Trans. Ind. Electron.* **2013**, *61*, 4780–4787. [CrossRef]
9. Wang, R.; Wang, S.; Wang, Y.; Tan, M.; Yu, J. A paradigm for path following control of a ribbon-fin propelled biomimetic underwater vehicle. *IEEE Trans. Syst. Man Cybern. Syst.* **2017**, *49*, 482–493. [CrossRef]

10. Cui, R.; Chen, L.; Yang, C.; Chen, M. Extended state observer-based integral sliding mode control for an underwater robot with unknown disturbances and uncertain nonlinearities. *IEEE Trans. Ind. Electron.* **2017**, *64*, 6785–6795. [CrossRef]
11. Fernandes, D.D.A.; Sørensen, A.J.; Pettersen, K.Y.; Donha, D.C. Output feedback motion control system for observation class ROVs based on a high-gain state observer: Theoretical and experimental result. *Control. Eng. Pract.* **2015**, *39*, 90–102. [CrossRef]
12. Wan, L.; Zeng, J.; Li, Y.; Qin, H.; Zhang, L.; Wang, J. Neural observer-based path following control for underactuated unmanned surface vessels with input saturation and time-varying disturbance. *Int. J. Adv. Robot. Syst.* **2019**, *16*, 1729881419878071. [CrossRef]
13. Liu, L.; Wang, D.; Peng, Z.; Li, T.; Chen, C.P. Cooperative path following ring-networked under-actuated autonomous surface vehicles: Algorithms and experimental results. *IEEE Trans. Cybern.* **2018**, *50*, 1519–1529. [CrossRef]
14. Lee, J.; Chang, H.J. Analysis of explicit model predictive control for path-following control. *PLoS ONE* **2018**, *13*, e0194110. [CrossRef]
15. Li, G.; Zhang, J.; Liu, Z.; Wang, L.; Sun, T. Predictive control for straight path following of underactuated surface vessels with roll constraints. In Proceedings of the 2016 Chinese Control and Decision Conference (CCDC), Yinchuan, China, 28–30 May 2016; pp. 583–588.
16. Helling, S.; Roduner, C.; Meurer, T. On the dual implementation of collision-avoidance constraints in path-following MPC for underactuated surface vessels. In Proceedings of the 2021 American Control Conference (ACC), New Orleans, LA, USA, 25–28 May 2021; pp. 3366–3371.
17. Wang, Y.; Wang, R.; Wang, S.; Tan, M.; Yu, J. Underwater bioinspired propulsion: From inspection to manipulation. *IEEE Trans. Ind. Electron.* **2019**, *67*, 7629–7638. [CrossRef]
18. Fossen, T.I. Guidance and Control of Ocean Vehicles. Ph.D. Thesis, University of Trondheim, Trondheim, Norway, 1999.
19. Zhang, Y.; Liu, X.; Luo, M.; Yang, C. MPC-based 3-D trajectory tracking for an autonomous underwater vehicle with constraints in complex ocean environments. *Ocean. Eng.* **2019**, *189*, 106309. [CrossRef]
20. Long, C.; Qin, X.; Bian, Y.; Hu, M. Trajectory tracking control of ROVs considering external disturbances and measurement noises using ESKF-based MPC. *Ocean. Eng.* **2021**, *241*, 109991. [CrossRef]
21. Zhao, C.; Wang, D.; Hu, J.; Pan, Q. Nonlinear model predictive control-based guidance algorithm for quadrotor trajectory tracking with obstacle avoidance. *J. Syst. Sci. Complex.* **2021**, *34*, 1379–1400. [CrossRef]
22. Yang, X.; Huang, Y. Capabilities of extended state observer for estimating uncertainties. In Proceedings of the 2009 American Control Conference, St. Louis, MO, USA, 10–12 June 2009; pp. 3700–3705.
23. Gu, N.; Wang, D.; Peng, Z.; Wang, J.; Han, Q.L. Disturbance observers and extended state observers for marine vehicles: A survey. *Control. Eng. Pract.* **2022**, *123*, 105158. [CrossRef]
24. Han, J. From PID to active disturbance rejection control. *IEEE Trans. Ind. Electron.* **2009**, *56*, 900–906. [CrossRef]
25. Gao, Z. Scaling and bandwidth-parameterization based controller tuning. In Proceedings of the 2003 American Control Conference, Denver, CO, USA, 4–6 June 2003; pp. 4989–4996.
26. Li, Z.; Li, R.; Bu, R. Path following of under-actuated ships based on model predictive control with state observer. *J. Mar. Sci. Technol.* **2021**, *26*, 408–418. [CrossRef]
27. Li, Z.; Bu, R. Trajectory tracking of under-actuated ships based on optimal sliding mode control with state observer. *Ocean. Eng.* **2021**, *233*, 109186. [CrossRef]
28. Kamel, M.; Alonso-Mora, J.; Siegwart, R.; Nieto, J. Robust collision avoidance for multiple micro aerial vehicles using nonlinear model predictive control. In Proceedings of the 2017 IEEE/RSJ International Conference on Intelligent Robots and Systems (IROS), Vancouver, BC, Canada, 24–28 September 2017; pp. 236–243.
29. Mousazadeh, H.; Jafarbiglu, H.; Abdolmaleki, H.; Omrani, E.; Monhaseri, F.; Abdollahzadeh, M.R.; Mohammadi-Aghdam, A.; Kiapei, A.; Salmani-Zakaria, Y.; Makhsoos, A. Developing a navigation, guidance and obstacle avoidance algorithm for an Unmanned Surface Vehicle (USV) by algorithms fusion. *Ocean. Eng.* **2018**, *159*, 56–65. [CrossRef]
30. Rao, A.V. A survey of numerical methods for optimal control. *Adv. Astronaut. Sci.* **2009**, *135*, 497–528.
31. Andersson, J.A.; Gillis, J.; Horn, G.; Rawlings, J.B.; Diehl, M. CasADi: A software framework for nonlinear optimization and optimal control. *Math. Program. Comput.* **2019**, *11*, 1–36. [CrossRef]

sensors MDPI

Article

Unmanned Ariel Vehicle (UAV) Path Planning for Area Segmentation in Intelligent Landmine Detection Systems

Ahmed Barnawi [1,*], Krishan Kumar [2], Neeraj Kumar [1], Nisha Thakur [2], Bander Alzahrani [1] and Amal Almansour [1]

1 Faculty of Computing and Information Technology, King Abdulaziz University, Jeddah 21589, Saudi Arabia; neeraj.kumar@thapar.edu (N.K.); aalmansour@kau.edu.sa (A.A.)
2 Thapar Institute of Engineering and Technology, Deemed to be University, Patiala 147004, India; kkumar_phd19@thapar.edu (K.K.); nthakur_phd20@thapar.edu (N.T.)
* Correspondence: ambarnawi@kau.edu.sa

Abstract: Landmine contamination is a significant problem that has devastating consequences worldwide. Unmanned aerial vehicles (UAVs) can play an important role in solving this problem. The technology has the potential to expedite, simplify, and improve the safety and efficacy of the landmine detection process prior to physical intervention. Although the process of detecting landmines in contaminated environments is systematic, it is proven to be rather costly and overwhelming, especially if prior information about the location of the lethal objects is unknown. Therefore, automation of the process to orchestrate the search for landmines has become necessary to utilize the full potential of system components, particularly the UAV, which is the enabling technology used to airborne the sensors required in the discovery stage. UAVs have a limited amount of power at their disposal. Due to the complexity of target locations, the coverage route for UAV-based surveys must be meticulously designed to optimize resource usage and accomplish complete coverage. This study presents a framework for autonomous UAV-based landmine detection to determine the coverage route for scanning the target area. It is performed by extracting the area of interest using segmentation based on deep learning and then constructing the coverage route plan for the aerial survey. Multiple coverage path patterns are used to identify the ideal UAV route. The effectiveness of the suggested framework is evaluated using several target areas of differing sizes and complexities.

Keywords: landmine detection; segmentation; coverage path planning; deep learning

Citation: Barnawi, A.; Kumar, K.; Kumar, N.; Thakur, N.; Alzahrani, B.; Almansour, A. Unmanned Ariel Vehicle (UAV) Path Planning for Area Segmentation in Intelligent Landmine Detection Systems. *Sensors* **2023**, 23, 7264. https://doi.org/10.3390/s23167264

Academic Editors: Zhun Fan and Gang Peng

Received: 15 July 2023
Revised: 11 August 2023
Accepted: 17 August 2023
Published: 18 August 2023

1. Introduction

Landmines are devices with explosive properties that are strategically placed on the ground or buried slightly beneath the surface. Their purpose is to harm or incapacitate adversary assets, including troops, vehicles, or tanks, as they approach or pass over the mine. The mines can be categorized into two types: anti-personnel and anti-vehicle, based on their intended targets. The detonation is triggered by the pressure, presence, or proximity of a person or vehicle [1]. Landmine explosions can cause severe consequences and physical damage.

Landmines are indiscriminate weapons with unfortunate far-reaching consequences even after conflicts cease. They bear significant social, economic, and environmental implications, serving as area-denial munitions and tactical barriers [2]. While their deployment is relatively cheap and easy, removing them is time-consuming, hazardous, and costly. Figure 1 displays the casualties from landmines or Explosive Remnants of War (ERW) recorded by Landmine Monitor for 2015–2021. According to a 2022 report [3], sixty countries and territories remain contaminated by anti-personnel landmines, with fifty casualties reported in 2021. Of the 5544 fatalities in 2021, 4200 were civilians.

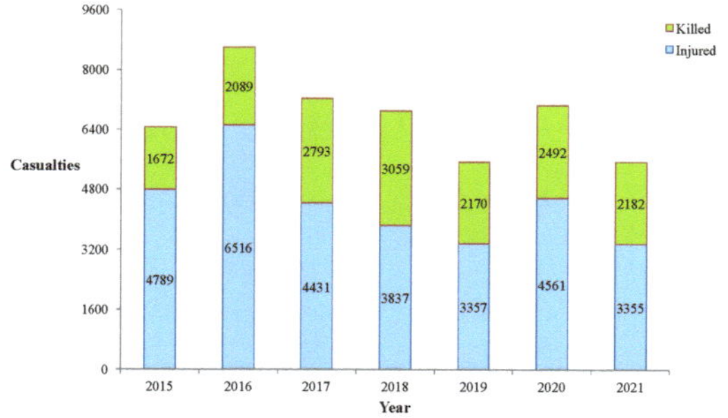

Figure 1. Casualties due to landmines/ERW during 2015–2021.

Landmine clearance or demining refers to removing landmines from a particular location. There are both military and humanitarian demining operations. Military demining aims to quickly clear a passage through a minefield, whereas humanitarian demining removes all landmines to a set depth and makes the area safe for human usage. Considering the catastrophic effects of landmines, measures must be undertaken to clear the minefields. Landmine clearance is a challenging and time-consuming task. It is a high-risk operation requiring specific equipment and specialized human resources to locate and remove landmines. In 2021, there were a total of 27 reported casualties of deminers [3].

As unmanned aerial vehicles (UAVs) and sensor technologies advance, they are finding more valuable use cases. The aerial survey using drone technology has the potential to save many lives by making the clearance of landmines a safer, quicker, and more affordable process. UAVs also make it feasible to survey otherwise inaccessible and complex geographies. The minefield can be inspected using a UAV mounted with suitable sensor devices for detecting landmines. The UAVs' resources, such as their batteries, are limited. Additionally, the operation involves a complicated environment with obstructions. It necessitates determining the best UAV traversal route that covers the whole target area.

The detection of landmines poses a significant challenge since numerous elements influence the accuracy and efficacy of this operation.

- Sensor technology: The selection of sensor technology plays a crucial role in the detection of landmines. Various types of sensors, including metal detectors, ground-penetrating radars (GPRs), and magnetometers, exhibit diverse capabilities and sensitivities toward certain categories of landmines [4]. Every type of sensor possesses distinct advantages and limits when detecting particular landmine materials and depths of burial.
- Landmine type: Landmine detection is influenced by their size and material composition. Detecting larger or metallic landmines is typically more feasible than smaller or non-metallic landmines. The capacity of a sensor to identify landmines with varied compositions is contingent upon its sensitivity to various materials.
- Soil type: The detection of landmines is influenced by several soil conditions, including but not limited to moisture content, conductivity, and magnetic susceptibility, as these soil properties play a significant role in the detection process. Variations can influence the electrical conductivity of the soil in moisture levels, impacting the sensor's performance. The presence of diverse soil compositions and changes in soil types can potentially lead to erroneous alerts or instances of undetected occurrences.
- Environmental factors: Environmental factors, including nearby structures or electronic devices emitting electromagnetic interference, can introduce unwanted signals into the sensor data, impacting its accuracy and dependability. Precise calibration and signal processing techniques must be meticulously applied to counteract this interference.

- Spatial resolution: The spatial resolution of the sensor system, which is determined by the sensor's characteristics and flight parameters, influences its ability to detect small or closely spaced landmines. A higher spatial resolution can increase detection accuracy but may reduce the flight's coverage area.
- Flight altitude and speed: The UAV's altitude and speed influence the efficacy of the sensor. Higher altitudes may reduce spatial resolution, while lower altitudes may provide greater sensitivity but increase the risk to the UAV. The flight pace influences the time available for data collection and may affect the data quality.
- Terrain and topography: The UAV's stability and the sensor's proximity to the ground might be influenced by the topography and roughness of the terrain. Variations in readings may occur due to the influence of uneven terrain on the orientation of the sensor.

Utilizing UAV technology to conduct low-altitude, low-speed surveys for landmine detection is of the utmost importance. It takes advantage of the UAV's ability to operate close to the ground, thereby facilitating the acquisition of detailed data and enhancing the detection of smaller or more deeply concealed landmines. The method with reduced survey speeds ensures comprehensive data acquisition, improving target identification and discrimination capabilities.

This research focuses on the challenge of automated area segmentation and route planning for UAV-based landmine detection. Initially, deep learning-based extraction isolates the region of interest (ROI) from the satellite imagery. Then, a route determination strategy for UAV-based coverage is formulated. It helps find the most efficient route considering both time and power consumption. UAV uses the specified course to traverse the target region. When equipped with suitable sensing technologies, the signatures of landmines can be recognized and localized. It paves the way for automated, remote landmine-detecting operations. In our study, we leverage offline path planning as a fundamental component of the proposed approach for UAV-based aerial surveys for landmine detection. Its primary objective is to generate an optimized coverage path that the UAV will follow during its survey mission. Once the path planning is completed, the optimized coverage path is transferred to the UAV's onboard system. The generated path guides the UAV's autonomous execution during the aerial survey. It enables the UAV to collect sensor data systematically and efficiently, improving data quality. Moreover, the streamlined path minimizes energy consumption, allowing extended flight endurance and maximizing the survey area coverage. The onboard computations primarily pertain to real-time navigation and control, ensuring that the UAV accurately follows the predetermined path.

1.1. Motivation and Contribution

As a result of this paper, the following notable contributions are made.

1. Deep learning-based segmentation is used to extract the region of interest (ROI) from satellite imagery of the geographical area.
2. The path planning strategy is formulated to optimize the coverage route of the UAV-based region survey.
3. The workflow is designed for UAV-based minefield surveys.
4. The efficacy of the proposed strategy is evaluated in terms of energy and time constraints.

1.2. Organization

The general outline of the paper is as follows. Section 2 provides an overview of segmentation and path planning concepts. In Section 3, we dive deeper into the details of the proposed framework. In Section 4, the results and their interpretation are presented. Section 5 provides a conclusion of the work and suggests some avenues for further study.

2. Background

Unmanned aerial vehicles (UAVs), sometimes known as "drones", are unmanned aircraft that can be flown without a pilot on board. Aircraft, ground control stations, and communications systems all fall under the umbrella term unmanned aircraft systems (UAS), which describes the infrastructure necessary for sophisticated drone operations. An autonomous drone is a UAV that can fly missions independently of a human pilot. It can take off, execute its task, and return to base without human assistance. Rather than relying on a human pilot, communications management software handles mission planning and flight control for autonomous drones.

2.1. UAV-Based Landmine Detection

Major application areas for drone technology include agriculture, disaster management, remote surveying, network coverage, and product delivery. In addition to the developments in network technology, UAVs play a greater role in maintaining and enhancing network capabilities. The UAVs play a crucial role in topographical surveying by offering a quick, simple, inexpensive, and safe approach with decreased human participation and improved access to challenging terrain. Recent research has envisioned a similar system for landmine detection [4]. Various sensing devices can detect the landmine signature, including metal detectors, ground penetration radar (GPR), infrared, multispectral, acoustic, magnetometer, and others [2,5]. More specialized sensors are now compatible with UAVs, allowing for more precise and thorough airborne surveys as technology advances.

Figure 2 depicts the process of landmine detection where the minefield survey utilizes a UAV equipped with detection sensors. The area is first chosen for the UAV-based investigation and its geographical map serves as the input. The required portion is then retrieved to determine the region of interest. A route or path plan is established to produce the coverage path in the extracted area. UAVs utilize the created course to explore all geographical places and gather sensor-based information. Information collected may subsequently be processed using artificial intelligence (AI) techniques to help locate and identify landmines. Additionally, the minefield layout may be recognized by post-processing. The demining team can use the detection results to inspect the specified locations and remove or destroy the landmines.

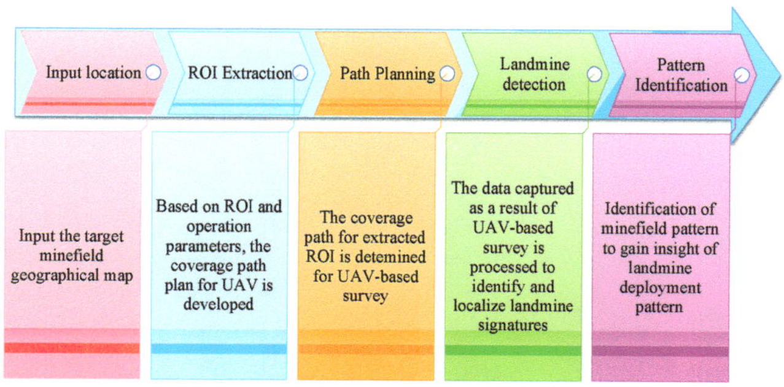

Figure 2. UAV-based landmine detection.

The use of UAVs in landmine detection provides several benefits, including a reduction in survey time, improved accessibility to difficult terrain, and risk-free operation. Consequently, several researchers have investigated the possibility of improving the detection procedure by using UAVs. Using low-cost drones, the visual detection system was proposed in [6] to detect fully and partially visible landmines. The system provides good efficiency in a low-altitude survey and low flight speed. The UAV equipped with a metal detector was considered in [7] to carry out aerial detection in rough terrains. The developed system can identify metals, mines, and bare explosives buried underground. A metal detector requires a certain amount of metal content to generate the landmine signature. It makes the detection of low-metal landmines difficult using metal detectors. Colorado et al. [8]

integrated a software-defined radio (SDR)-based GPR system with a UAV that was capable of detecting landmines in variable terrains. The flight control system was developed to enable the steady operation of the UAV. The ground-penetrating synthetic aperture radar (GPSAR) was used with UAV in [9] to accelerate the demining process. The strong clutter generated at the surface–air interface affected the GPR-based landmine detection. The time-gating and average subtraction technique was used for clutter removal in [10], and SAR processing was used to obtain high-resolution images.

A lightweight and low-power GPR was developed in [11], based on the stepped frequency continuous wave (SCFW) radar. The developed system efficiently detects metallic and plastic landmines while flying at low altitudes. But the detection system has limited scanning speed and suffers from the moisture content of the soil. In [12], an improved SAR-based GPR detection system was used with UAV for landmine and improvised explosive device (IED) detection. The system provides a better signal-to-clutter ratio in GPR images using clutter filtering based on singular value decomposition (SVD).

The UAV equipped with a thermal and multispectral sensor was employed for scatterable landmine detection in [13]. It is based on a convolution neural network (CNN)-based identification and localization of landmine signatures in the captured sensor images. The magnetometer-based survey using UAV was presented in [14] to detect different types of mines. The study has taken into account a sensor-equipped UAV with a height of 1 m and a survey line spacing of 3 m. It was improved in [15] by altering the sensor position to enable low-altitude operation. The low-pass filtering and moving average methods were used to eliminate the magnetic noise and improve detection.

Various sensing technologies can be used with UAVs for the remote survey of minefields, each serving some advantages and disadvantages in different use cases. In most studies, the survey was remotely controlled in a simple environment. It poses a challenge for UAV operation in complex terrains where the optimal utilization of resources and complete coverage are critical. It requires automation of trajectory formulation for UAV-based surveys.

2.2. Segmentation

Segmentation is the process of breaking down a picture into multiple pieces or regions [16]. It entails dividing a picture into smaller, more manageable regions, with each section being represented by a mask or a set of labels. Simply said, it is the process of assigning labels to the image pixels. The pixels that share the same attributes are labeled alike. It helps in the identification of different image components, and instead of processing the complete image, just the relevant parts need to be processed. The segmentation can be categorized as semantic segmentation, instance segmentation, and panoptic segmentation.

In this study, K-means clustering and normalized RGB color space are used to segment the green terrain in the provided image. Using K-means clustering and normalized RGB color space, the results demonstrate that the shared Google Earth photographs can be divided into segments that are roughly 40.50% and 47.01% of the total image pixels, respectively [17].

When image segmentation methods are developed, they could enter the medical area and help with disease diagnosis. Using pathology images to expedite clinical diagnosis and automate image analysis with reliability and effectiveness is still highly challenging. For several machine learning methods, including convolutional neural networks, automatic pathological picture segmentation was proposed. Accuracy and processing speed are provided for this segmentation using fully convolutional networks and other deep learning techniques [18].

The segmentation process plays a crucial role in enhancing the efficacy of landmine detection procedures. By precisely identifying and isolating areas with a high probability of containing landmines, segmentation lays the groundwork for targeted and efficient UAV-based aerial surveys. Moreover, the segmentation process assists in overcoming the obstacles posed by complex and diverse terrains. It enables the UAV to prioritize survey activities, concentrating on pre-identified regions more likely to contain landmines. This targeted strategy not only increases survey efficiency but also improves operational safety. By incorporating the segmentation process into the landmine detection workflow, the accuracy, efficiency, and overall effectiveness of UAV-based aerial surveys can be significantly improved. By intelligently identifying regions of interest, the UAV can implement survey missions with greater precision, resulting in enhanced landmine detection results and the establishment of safer environments. As part of our investigation, we collected satellite images from Google Earth for our intended use. The objective is to determine the precise target area for UAV aerial surveys. This requires the initial collection of datasets, followed by the precise definition of the intended area of interest. Then, a deep learning model is utilized to effectively recognize and categorize the distinct regions.

2.3. UAV Path Planning

In recent years, various studies are focused on automating the UAV flight. One such component examined in the literature to automate the UAV-based aerial survey is autonomous route determination, also known as path planning. It covers the estimation of the best route between a source and a destination [19]. In some applications, like archaeological surveys, the objective is to cover every location within a region. It requires the determination of the optimal coverage path, denoted as coverage path planning [20]. Figure 3 illustrates the execution of an aerial survey, wherein the UAV systematically traverses a predetermined coverage path to collect data, covering the whole target region.

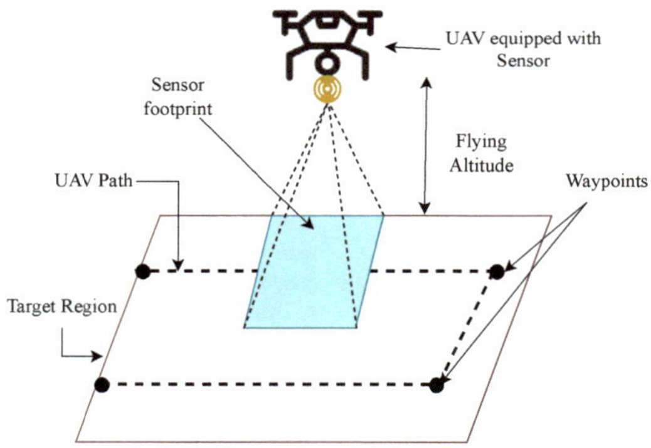

Figure 3. Representation of the UAV-based survey of the target area.

Several recent works have investigated the coverage path planning of UAVs. Surveys involving UAVs encompass operations under a range of conditions, accounting for varying complexities and the number of regions Accordingly, different strategies have been formulated to automate the coverage path determination. The target region having complex geometry often requires partitioning the free space into cells to simplify the survey path, termed cellular decomposition [21]. It can be categorized as the exact and approximate cellular decomposition or grid decomposition. In the exact decomposition method, the region is partitioned into smaller sub-areas, the combination of which forms the same region. The path for each sub-region is then determined, and the integration of all the paths generates the coverage path. On the other hand, the approximate decomposition discretizes the area into a set of regular cells at some resolution. The traversal of all the cells in some order generates the coverage path.

The authors of [22] focused on the exact cellular decomposition of the region for a UAV-based survey using a greedy strategy. The back-and-forth motion was used to obtain the coverage path for each sub-region. The minimum spanning tree (MST)-based traversal was used to obtain the complete coverage path for the undirected graph of sub-regions. Torres et al. considered the line sweep direction computation for the polygon, and then the coverage path was generated perpendicular to it. In [23], the work focused on UAV operations in the presence of wind, and a dynamic programming-based approach was used for polygon decomposition. The selected direction of the coverage path line was perpendicular to the wind direction to consume the minimum flight time.

The approximate cellular decomposition and a gradient-based approach were used in [24] to obtain a coverage path with the minimum number of turns. The energy model for UAV was derived in [25], which was used to determine the optimal coverage path along with safety mechanisms. The authors of [26] examined the coverage path formulation based on exact and grid-based decomposition methods for various CPP widths.

In some applications, the target area to be surveyed can be considerably large, making it difficult to be covered with a single UAV. Also, certain applications, like disaster management, are time-critical and require the survey to be completed as quickly as possible [27]. The survey task can be enhanced by employing multiple UAVs leading to cooperative or multi-UAV coverage [28]. The region can be decomposed into smaller fragments that are assigned to different UAVs. It requires careful partitioning and sub-region allocation to optimize the complete coverage [29]. The hexagonal decomposition of the target region and clustering-based allocation to multiple UAVs was proposed in [30]. Similarly, Morse-based decomposition was used in [31] where sub-region assignment considered the UAV start and end positions. Choi et al. [32] proposed an optimization model with column generation to track energy consumption during an aerial survey. The work focused on minimizing the number of UAVs and energy requirements for aerial surveys.

Researchers have investigated the cooperative utilization of several UAVs to achieve sustained coverage, with a focus on grids that are visited less frequently. Furthermore, specific strategies have been implemented to address emergency situations in particular regions. In some scenarios, multiple UAVs can operate in a certain formation, and the coverage path can be determined accordingly [33]. The authors of [34] included mixed-integer linear programming (MILP) modeling of the coverage task and proposed a randomized search-based algorithm to optimize the coverage path for heterogeneous UAVs.

In certain applications, UAVs need to survey multiple separated regions instead of a single area. It requires proper planning and sequence determination if a single UAV is used to cover all the regions to determine an efficient coverage path. Xie et al. [35] formulated the problem as the integration of the traveling salesperson problem (TSP) and CPP. Grid-based and dynamic programming (DP)-based approaches have been used to determine each region's entry and exit point and optimize intra and inter-regional coverage paths. The work was extended in [36], where a heuristic method was employed. The exact methods have limited scalability, where the time complexity increases exponentially with the problem scale. The heuristic approaches can provide appropriate solutions in an acceptable time and improved scalability.

In UAV-based landmine detection, the minefield survey often includes complex terrain, and the UAV resources are also constrained. The survey also needs to consider several factors to determine the gap between sweep lines. It, thus, requires autonomous and efficient path planning that determines the route for UAV passing through each location of the target region with minimum energy and time requirements. The proposed work is focused on addressing the CPP problem for landmine detection.

3. Methodology

The determination of the coverage path for a UAV-based survey includes two phases: ROI segmentation and path planning. Figure 4 depicts the various steps involved in the whole process. Initially, a geographical map serves as input for selecting a target area. The region is segmented using a deep learning-based technique to extract the area of interest. The UAV route is created as a series of geographical waypoints in the subsequent path planning stage using the coverage path determination and coordinate conversion strategies.

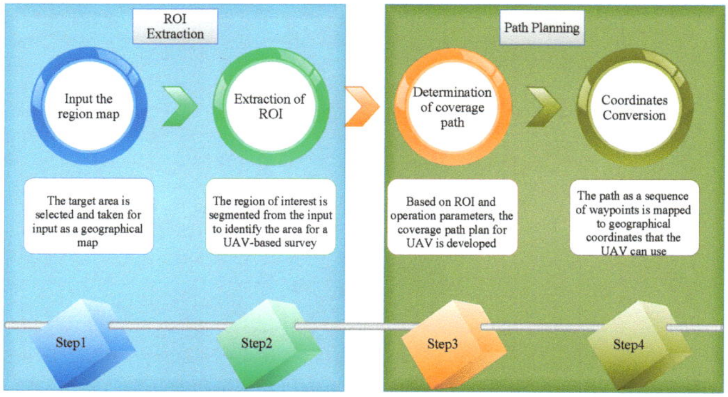

Figure 4. ROI segmentation and coverage path panning for UAV-based survey.

3.1. ROI Segmentation

The first phase involves extracting the region to be surveyed from the input region map. Landmine detection is often carried out in terrains where every part or region, like rocky regions, does not form a part of detection. So, it becomes essential to identify the region of interest so that only desired area is considered for the UAV-based survey. The input to the first phase is in the form of a satellite map image of the location where landmine detection is to be carried out. The ROI is segmented from the image, acting as an input for the path planning phase.

The U-Net model is used for image segmentation that segments the input satellite image into ROI and non-ROI regions. The ROI region is the intended ROI that the UAV surveys for detecting landmines. U-Net is a classic image segmentation model used in a wide range of applications. It has two networks: an encoder and a decoder. U-Net conducts semantic segmentation that is based on pixel-level classification. Due to several factors, U-Net has increased in popularity and is seen as superior to several other semantic segmentation techniques.

- Satellite images: U-Net can be used to segment various land cover types, including forests, aquatic bodies, urban areas, and agricultural land, in satellite imagery. The model will produce masks with separate labels for each land cover type. And here, our data takes satellite images of a particular region.
- The U-Net architecture is made to handle small amounts of data: Obtaining big annotated datasets for training deep learning models can be difficult in many real-world applications. Due to its symmetric design and skip connections, U-Net can function well even with little training data. Here, we captured satellite images that are limited in number, so we used the U-Net model for image segmentation.

Figure 5 shows the encoder in its first half. It involves the usage of a pre-trained classification network like VGG/ResNet. Convolution blocks and max pool downsampling encode the input image into feature representations at different levels. The second element of the architecture is the decoder. It semantically projects the encoder's learned and low-resolution selected features onto the higher-resolution pixel space to provide a dense categorization. Upsampling, concatenation, and conventional convolution procedures make up the decoder. It consists of a route that expands to the right and contracts to the left. Here we took the input size of an image (256, 255, 3). The contracting path follows the common convolutional network architecture. Repeatedly, a rectified linear unit (ReLU) and a 2×2 max pooling operation with stride 2 are used for downsampling, followed by two 3×3 convolutions (unpadded convolutions). With each downsampling step, we double the number of feature channels by four. A 2×2 convolution ("up-convolution"), which reduces the number of feature channels in half, an association with the correspondingly cropped feature map from the contracting path, and two 3 × 3 convolutions, each followed by a ReLU, at each stage of the expansive path are all performed after upsampling the feature map. Cropping is necessary because each convolution loses boundary pixels. Using a 1 × 1 convolution, each 64-component feature vector is translated to the necessary number of classes at the top layer. There are a total of 23 convolutional layers in the network. An associated multi-channel feature map is shown by each blue box. On the top of the box, there is a channel count indication. In this case, brown squares signify replicated feature maps. The various operations are shown by the arrows.

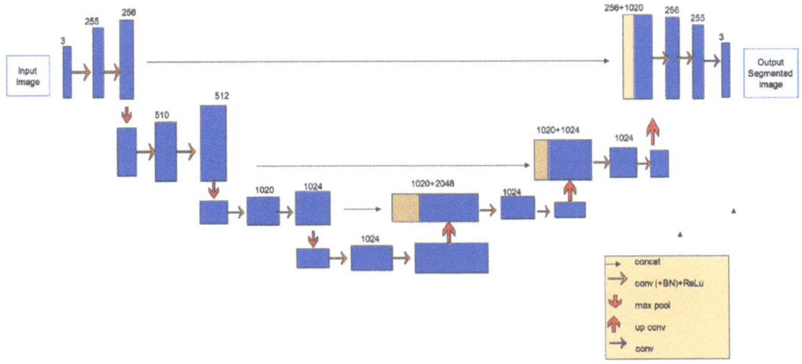

Figure 5. U-net architecture (example for 32 × 32 pixels in the lowest resolution).

3.2. Path Planning

3.2.1. Coverage Path for ROI

The second phase includes the formation of a UAV trajectory that covers all the points of the segmented region obtained from the previous stage. This paper examines the use of spiral and zig-zag algorithms to determine an optimal coverage path for the UAV employed in aerial surveys for landmine detection. The spiral path planning algorithm was selected for its ability to efficiently cover extensive areas while maintaining a central focus, which corresponds well with the requirement for comprehensive coverage in landmine detection. In contrast, the zig-zag path planning algorithm provides a structured approach for systematic coverage. It is selected due to its unique ability to minimize the number of turns and overall path length, which are essential for optimizing the efficacy of UAV-based aerial surveys. The coverage path is affected by region shape, sensor footprint, UAV height, resolution requirement, and other factors. All the parameters determine the waypoints and distance between survey lines. For the sake of simplicity, our proposed approach entails generating a generalized path that adheres to the predetermined line gap and UAV height specifications. This streamlined approach facilitates ease of implementation and serves as a foundation for more complex adaptations in the future.

Energy efficiency is vital for UAV missions, as it directly impacts the UAV's endurance and operational range. Minimizing the path length helps conserve energy and extend the mission's flight time, enabling the UAV to cover larger areas or stay airborne for extended periods. Turning points in the flight path of a UAV during aerial surveys also play an essential role in determining the mission's efficacy and energy consumption. A turning point is where the UAV alters direction or bearing to navigate the survey area. Depending on the path planning strategy, these locations may be abrupt turns or incremental course adjustments. Turning points require the UAV to adjust its orientation and alter its flight trajectory. Each turn consumes additional energy because the UAV must surmount aerodynamic drag and inertial forces during the direction change. Frequent reversals can increase energy consumption, reducing the mission's overall endurance and flight range. The time the UAV takes to complete its mission increases with every turn. The length of a mission is affected by factors like the time needed to complete a turn and the time needed to stabilize the UAV's flight after the turn. Keeping the number of turns to a minimum improves trip time and data-gathering efficiency.

The primary objective of path planning is to conserve energy and time by shortening the path and minimizing turns. It is accomplished by determining the most feasible path that avoids needless detours and abrupt turns. With this optimized route, the UAV can efficiently survey the area, consuming less energy and concluding the mission faster. It makes UAV missions more cost-effective and sustainable while maximizing resource utilization and enhancing overall performance.

The spiral and zig-zag path planning methods reduce the turning radius. They frequently provide tighter turns and lower turning radii, which might be essential in confined spaces. It may result in more significant space usage and more effective navigation. Our dataset contains satellite images of particular regions, with ROI of different polygon shapes with tighter turns and less turning radii. These path planning methods can help minimize the risk of collision. These algorithms also provide smooth paths that can increase energy efficiency since unexpected shifts in direction can increase energy consumption, particularly for mobile robots or vehicles.

Algorithm 1 describes the procedure for spiral path planning. First, we initialize the inputs with S_I representing the starting point and R_I representing the region of interest. It then starts with the spiral path planning method, which divides the ROI into grid form. Here, we have considered the width of size w for each grid.

Algorithm 1 Spiral path planning (Spp) algorithm

```
Input : S_I = Starting point,
R_I = Region of interest
w = Linewidth
Output : Waypoints : N Procedure:
N ← S_I
r_s = 0, r_e = m − 1
c_s = 0, c_e = n − 1
while ((r_s <= r_e) && (c_s <= c_e)) do
    for (i = c_s; i <= c_e; i = i++) do
        N ← (i, r_s)
        r_s = r_s + 1
        for (j = r_s; j <= r_c; j = j++) do
            N ← (j, i)
            if (r_s ⩽= r_e) then
                for (k = c_e; k >= c_s; k = k−−) do
                    N ← (k, j)
                    r_e = r_e − 1
                    if (c_s <= c_e) then
                        for (l = r_e; l >= r_s; l = l−−) do
                            N ← (l, k)
                            c_s = c_s + 1
                        end for
                    end if
                end for
            end if
        end for
    end for
end while
return N
```

The explanation of the algorithm is given below:

- Lines 2, 3 initialize variables r_s, r_e, c_s, and c_e, which represent the start and end indices of the rows and columns, respectively, stored in the current iteration.
- In line 4, it enters an iteration that continues as long as the start index of the rows is less than or equal to the end index of the rows and the start index of the columns is less than or equal to the end index of the columns.
- Lines 6, 7 use iteration to store the elements in the top row from left to right. The start index of the rows is incremented by 1 to indicate that the top row has been stored.
- In line 8, it again goes into another iteration to store the elements in the rightmost column from top to bottom. The end index of the columns is decremented by 1 to indicate that the rightmost column has been stored.
- In line 10, it checks if there are any rows left. If so, in line 11, it again goes into iteration for the elements in the bottom row from right to left. The end index of the rows is decremented by 1 to indicate that the bottom row has been stored.
- Line 14 checks if there are any columns left. If so, it uses an iteration again for the elements in the leftmost column from bottom to top.
- The start index of the columns is incremented by 1. The iteration continues with the updated indices, and the process is repeated until all elements are stored in ordered nodes (N).
- It then returns the coordinates in an ordered node N.

Figure 6 shows the spiral path planning on the segmented image. The complexity of the algorithm is O(r*c). It is explained below:

- The algorithm uses nested loops. The outer while loop continues until both r_s are less than or equal to r_e and c_s are less than or equal to c_e.
- Inside the while loop, there are two nested for loops:
  - The first nested for loop iterates over rows within the current column (from r_s to r_e).
  - The second nested for loop iterates over columns within the current row (from c_s to c_e).
- Additionally, there are two more nested for loops within conditions:
  - If $r_s <= r_e$, a loop iterates over a row in reverse (from c_e to c_s).
  - If $c_s <= c_e$, a loop iterates over a column in reverse (from r_e to r_s).

  In terms of time complexity:

  - The first while loop runs at most $\min(r, c)$ times, where r is the number of rows and c is the number of columns.
  - The nested loops inside the while loop are responsible for iterating through sub-matrices with decreasing dimensions. The total number of iterations performed by these loops can be roughly approximated as O(r*c).

  Therefore, the overall time complexity of the algorithm is approximately O(r*c).

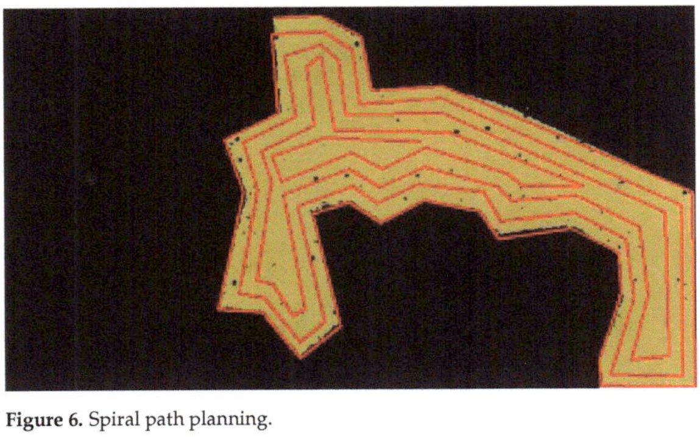

Figure 6. Spiral path planning.

Algorithm 2 depicts zig-zag path planning. It takes starting point S_I, R is the ROI of an input image, r_e is the even row, r_o is the odd row, c is the column, and in N, we store the waypoints. It outputs a list of ordered nodes (N), which represents a path of points in the ROI R.

Algorithm 2 Zig-zag path planning

Input: S_I: Starting point R:ROI Linewidth:w **Output**: Waypoints N **Procedure**:

```
1:  N ← S_I
2:  r_e = 0
3:  r_o = 1
4:  while r_e < r do
5:      for i in range c do
6:          N ← (i, r_e)
7:          r_e = r_e + 2
8:          if r_o < r then
9:              for j in range (c − 1, −1, −1) do
10:                 N ← (j, i)
11:                 r_o = r_o + 2
12:             end for
13:         end if
14:     end for
15: end while
    return N
```

It then starts with the zig-zag path planning method, which divides the ROI into grid form. We took the linewidth of size w for each grid. Here is a step-by-step explanation of the procedure:

- Line 1 points to the starting point and stores it in N. It then initializes variables r_e, and r_o to 0 and 1 respectively (lines 2,3).
- Then it starts with the first iteration (lines 4,5), where it starts with the even row and stores the waypoints of the path in the same direction in N.
- We skip the next row (oddRow) and move to the next evenRow.
- Similarly, it again iterates (line 8) for the odd rows and stores the path waypoints in the opposite direction in N.
- We skip the next row and move to the next even row.
- Finally, we obtain the output in the form of a "path" list as the ordered nodes(N) of the path for the ROI R. Note that this procedure traverses the ROI, R, in a zig-zag pattern, alternating the direction of traversal in each row.

Figure 7 shows the zig-zag path planning on the segmented image. The computation complexity of the above algorithm is explained below:

- Initializing N with the values of S_1 takes constant time, O(1).
- Initializing variables r_e and r_o takes constant time, O(1).
- The while loop continues as long as r_e is less than r. Each iteration involves a constant number of operations (incrementing r_e and potentially entering the inner loop). The number of iterations is $r/2$, so the complexity of the while loop is O(r).
- The first for loop iterates over the indices of array r_e, which contains c elements. Each iteration involves a constant number of operations (assigning a value to N and incrementing r_e). So, the complexity of this loop is O(c).
- The second for loop (inside the first loop) iterates in reverse over the indices of array r_o, which contains c elements. Similar to the previous loop, each iteration involves a constant number of operations. Since this loop only occurs if r_o < r, we can ignore its complexity if r_o is small compared to r.

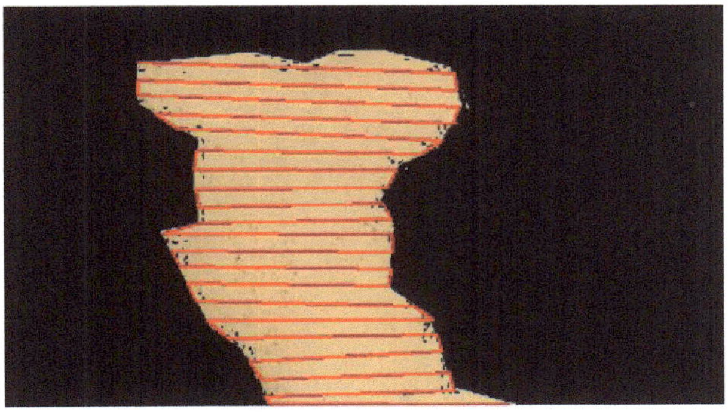

Figure 7. Zig-zag path planning.

The overall time complexity is determined by the most time-consuming part of the code, which is the while loop. Therefore, the total time complexity of the provided pseudo-code is approximately O(r * c), where r is the predefined value and c is the number of elements in arrays r_e and r_o.

3.2.2. Coordinates Transformation

The result of the path planning is a sequence of waypoints representing the order of locations in the path to be followed by UAVs. Since the segmentation's input and output are in the form of an image, the waypoints are also obtained in pixel coordinates. The transformation of pixel coordinates to geographical coordinates is required for UAV operation. To achieve it, the geographical coordinates of some points, like corner pixels, are maintained and later used for mapping pixel coordinates with geographical ones. In this way, the final coverage path as a sequence of geographical coordinates is obtained.

4. Experimentation and Results

The proposed method, including segmentation and path planning, was simulated using Python and the ArduPilot-based Mission planner simulation. The goal is to obtain the segmentation accuracy and efficiency of generated UAV coverage path.

4.1. Segmentation

4.1.1. Dataset

The dataset is obtained from Google Earth, where 100 samples are taken from various places. The Label box tool is employed to assign labels to the pixels of each instance, categorizing them into two distinct classes: region of interest (ROI) and non-ROI. Figure 8a,b contains the original image, and Figure 8c,d contains the segmented images where the ROI and non-ROI parts are segmented from the original image. The ROI part represents the region more likely to contain landmines, whereas non-ROI is the less significant region.

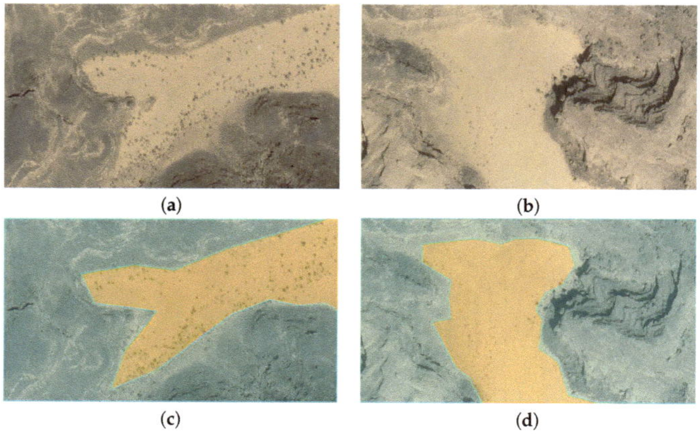

Figure 8. (**a**,**b**) Images taken from Google Earth; (**c**,**d**) are labeled images.

4.1.2. Results

This work utilized the U-Net segmentation model to improve the identification of landmines by analyzing satellite images. A dataset of 100 satellite photos from Google Earth was carefully selected, focusing on specific regions of interest (ROIs) intended for aerial surveys conducted using UAVs. The U-Net architecture was specifically designed for binary segmentation, allowing it to accurately determine the categorization of each pixel as either belonging to the ROI or the non-ROI category. To mitigate the limitations of the restricted dataset, the researchers implemented data augmentation techniques like scaling, rotation, shearing, flipping, and zooming, enhancing the model's capacity to generalize well over a wide range of circumstances. During a comprehensive training and validation process, essential performance measures such as loss and accuracy provided vital insights into the model's competency, depicted in Figure 9. The segmentation methodology has been developed to enhance the accuracy of UAV survey operations, enabling focused examination of areas that are more likely to harbor landmines.

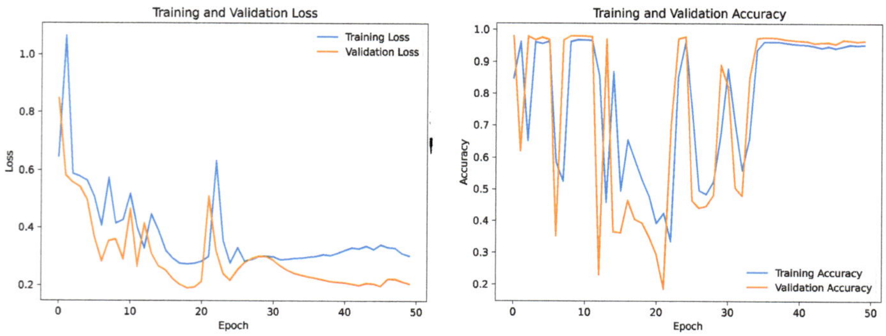

Figure 9. Graph showing the performance of the U-net model in terms of loss and accuracy on the given dataset.

4.2. Path Planning

Experiments for UAV-based surveys for different input locations have been performed using Python and ArduPilot-based simulation. It helps to check the efficiency of the obtained coverage path using the proposed method regarding path length, time, and energy consumption. In conjunction with the region's geometry, certain UAV parameters govern the generation of coverage routes. Table 1 lists UAV parameters and their corresponding values. The drone speed determines the maximum speed of a drone to the ground. At the turns, the velocity can vary, based on the trajectory. Regarding coverage path generation, the UAV's height is also crucial. The UAVs are outfitted with various sensors for data collection during aerial surveys. With an increase in height, the sensor footprint also changes, resulting in a modification of the required line spacing along the coverage path. The percentage of overlap can be selected for the coverage depending on the required resolution. The importance of performing surveys at lower altitudes is in their capacity to provide enhanced resolution and sensitivity, which are essential for the precise identification and localization of landmines. By operating at lower altitudes, UAVs can acquire more detailed data, mitigating the potential for erroneous negative results and augmenting the overall detection precision. This enhances the identification of tiny or deeply concealed landmines that may be disregarded in surveys conducted at higher altitudes. Several aspects, including altitude, sensor field of view, spatial resolution, and overlap percentage impact the survey line spacing in coverage route planning using UAVs. To ensure simplicity and facilitate implementation, the present research has used some fixed values listed in Table 1. Two types of coverage paths were considered. The first one with line space of 10 m, a UAV height of 5 m, and a maximum speed of 10 m/s. The second one is based on magnetometer-based sensing as employed in [14]. It has a line space of 3 m, with a UAV height of 1 m and a maximum UAV speed of 5 m/s.

Table 1. Parameters considered for testing the coverage paths for UAVs.

Parameter	Value
UAV Max speed	5 m/s, 10 m/s
Altitude	1 m, 5 m
Line Gap	3 m, 10 m
Overlap	0.5, 0.5

The integration of digital elevation models (DEMs) plays a crucial role in the aerial operations of UAVs. The provided datasets enable a thorough depiction of terrain elevation, enabling an enhanced understanding of the topographical differences within the region. By incorporating DEMs into the UAV's navigation system, the aircraft can make informed decisions in real time, adjusting its height and route to accommodate the current terrain conditions. In the context of our simulation configuration, we used the functionalities offered by the integrated DEM given by Mission Planner. It enables a dynamic representation of the terrain's elevation data, revealing the topographical characteristics of the surveyed area. By incorporating the DEM data into our simulation, we can simulate real-world conditions and allow the UAV to alter its flight parameters in response to variations in terrain elevation.

The time a UAV requires to cover a given area is critical in many applications, such as search and rescue operations or disaster response efforts. The faster the UAV can cover the area, the quicker it can provide crucial information or aid to those in need. Minimizing energy usage can increase the flight time and distance a UAV can cover before it returns for recharging or refueling. It can improve the overall efficiency of the UAV mission and reduce costs. The energy consumption of a UAV increases with the distance covered. A longer coverage path will require the UAV to fly for a longer period, reducing the available flight time. Figure 10 depicts the total length of the path generated by various coverage path planning techniques.

The spiral pattern generates a short path for simple regions, but with increased complexity, the path length becomes higher than the zig-zag. The proposed coverage pattern in the study was compared with the minTraversal strategy [22]. The minTraversal strategy produces a coverage trajectory that is longer than that generated by the zig-zag strategy.

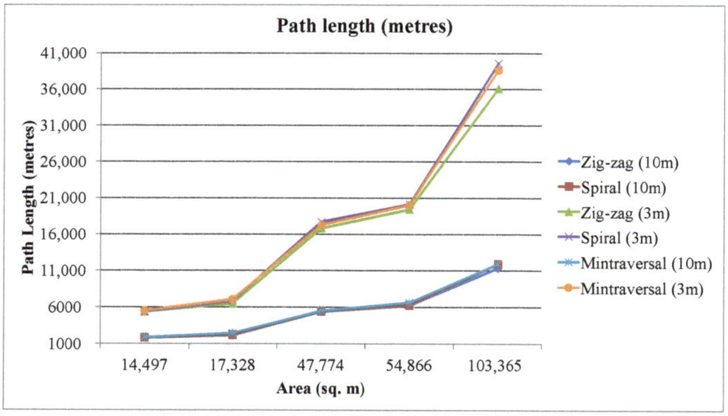

Figure 10. Total distance or path length (in meters) for different coverage paths.

The number of turning points in a UAV coverage path can significantly impact the energy consumption and flight time of the UAV. The turning point requires the UAV to slow down, change direction, and accelerate again, which consumes additional energy. The comparison of the number of turning points in different patterns is presented in Figure 11. The zig-zag pattern generally has fewer turning points but requires more sharp turns if the region is complex, and remains relatively smooth in the spiral pattern. Similarly, the time consumption is depicted in Figure 12. The zig-zag pattern generates the trajectory requiring lesser time to cover the entire region as compared to other methods.

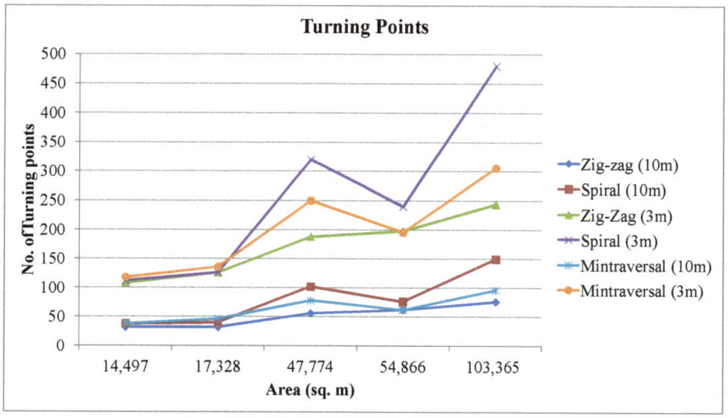

Figure 11. Number of turning points in different coverage paths.

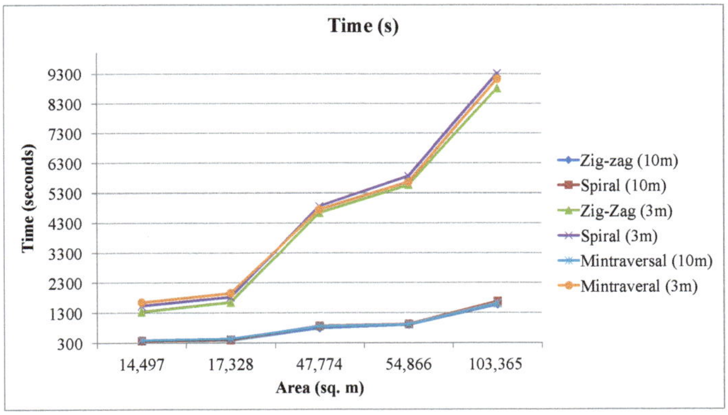

Figure 12. Total time consumption (in seconds) by UAV to traverse different coverage paths.

Figure 13 shows the relative energy requirements of various coverage patterns. Similarly, the energy consumption per kilometer is depicted in Figure 14. Table 2 summarizes the coverage results obtained for different strategies. The zig-zag path planning approach has demonstrated superior performance, showcasing its potential to enhance efficiency and optimize UAV-based surveys. Through its innovative navigation strategy, the zig-zag pattern significantly reduces the number of turns required during the survey, resulting in a streamlined flight trajectory. This reduction in the number of turns translates to both time and energy savings, allowing the UAV to cover the designated area more swiftly and with reduced battery consumption. In addition, the zig-zag path planning method generates shorter survey routes than other methods. By strategically alternating flight directions, the zig-zag pattern ensures that adjacent survey lines are closely spaced, effectively reducing coverage gaps and enhancing the overall quality of data collection. The compact path design provides extensive coverage without compromising data precision.

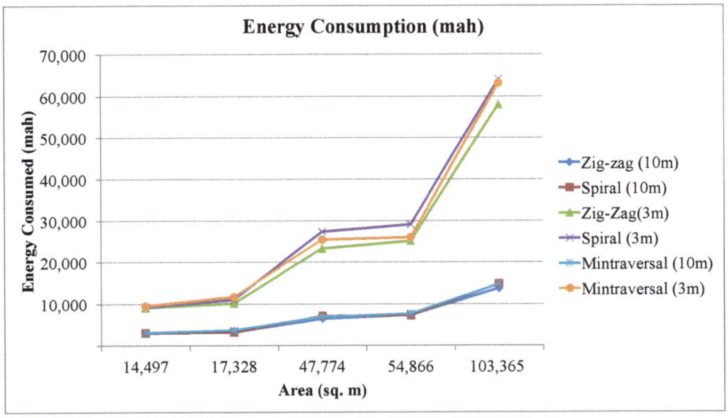

Figure 13. Energy consumption in milliampere-hour (mAh) for different coverage paths.

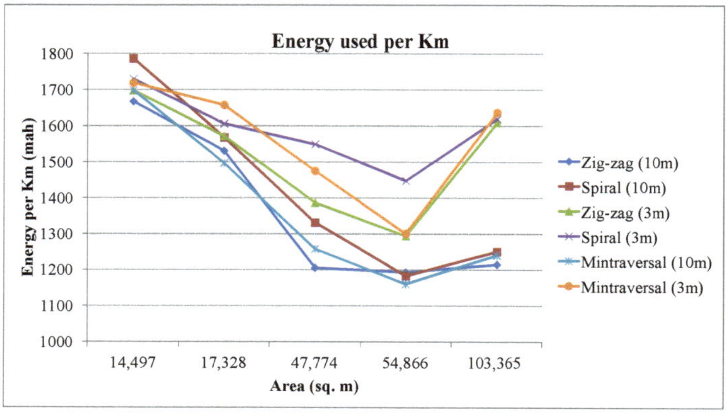

Figure 14. Energy consumed per kilometer for different coverage paths.

Table 2. Comparative analysis of different coverage paths for various sample regions with line gap 10 m.

Sample	Area	Path	Path Length	Time(s)	Energy Consumed	Energy mAh/Km	No. of Turning Points	Average Speed
S1	14,497	Zig-zag	1849.64	385	3083	1666.81	32	4.8
S2	17,328	Zig-zag	2154	410	3298	1530.78	32	5.32
S3	47,774	Zig-zag	5405.41	801	6514	1205.09	56	6.75
S4	54,866	Zig-zag	6235.49	915	7435	1192.37	62	6.81
S5	103,365	Zig-zag	11,350.54	1575	13,785	1214.47	76	7.21
S1	14,497	Spiral	1750.99	385	3128	1786.42	38	4.55
S2	17,328	Spiral	2223.95	405	3264	1467.66	40	5.42
S3	47,774	Spiral	5334.95	872	7103	1331.41	102	6.12
S4	54,866	Spiral	6295.56	900	7317	1162.25	76	6.99
S5	103,365	Spiral	11915	1675	14,898	1250.36	150	7.11
S1	14,497	minTraversal	1876.74	405	3187	1698.15	38	4.63
S2	17,328	minTraversal	2473.92	445	3703	1496.81	46	5.56
S3	47,774	minTraversal	5522.08	865	6945	1257.67	78	6.38
S4	54,866	minTraversal	6645.74	925	7710	1160.14	62	7.18
S5	103,365	minTraversal	11,901.89	1615	14,756	1239.80	96	7.36

5. Conclusions

The landmines continue to create problems for people residing in contaminated areas, and manual detection methods are time-consuming and risky. The automation of the detection process using UAVs can greatly help in the efforts of humanitarian demining. The work conducted is focused on the task of automating the path determination for UAV-based minefield surveys. It works by segmenting the required area and then generating the coverage path. It can help optimize the UAV performance while operating in complex terrains. The study compared the time and energy consumption of different coverage path planning algorithms. The results showed that a carefully planned zig-zag path was able to cover a larger area in a smaller amount of time with lesser energy consumption compared to other methods.

The study can be extended with real-time experiments where various factors for controlled and safe UAV operation can be explored under different environments. The use of multiple UAVs to further enhance the minefield coverage in various regions can be considered in future work. Moreover, dynamic UAV operation in the presence of obstacles and other environmental factors needs to be examined to ascertain robustness in UAV-based landmine detection.

Author Contributions: Software, K.K. and N.T.; Validation, B.A.; Formal analysis, A.A.; Investigation, A.A.; Supervision, A.B. and N.K.; Project administration, A.B. All authors have read and agreed to the published version of the manuscript.

Funding: This research work was funded by Institutional Fund Projects under grant no (IFPNC-001-611-2020). Therefore, the authors gratefully acknowledge technical and financial support from the Ministry of Education and King Abdelaziz University, Jeddah, Saudi Arabia.

Institutional Review Board Statement: Not applicable.

Informed Consent Statement: Not applicable.

Data Availability Statement: The data are contained within the article.

Conflicts of Interest: The authors declare no conflict of interest.

References

1. International Campaign to Ban Landmines. Available online: http://icbl.org/en-gb/problem/what-is-a-landmine.aspx (accessed on 30 December 2022).
2. Bello, R. Literature review on landmines and detection methods. *Front. Sci.* **2013**, *3*, 27–42.
3. Landmine Monitor 2022. Available online: http://www.the-monitor.org/en-gb/reports/2022/landmine-monitor-2022.aspx (accessed on 30 December 2022).
4. Barnawi, A.; Budhiraja, I.; Kumar, K.; Kumar, N.; Alzahrani, B.; Almansour, A.; Noor, A. A comprehensive review on landmine detection using deep learning techniques in 5G environment: Open issues and challenges. *Neural Comput. Appl.* **2022**, *34*, 21657–21676. [CrossRef]
5. Kasban, H.; Zahran, O.; Elaraby, S.M.; El-Kordy, M.; Abd El-Samie, F.E. A comparative study of landmine detection techniques. *Sens. Imaging Int. J.* **2010**, *11*, 89–112. [CrossRef]
6. Castiblanco, C.; Rodriguez, J.; Mondragon, I.; Parra, C.; Colorado, J. Air drones for explosive landmines detection. In Proceedings of the ROBOT2013: First Iberian Robotics Conference, Madrid, Spain, 28–29 November 2013; Springer: Berlin/Heidelberg, Germany, 2014; pp. 107–114.
7. Fattah, S.A.; Haider, M.Z.; Chowdhury, D.; Sarkar, M.; Chowdhury, R.I.; Islam, M.S.; Karim, R.; Rahi, A.; Shahnaz, C. An aerial landmine detection system with dynamic path and explosion mode identification features. In Proceedings of the 2016 IEEE Global Humanitarian Technology Conference (GHTC), Seattle, WA, USA, 13–16 October 2016; pp. 745–752.
8. Colorado, J.; Perez, M.; Mondragon, I.; Mendez, D.; Parra, C.; Devia, C.; Martinez-Moritz, J.; Neira, L. An integrated aerial system for landmine detection: SDR-based Ground Penetrating Radar onboard an autonomous drone. *Adv. Robot.* **2017**, *31*, 791–808. [CrossRef]
9. Schartel, M.; Burr, R.; Mayer, W.; Docci, N.; Waldschmidt, C. UAV-based ground penetrating synthetic aperture radar. In Proceedings of the 2018 IEEE MTT-S International Conference on Microwaves for Intelligent Mobility (ICMIM), Munich, Germany, 15–17 April 2018; pp. 1–4.
10. Fernández, M.G.; López, Y.Á.; Arboleya, A.A.; Valdés, B.G.; Vaqueiro, Y.R.; Andrés, F.L.H.; García, A.P. Synthetic aperture radar imaging system for landmine detection using a ground penetrating radar on board a unmanned aerial vehicle. *IEEE Access* **2018**, *6*, 45100–45112. [CrossRef]
11. Šipoš, D.; Gleich, D. A lightweight and low-power UAV-borne ground penetrating radar design for landmine detection. *Sensors* **2020**, *20*, 2234. [CrossRef] [PubMed]
12. Garcia-Fernandez, M.; López, Y.Á.; Andrés, F.L.H. Airborne multi-channel ground penetrating radar for improvised explosive devices and landmine detection. *IEEE Access* **2020**, *8*, 165927–165943. [CrossRef]
13. Baur, J.; Steinberg, G.; Nikulin, A.; Chiu, K.; de Smet, T.S. Applying deep learning to automate UAV-based detection of scatterable landmines. *Remote Sens.* **2020**, *12*, 859. [CrossRef]
14. Yoo, L.S.; Lee, J.H.; Ko, S.H.; Jung, S.K.; Lee, S.H.; Lee, Y.K. A drone fitted with a magnetometer detects landmines. *IEEE Geosci. Remote. Sens. Lett.* **2020**, *17*, 2035–2039. [CrossRef]
15. Yoo, L.S.; Lee, J.H.; Lee, Y.K.; Jung, S.K.; Choi, Y. Application of a drone magnetometer system to military mine detection in the demilitarized zone. *Sensors* **2021**, *21*, 3175. [CrossRef]

16. Sultana, F.; Sufian, A.; Dutta, P. Evolution of image segmentation using deep convolutional neural network: A survey. *Knowl.-Based Syst.* **2020**, *201*, 106062. [CrossRef]
17. Rozanda, N.E.; Ismail, M.; Permana, I. Segmentation Google Earth imagery using k-means clustering and normalized RGB color space. In *Computational Intelligence in Data Mining*; Springer: Berlin/Heidelberg, Germany, 2015; Volume 1, pp. 375–386.
18. Wang, S.; Yang, D.M.; Rong, R.; Zhan, X.; Xiao, G. Pathology image analysis using segmentation deep learning algorithms. *Am. J. Pathol.* **2019**, *189*, 1686–1698. [CrossRef]
19. Aggarwal, S.; Kumar, N. Path planning techniques for unmanned aerial vehicles: A review, solutions, and challenges. *Comput. Commun.* **2020**, *149*, 270–299. [CrossRef]
20. Galceran, E.; Carreras, M. A survey on coverage path planning for robotics. *Robot. Auton. Syst.* **2013**, *61*, 1258–1276. [CrossRef]
21. Cabreira, T.M.; Brisolara, L.B.; Paulo R, F.J. Survey on coverage path planning with unmanned aerial vehicles. *Drones* **2019**, *3*, 4. [CrossRef]
22. Li, Y.; Chen, H.; Er, M.J.; Wang, X. Coverage path planning for UAVs based on enhanced exact cellular decomposition method. *Mechatronics* **2011**, *21*, 876–885. [CrossRef]
23. Coombes, M.; Fletcher, T.; Chen, W.H.; Liu, C. Optimal polygon decomposition for UAV survey coverage path planning in wind. *Sensors* **2018**, *18*, 2132. [CrossRef]
24. Valente, J.; Sanz, D.; Del Cerro, J.; Barrientos, A.; de Frutos, M.Á. Near-optimal coverage trajectories for image mosaicing using a mini quad-rotor over irregular-shaped fields. *Precis. Agric.* **2013**, *14*, 115–132. [CrossRef]
25. Di Franco, C.; Buttazzo, G. Coverage path planning for UAVs photogrammetry with energy and resolution constraints. *J. Intell. Robot. Syst.* **2016**, *83*, 445–462. [CrossRef]
26. Pérez-González, A.; Benítez-Montoya, N.; Jaramillo-Duque, Á.; Cano-Quintero, J.B. Coverage path planning with semantic segmentation for UAV in PV plants. *Appl. Sci.* **2021**, *11*, 12093. [CrossRef]
27. Nedjati, A.; Izbirak, G.; Vizvari, B.; Arkat, J. Complete coverage path planning for a multi-UAV response system in post-earthquake assessment. *Robotics* **2016**, *5*, 26. [CrossRef]
28. Zhang, H.; Xin, B.; Dou, L.h.; Chen, J.; Hirota, K. A review of cooperative path planning of an unmanned aerial vehicle group. *Front. Inf. Technol. Electron. Eng.* **2020**, *21*, 1671–1694. [CrossRef]
29. Araujo, J.; Sujit, P.; Sousa, J.B. Multiple UAV area decomposition and coverage. In Proceedings of the 2013 IEEE Symposium on Computational Intelligence for Security and Defense Applications (CISDA), Singapore, 16–19 April 2013; pp. 30–37.
30. Azpúrua, H.; Freitas, G.M.; Macharet, D.G.; Campos, M.F. Multi-robot coverage path planning using hexagonal segmentation for geophysical surveys. *Robotica* **2018**, *36*, 1144–1166. [CrossRef]
31. Guastella, D.C.; Cantelli, L.; Giammello, G.; Melita, C.D.; Spatino, G.; Muscato, G. Complete coverage path planning for aerial vehicle flocks deployed in outdoor environments. *Comput. Electr. Eng.* **2019**, *75*, 189–201. [CrossRef]
32. Choi, Y.; Choi, Y.; Briceno, S.; Mavris, D. Energy-constrained multi-UAV coverage path planning for an aerial imagery mission using column generation. *J. Intell. Robot. Syst.* **2020**, *97*, 125–139. [CrossRef]
33. Cao, Y.; Cheng, X.; Mu, J. Concentrated Coverage Path Planning Algorithm of UAV Formation for Aerial Photography. *IEEE Sensors J.* **2022**, *22*, 11098–11111. [CrossRef]
34. Cho, S.W.; Park, H.J.; Lee, H.; Shim, D.H.; Kim, S.Y. Coverage path planning for multiple unmanned aerial vehicles in maritime search and rescue operations. *Comput. Ind. Eng.* **2021**, *161*, 107612. [CrossRef]
35. Xie, J.; Carrillo, L.R.G.; Jin, L. An integrated traveling salesman and coverage path planning problem for unmanned aircraft systems. *IEEE Control. Syst. Lett.* **2018**, *3*, 67–72. [CrossRef]
36. Xie, J.; Carrillo, L.R.G.; Jin, L. Path planning for UAV to cover multiple separated convex polygonal regions. *IEEE Access* **2020**, *8*, 51770–51785. [CrossRef]

sensors

MDPI

Article

Research on Autonomous and Collaborative Deployment of Massive Mobile Base Stations in High-Rise Building Fire Field

Ke Li [1], Chen Huang [1], Jiaping Liang [1], Yanbin Zou [1], Biao Xu [1], Yao Yao [1,*], Yang Zhang [2] and Dandan Liu [3]

[1] Department of Electronic Engineering, Shantou University, Shantou 515041, China; ericlee@stu.edu.cn (K.L.); 21chuang@stu.edu.cn (C.H.); 21jpliang@stu.edu.cn (J.L.); ybzou@stu.edu.cn (Y.Z.); xubiao@stu.edu.cn (B.X.)

[2] Naval Research Academy, Beijing 100161, China

[3] Yancheng Institute of Technology, Yancheng 224051, China; ldd@ycit.edu.cn

* Correspondence: yaoyao@stu.edu.cn; Tel.: +86-13917385576

Abstract: High-rise building fires pose a serious threat to the lives and property safety of people. The lack of reliable and accurate positioning means is one of the main difficulties faced by rescuers. In the absence of prior knowledge of the high-rise building fire environment, the coverage deployment of mobile base stations is a challenging problem that has not received much attention in the literature. This paper studies the problem of the autonomous optimal deployment of base stations in high-rise building fire environments based on a UAV group. A novel problem formulation is proposed that solves the non-line-of-sight (NLOS) positioning problem in complex and unknown environments. The purpose of this paper is to realize the coverage and deployment of mobile base stations in complex and unknown fire environments. The NLOS positioning problem in the fire field environment is turned into the line-of-sight (LOS) positioning problem through the optimization algorithm. And there are more than three LOS base stations nearby at any point in the fire field. A control law which is formulated in a mathematically precise problem statement is developed that guarantees to meet mobile base stations' deployment goals and to avoid collision. Finally, the positioning accuracy of our method and that of the common method were compared under many different cases. The simulation result showed that the positioning error of a simulated firefighter in the fire field environment was improved from more than 10 m (the positioning error of the traditional method) to less than 1 m.

Keywords: swarm intelligence; trajectory planning; fire rescue; autonomous deployment; collaborative positioning

Citation: Li, K.; Huang, C.; Liang, J.; Zou, Y.; Xu, B.; Yao, Y.; Zhang, Y.; Liu, D. Research on Autonomous and Collaborative Deployment of Massive Mobile Base Stations in High-Rise Building Fire Field. *Sensors* **2023**, 23, 7664. https://doi.org/10.3390/s23187664

Academic Editor: Gang Peng

Received: 24 July 2023
Revised: 16 August 2023
Accepted: 29 August 2023
Published: 5 September 2023

1. Introduction

Due to the high land price, limited land resources, and the rapidly growing economy's demand for landmark buildings, high-rise and super high-rise buildings have become one of the effective means to address the space demand of urbanization and economic growth [1–3]. China is home to 5 of the world's top 10 skyscrapers [4]. Statistics show that the number of high-rise and super high-rise buildings in China have been in the forefront in the world for eight consecutive years [5–7]. Although super high-rise buildings bring people enjoyment in life and greatly save space, they also bring people many problems. Fire is one of the most serious problems. According to statistics, urban high-rise building fire accidents can account for more than 90% of all fire accidents [8,9], and it is difficult to rescue and safely evacuate, which has brought huge losses to the national economy and people's lives and property.

The biggest problem in the high-rise building fire rescue and evacuation of people in distress is positioning. Because the global positioning system (GPS) cannot be used in the indoor environment and the positioning base station cannot be deployed in advance, the positioning system in a high-rise building fire rescue has to face the impact of the non-line-of-sight environment, which greatly affects the positioning accuracy and real time.

Firefighters and people in distress cannot obtain their exact positions in real time, which leads to the very low efficiency of rescue and evacuation. Therefore, the ultra-wide band (UWB) positioning system is the best choice in this case. The accuracy of the UWB base station coordinates affects the positioning accuracy of firefighters directly. So, obtaining the accurate position of the UWB base stations is very significant for the firefighters' safety. But traditional methods are very difficult to obtain the accurate position of the base station in an environment with unknown obstacles [10]. It cannot meet the needs for the rapid deployment of UWB base stations.

Non-line-of-sight (NLOS) positioning means that there is at least one obstacle on the straight line between the base station and the positioning objective. Whereas line-of-sight (NLOS) positioning means that there is no obstacle on the straight line between the base station and the positioning objective. Non-line-of-sight (NLOS) positioning is much more difficult than LOS positioning so it has always been a bottleneck problem to improve the positioning accuracy in complex environments [11]. Obstacle obstruction will cause multipath effects, signal hysteresis, and other negative effects [12,13]. Up to now, no ideal solution has been found [14,15]. Targeting this bottleneck problem, a method for the autonomous deployment of positioning base stations based on swarm intelligence is proposed. The cooperation among unmanned aerial vehicles (UAVs) is used to achieve the autonomous coverage and deployment of the fire field, and the non-line-of-sight environment is converted into the line-of-sight environment, so as to achieve the accurate positioning of targets in the fire field environment.

Given a known environment without obstacles, the algorithms for the deployment of base stations are well studied [16–18]. The main practical motivation and application for the deployment of base stations in an ideal environment is to construct a communication network in the air to provide wireless signal coverage for the fire field [19,20]. In recent years, many scholars have applied machine-learning methods to solve control and path planning problems in multi-agent systems, such as deep learning, reinforcement learning, and deep reinforcement learning, and have achieved some success [21–23]. However, there are still many big challenges in solving the control problem of multi-agent systems in complex unknown closed environments, such as high-rise building fire fields. The multi-agent system used for fire environment detection has extremely high requirements for stability, reliability, environmental adaptability, lightweight, and so on. Machine learning cannot meet the application requirements for the detection of a high-rise building fire field because of the lack of interpretability and generalization.

Therefore, there has been very little work in the literature on how multi-UAV systems solve the positioning problem of NLOS in an unknown complex closed and GPS-denied environment with swarm UAVs. Existing works in the literature require robots equipped with sensors that are able to localize themselves precisely. The literature [24,25] focuses on using object-level features with both semantic and geometric information to model landmarks in the environment not need a prior constructed precise geometric map, which greatly releases the storage burden, especially for large-scale navigation. The algorithm is effective, but the deployment needs the accurate positioning information of UAVs that cannot meet the requirements for a high-rise building fire rescue. To the best of our knowledge, this is the most feasible approach for solving the precise positioning problem of a high-rise building fire rescue using GPS-denied swarm UAVs.

The main contribution and novel part in this paper is that it provides the approach for transforming the NLOS positioning problem into the LOS positioning problem in a 2D unknown complex environment with swarm UAVs and a host computer. There are two fundamental problems in the deployment of mobile base stations for coverage with communication constraints. The first problem is how to control the movement and behavior of multi-agent systems to realize a desired configuration. The second one is how to optimize the deployment in order to enhance the efficiency of the multi-agent system. A UAV lands and keeps static as a base station when UAV obtains the right position. Every UAV is equipped with a UWB positioning device to obtain the distance to adjacent UAVs and

Zigbee equipment to send its own relative position information to other adjacent base stations and the host computer.

The rest of this paper is organized as follows: In Section 2, we formulate the problem and then define the sensing area of the UAV and the function of the sensing ability. The optimization method for the autonomous cooperative coverage deployment of base stations is presented in Section 3. The control law of multi-UAVs is given to make the coverage deployment operation feasible in complex environments in the presence of unknown obstacles and to ensure it can turn a non-line-of-sight environment into a line-of-sight environment. Section 4 demonstrates the effectiveness and feasibility of the proposed approach via computer simulations. Section 5 summarizes the paper and draws some conclusions.

2. Problem Formulation

To achieve precise target positioning in the fire field, there are two main problems to be solved in the optimization algorithm for the autonomous deployment of base stations in complex unknown environments.

1. How to realize the coverage deployment of mobile base stations in the fire field, and ensure that the target at any point in the area will be surrounded by at least three base stations nearby that can locate the target with LOS;
2. How to avoid collision when a large-scale multi-UAV system moves in a complex and unknown environment.

This becomes more difficult in the absence of GPS for navigation and positions. Without the global information, agents can only obtain local localization via a limited information exchange and estimate their own position or neighbor's relative position based on the relative distance they measured.

2.1. Study Objective

In the process of solving the first problem, we set the different coverage values for the fire field according to the duration of the different area being detected and guided the UAV to move to the area where it is not effectively detected (the coverage value of the area is less than the effective coverage value).

2.2. Assumptions and Constrains

In this paper, a UAV is denoted by A. Let S be a region that it is required to cover. Let N be the number of UAVs and let p_i and v_i denote the position and the velocity of UAV A_i, respectively, $i \in I = \{1, 2, 3, \ldots, N\}$. Each UAV A_i, $i \in I$, satisfies the following such kinematic equations of motion:

$$\dot{p}_i = v_i,\ i \in I \tag{1}$$

Define the instantaneous coverage function D_i as a continuous map that describes how effective a UAV A_i senses a point $p \in S$. In this paper, we consider sensors with the following properties:

(1) The communication of a multi-UAV system is fully connected.

(2) The detection probability of the on-board sensor is 100%, that is, the detection probability of the sensor detection equipment has no effect on the results.

(3) Each UAV has a limited sensory domain $D_i(t)$ with a sensory range R_i. The sensory domain of each UAV is given by

$$D_i(t) = \{p \in S : ||p_i(t) - p|| \leq R_i\} \tag{2}$$

Let the union of all coverage regions be denoted by $D(t) = \cup_{i \in S} D_i(t)$.

(4) Without the loss of generality, we consider the vision-based sensor for aerial detection may be modeled by following mathematical Formula (3). Each UAV's sensor has a circular sensing symmetry about the position p_i, which is a practical property of a

sensor in real applications. Within the sensing range of the UAV, each UAV has a peak sensing capacity of M_i exactly at the position p_i of UAV A_i. That is, we have

$$D_i(q_i, q_i) = M_i > D_i(q, q_i),\ \forall q \neq q_i \tag{3}$$

Hence, the divergence angle of the detector is θ, the detection height is h_i, then, the sensing radius is R_i, as shown in Figure 1.

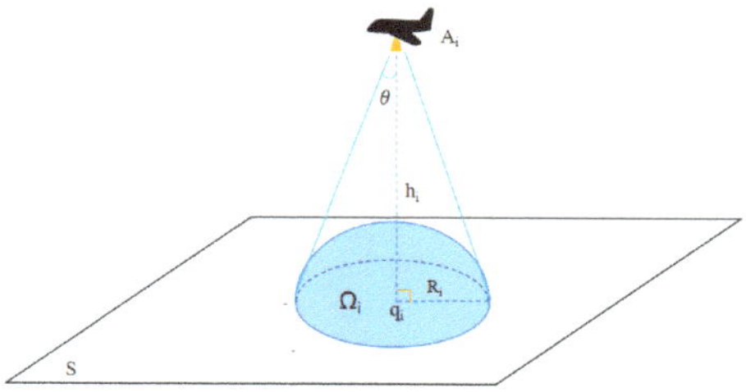

Figure 1. Diagram of the sensing region.

(5) Mathematically, such a sensor function is a second-order polynomial function of $d = |q_i - q|$. Hence, the perceptual ability at the different position q of the UAV can be given by

$$G_i(d) = \begin{cases} \frac{M_i}{R_i^4}(d^2 - R_i^2)^2 & ,d \leq R_i \\ 0 & ,d > R_i \end{cases} \tag{4}$$

The information accumulation of a point in the fire field obtained by a UAV using the sensor increases with time, that is, the coverage value of a point in the area increases with time. The coverage level of any point $q \in \Omega_i$ from the initial time $t = 0$ to t is defined as follows:

$$\gamma_i(q, t) = \int_0^t G_i(|q_i - q|)d\tau \tag{5}$$

Let $k \subseteq I$ be the subset of UAVs that covered q and the effective coverage by a subset of UAVs $A_k = \{A_i | i \in k \subseteq I\}$ in surveying q is then given by [26,27]

$$\gamma_k(q, t) = \int_0^t \sum_{i \in k} G_i(|q_i(\tau) - q|_2)d\tau \tag{6}$$

It can be obviously seen that $\gamma_k(q, t)$ is a non-decreasing function of time. That is,

$$\frac{\partial}{\partial t}\gamma_k(q, t) = \sum_{i \in k} G_i(|q_i(\tau) - q|) \geq 0 \tag{7}$$

3. Adaptive Control Law

In the process of solving the second problem, the traditional artificial potential field method was abandoned. Because the acceleration of UAVs is limited, they need time and space to brake and avoid collisions. Therefore, the alignment interaction range was determined based on the expected optimal relation between the velocity difference and distance. Consequently, the movement state of each UAV itself can be adjusted in real time to effectively avoid collisions according to the relative distance and relative speed of the surrounding objects.

The rationale of adaptive control law is to use the flocking aggregation morphology control algorithm to enable the UAV group to construct a local network. The perception range is expanded to avoid collisions through the information interaction between adjacent

UAVs. In addition, the gradient descent method is used to optimize the navigation path in real time to achieve the autonomous coverage and deployment of the UAV group to the fire field.

When an agent moves in the fire field, it will be affected by the driving force from the unknown area, the repulsive force from the obstacles or other adjacent agents, and the viscous force to reduce the difference in the velocity vectors of nearby UAVs or obstacles. So, the optimal velocity of an agent is the sum of the vectors of the optimum coverage velocity, repulsive velocity, and viscous velocity. The specific design process and implementation steps are as follows:

3.1. Optimum Coverage Velocity

The rationale of the optimum coverage velocity is to make each UAV move toward the direction of a low coverage level by establishing the potential energy function regarding the error between the expected coverage level and the actual coverage level, so that UAVs can autonomously and intelligently complete the coverage deployment task of mobile base stations.

Let C_0 be the desired attained effective coverage at all points in the S. The goal is to attain a network coverage of $\gamma_k(q,t) = C_0$ for all points $\boldsymbol{q} \in S$ at some time t. Consider the following error function:

$$J(t) = \int_{\Omega} h(C_0 - \gamma_I(q,t))dq \tag{8}$$

where $h(x)$ is a penalty function, which is strictly convex in the interval $(0,C_0]$, twice differentiable, and positive definite, that is, $h(x) > 0, h'(x) > 0, h''(x) > 0, \forall x \in (0, C_0]$. And that satisfies $h(x) = h'(x) = h''(x) = 0$ for all $x \leq 0$. When $J(t) = 0$, it means that the coverage task in the sensing area Ω is completed.

To guarantee the coverage of the entire domain S with an effective coverage of C_0 under appropriate assumptions, the UAV has to move toward the area where $\gamma_I(q,t) < C_0$. In order to improve the coverage efficiency, the area coverage problem can be described as the optimization problem of seeking the minimum $J(t)$ in Formula (9) [28,29]:

$$\text{Minimize } J(t) = \int_{S} h(C_0 - \gamma_I(q,t))dq,\ q_i \in S \tag{9}$$

The gradient descent method can be used to solve (9) to obtain the optimal coverage speed of the UAV. The accumulated error will generate an attractive "force" on a UAV. Once $\gamma_I(q,t) = C_0$ at a point, the error at this point is zero no matter how much additional time UAVs spend detecting that point, that is, the excessive coverage has no effect on the motion. Therefore, the optimal coverage speed of the UAV can be designed as follows:

$$\begin{aligned} \vec{v}_{i1}(t) &= -k_c \cdot \tfrac{dJ(t)}{dt} \\ &= -k_c \cdot \textstyle\int_S h'(C_0 - \gamma_I(\boldsymbol{q},t)) \sum\limits_{i \in I} G(|\boldsymbol{q}_i - \boldsymbol{q}|)d\boldsymbol{q} \end{aligned} \tag{10}$$

where $k_c > 0$ is the fixed coverage velocity adjustment gain.

Since the optimal coverage speed of the UAV is based on local information, when the points in the perception area of the UAV are effectively covered, the UAV will tend to be stationary. If the whole area has not been covered effectively at this point, a traction speed is required for the UAV to avoid becoming stuck in a local minimum.

Firstly, set $P(t)$ represents a set of points that are not effectively covered in the area S at time t. That is

$$\boldsymbol{P}(t) = \{\boldsymbol{q} \in \mathrm{S} : 0 < \gamma_I(\boldsymbol{q},t) < C_0\} \tag{11}$$

For UAV A_i, let $P_i(t)$ be the set of points with the shortest distance from $\boldsymbol{q}_i(t)$ in set $P(t)$, that is

$$P_i(t) = \left\{\boldsymbol{q}^* \in P(t) : \boldsymbol{q}^* = \mathrm{argmin}_{q \in P(t)} \left|\boldsymbol{q}_i - \boldsymbol{q}\right|\right\} \tag{12}$$

A rule can be set to choose a unique point from the set to calculate the speed of the UAV at the next moment. The traction speed $\vec{v}_s$ of the UAV can be designed according to the artificial potential field method [30] as follows:

$$\vec{v}_s = -k_s \cdot (\vec{q}_i - \vec{q}_m) \tag{13}$$

where $k_s > 0$ is the traction velocity adjustment gain, $\vec{q}_i$ is the position vector of the UAV projected on the OXY plane, and $\vec{q}_m$ is the position vector of point q_m on the OXY plane.

In summary, the expected optimal coverage velocity of each UAV is:

$$\vec{v}_i^{\,c}(t) = \begin{cases} -k_c \cdot \frac{dJ(t)}{dt} & , J(t) \neq 0 \\ -k_s \cdot (\vec{q}_i - \vec{q}_m) & , J(t) = C_0 \neq 0 \end{cases} \tag{14}$$

3.2. Repulsion

A linear velocity term was chosen for the local repulsion. Let r_0 be the maximum interaction range under which the UAV starts to repulse other UAVs and obstacles:

$$\vec{v}_{ij}^{\,r} = \begin{cases} p_0 \cdot (r_0 - r_{ij}) \cdot \frac{\vec{r}_i - \vec{r}_j}{r_{ij}} & , \; r_{ij} < r_0 \\ 0 & , \text{otherwise} \end{cases} \tag{15}$$

In the equation above, p_0 is the linear gain of the pairwise repulsion and $r_{ij} = \left| \vec{r}_i - \vec{r}_j \right|$ is the distance between UAV i and obstacle j. The total repulsion term calculated for UAV i with respect to the other UAVs or obstacles is

$$\vec{v}_i^{\,r} = \sum_{j \neq i} \vec{v}_{ij}^{\,r} \tag{16}$$

where j is iterated for all other UAVs and obstacles.

3.3. Viscous Velocity

The pairwise velocity alignment can be obtained with a velocity term that depends on the difference in the velocity vectors of nearby UAVs or obstacles. Previous works based on the artificial potential field method adjusted the velocity of the UAV attenuating asymptotically to zero at the expected position according to the distance relative to the obstacle at a different time [31]. These models work fine when UAVs move within a lower velocity regime. However, the UAVs have to fly through the narrow corridor at a relatively high velocity in order to reduce the rescue time for people in distress at the high-rise fire scene. In this case, the previous method caused the obvious self-excited oscillations of the UAV in the process of the repulsion. Therefore, the viscous velocity appended to the control law was useful to solve this problem. It not only synchronizes motion to achieve the collective behavior of UAVs, but it also has to serve as a damping medium, reducing the self-excited oscillations emerging due to the relatively high velocity of the UAV during the repulsion.

To fulfill the practical deployment requirements of mobile base stations in the high-rise fire field, we have chosen an ideal braking function in the space that is denoted by $D(.)$:

$$D(r, a, p) = \begin{cases} 0 & \text{if } r \leq 0 \\ rp & \text{if } 0 < rp < a/p \\ \sqrt{2ar - a^2/p^2} & \text{otherwise} \end{cases} \tag{17}$$

where r is the distance between a UAV and an expected stopping point, a is the preferred acceleration, and p is a linear gain also determining the crossover point between the two phases of deceleration.

The rationale behind our viscous velocity term is to prohibit two UAVs having a larger velocity difference at a given distance than what is allowed by this ideal braking function and to serve as a buffer medium to eliminate the self-excited oscillations of the UAV:

$$v_{ij}^{\mathrm{fmax}} = \max(v^{\mathrm{f}}, D(r_{ij} - r_0^{\mathrm{f}}, a^{\mathrm{f}}, p^{\mathrm{f}})), \tag{18}$$

$$\vec{v}_{ij}^{\mathrm{f}} = \begin{cases} F_s(v_{ij} - v_{ij}^{\mathrm{fmax}}) \cdot \frac{\vec{v}_i - \vec{v}_j}{v_{ij}} & \text{if } v_{ij} > v_{ij}^{\mathrm{fmax}} \\ 0 & \text{otherwise} \end{cases} \tag{19}$$

In the equations above, F_s is a linear coefficient of the velocity alignment error reduction, v^{f} is a velocity slack to allow for a certain amount of velocity difference independently of the inter-UAV distance, r_0^{f} is the distance of the stopping point for UAV i relative to and in front of UAV j, p^{f} and a^{f} are the linear gain and the acceleration parameters of the pairwise alignment, and $v_{ij} = \left|\vec{v}_i - \vec{v}_j\right|$ is the amplitude of the velocity difference between UAVs i and j. The total viscous velocity term calculated for UAV i with respect to the other UAVs—similarly to the repulsion term—is

$$\vec{v}_i^{\mathrm{f}} = \sum_{j \neq i} \vec{v}_{ij}^{\mathrm{f}} \tag{20}$$

where j is iterated for all other UAVs.

4. Simulation Results

In this section, we provide the simulation results for the fully connected network case. Assume that S is a square region with the side length of $l = 700$ dm. Let the original position of each UAV be distributed randomly in the zone so that $x \in [0, 150]$ and $y \in [550, 700]$. Let the original velocity of each UAV be $\left|\vec{v}_0\right| = 10$ dm/s and set the direction of the original velocity be set randomly. Set the original relay distance $d_r = 50$ dm. Let the effective coverage $C_0 = 40$ and the number of UAVs $N = 78$. Obstacles are marked with the black color. The UAV is marked with the blue color and the base station is marked with the red color. The red line between two base stations means the distance between them $d_b \leq d_r$.

Due to the fire field being an unknown complex environment, there is no way to plan the path for each UAV before the beginning of the mission. The only way is to depend on their cooperation. It can be seen from the trajectory of UAVs at different times in Figure 2 that each UAV can keep a safe distance from obstacles and other UAVs. There is no one UAV which collides with the obstacles and beyond the boundaries. It can be seen from Figure 2e that the base stations were deployed evenly in the unknown mission area with obstacles and can cover the whole mission area. There is no obstacle in the middle of the line between every two adjacent base stations so that the non-line-of-sight problem in the unknown complex environment can be turned into the line-of-sight problem. Therefore, the efficiency and accuracy of the target positioning will be greatly enhanced.

At present, the commonly used positioning method is to set a UWB positioning base station at the periphery of the fire field. We assume that eight UWB base stations were deployed evenly on the boundary of the fire field as shown by the green triangle mark in Figure 3 and the inter distance of adjacent UWB base stations is 35 m. There are more than three LOS positioning base stations nearby at any point in the fire field after the deployment of mobile base stations in the fire field is finished. The position of the firefighter is (550, 480) shown as the blue star mark in Figure 3 The distance between the firefighter and UWB base station can be obtained due to the UWB label on the firefighter. The closest three mobile base stations to the firefighter are marked by the yellow triangle in Figure 3 and were used to obtain the precise position information of the firefighter. It can be easily seen from Figure 4 that the problem for the common method is to solve the NLOS positioning, but the problem for the method in this paper is to solve the LOS positioning.

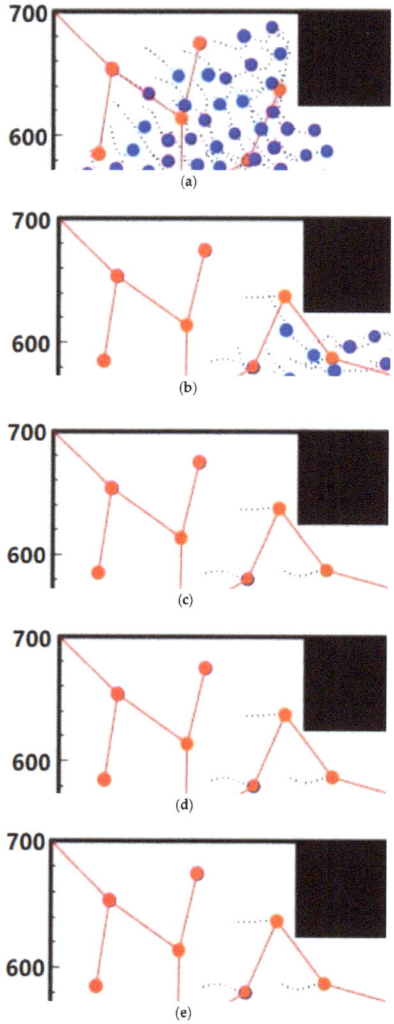

Figure 2. Coverage deployment of mobile base stations at different times: (**a**) coverage deployment at the time of 10 s; (**b**) coverage deployment at the time of 20 s; (**c**) coverage deployment at the time of 40 s; (**d**) coverage deployment at the time of 50 s; (**e**) coverage deployment at the time of 60 s.

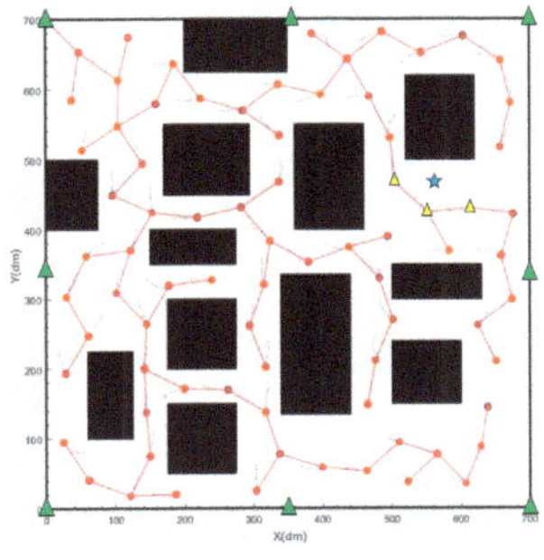

Figure 3. Diagram of base stations' deployment of two algorithms.

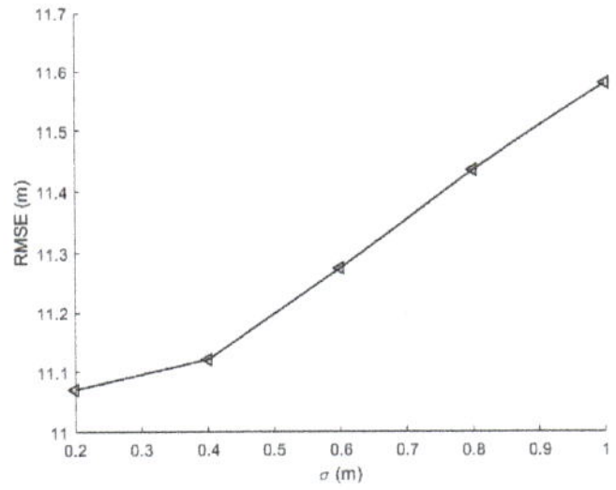

Figure 4. NLOS positioning error simulation results with eight base stations.

According to our experience, we assume that the system error of the UWB is 0.2 m, 0.4 m, 0.6 m, 0.8 m, and 1 m, respectively. The position of the firefighter was measured many times with the same UWB unit. The optimal position of the firefighter was estimated with the least square method [32,33]. The simulation results of the above two methods are shown in the following figures.

5. Discussion

The most important problem in the deployment of mobile base stations for coverage with connectivity constraints is how to control their movement and behavior to achieve a desired configuration and how to establish decentralized coordination among the multiple UAVs team. This becomes more difficult for the navigation and positions of UAVs in the absence of GPS. In the absence of GPS, robots can only perform local localization by estimating their adjacent UAV's relative locations and sending limited messages out with Zigbee. Especially in the fire field of a high-rise building, this mechanism is more significant to realize the fast deployment of UWB base stations with a UAV group.

In the simulation experiment, the system error of the UWB is set as 0.2 m, 0.4 m, 0.6 m, 0.8 m, and 1 m, respectively. It can be seen from Figure 5 that the positioning error of the algorithm presented in this paper is less than 1.2 m. It is much less than the positioning error of the traditional algorithm that is more than 10 m, as shown in Figure 4.

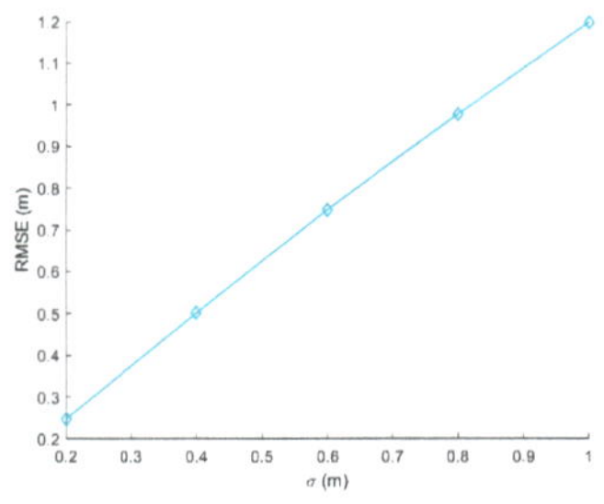

Figure 5. LOS positioning error simulation results with three base stations.

On the other hand, the large scale of a multi-agent system can effectively reduce the moving distance of each agent. The less moving distance not only reduces the accumulative positioning error but also means low energy a large scale of smaller and cheaper UAVs can be used to satisfy the requirement for the deployment of mobile UWB base stations in the fire field.

The real-world implementation of such a system could face potential challenges. The most important one is that the size, shape, and position of the obstacle would change randomly with the development of the fire. Therefore, the self-adaptive control law of the multi-agent system is a very challenging issue that is worth further research.

6. Conclusions

In this paper, we formulated a control problem that addresses the autonomous deployment of a massive multi-agent system in a complex unknown high-rise fire environment. The goal is to achieve the coverage deployment of mobile base stations to make sure that there are more than three base stations nearby to any one point in the fire field so that the NLOS positioning problem of the common positioning method can be transformed into the LOS positioning problem. The viscous velocity was appended to the control law. The simulation results show that the coverage deployment of the mobile base stations can be realized safely in the complex environment with unknown obstacles. Finally, we compared the simulation results of the common method and the method illustrated in this paper. The result of the comparison shows that this method can greatly improve the positioning accuracy and meet the practical positioning requirements of a high-rise fire rescue.

Author Contributions: Methodology, K.L. and Y.Z. (Yanbin Zou); Software, C.H. and J.L.; Validation, B.X.; Formal analysis, K.L.; Investigation, Y.Z. (Yang Zhang) and D.L.; Writing—original draft, K.L.; Writing—review & editing, Y.Y. All authors have read and agreed to the published version of the manuscript.

Funding: This research was funded by the Scientific Research Staring Foundation of Shantou University (NTF19028).

Institutional Review Board Statement: Not applicable.

Informed Consent Statement: Not applicable.

Data Availability Statement: Where no new data were created.

Conflicts of Interest: The authors declare no conflict of interest.

References

1. Liu, J.; Chow, K. Determination of fire load and heat release rate for high-rise residential buildings. *Procedia Eng.* **2014**, *84*, 491–497. [CrossRef]
2. Zhou, Q.; Li, Q.-S.; Han, X.-L.; Wan, J.-W.; Xu, K. Horizontal displacement estimation of high-rise structures by fusing strain and acceleration measurements. *Fire Saf. J.* **2022**, *57*, 104964. [CrossRef]

3. Xing, Z.-X.; Gao, W.-L.; Zhu, D.-Z.; Tang, Y. Research on the rescue force of high-rise building fire. *J. Saf. Sci. Technol.* **2012**, *08*, 35–40.
4. Ma, J.; Wang, J.; Zhou, X.-P.; Wei, Y.-Y. Research on Fire Risk Assessment of High-rise Buildings Based on Fuzzy Mathematics and Set-value Statistics Theory. *J. Comput.* **2018**, *29*, 145–160.
5. Li, Q.-S.; He, Y.-H.; Zhou, K.; Han, X.-L.; He, Y.-C.; Shu, Z.-R. Structural health monitoring for a 600 m high skyscraper. *Struct. Des. Tall Spéc. Build.* **2018**, *27*, e1543. [CrossRef]
6. Zhou, F.-B. Thinking on Fire Extinguishing and Rescue Measures for High-rise Buildings. *Constr. Des. Eng.* **2018**, *7*, 94–95, 98.
7. Murakami, S.; Murosaki, Y. Fire and Smoke Protection Measures for High-Rise Buildings. *J. Disaster Res.* **2011**, *6*, 629–643. [CrossRef]
8. Zmud, M. Public Perceptions of High-rise Building Emergency Evacuation Preparedness. *Fire Technol.* **2008**, *44*, 329–336. [CrossRef]
9. Hu, L.; Milke, J.A.; Merci, B. Special Issue on Fire Safety of High-Rise Buildings. *Fire Technol.* **2017**, *53*, 1–3. [CrossRef]
10. Kodur, V.K.R.; Venkatachari, S.; Naser, M.Z. Egress Parameters Influencing Emergency Evacuation in High-Rise Buildings. *Fire Technol.* **2020**, *56*, 2035–2057. [CrossRef]
11. Wei, J.-Y.; Wang, H.-W.; Su, S.-J.; Tang, Y.; Guo, X.; Sun, X. NLOS identification using parallel deep learning model and time-frequency information in UWB-based positioning system. *Measurement* **2022**, *195*, 111191. [CrossRef]
12. Tabaa, M.; Diou, C.; Saadane, R.; Dandache, A. LOS/NLOS Identification based on Stable Distribution Feature Extraction and SVM Classifier for UWB On-body Communications. *Procedia Comput. Sci.* **2014**, *32*, 882–887. [CrossRef]
13. Marano, S.; Gifford, W.M.; Wymeersch, H.; Win, M.Z. NLOS identification and mitigation for localization based on UWB experimental data. *IEEE J. Sel. Areas Commun.* **2010**, *28*, 1026–1035. [CrossRef]
14. Yang, X.; Zhao, F.; Chen, T. NLOS identification for UWB localization based on import vector machine. *AEU–Int. J. Electron. Commun.* **2018**, *87*, 128–133. [CrossRef]
15. Zhao, W.; He, R.; Ai, B.; Zhong, Z.; Zhang, H. Vehicle Localization Based on Hypothesis Test in NLOS Scenarios. *IEEE Trans. Veh. Technol.* **2022**, *71*, 2198–2203. [CrossRef]
16. Cwalina, K.K.; Rajchowski, P.; Blaszkiewicz, O.; Olejniczak, A.; Sadowski, J. Deep Learning-Based LOS and NLOS Identification in Wireless Body Area Networks. *Sensors* **2019**, *19*, 4229. [CrossRef]
17. Zeng, Y.; Zhang, R.; Lim, T.J. Throughput maximization for UAVenabled mobile relaying systems. *IEEE Trans. Commun.* **2016**, *64*, 4983–4996. [CrossRef]
18. Zhang, J.; Zeng, Y.; Zhang, R. Spectrum and energy efficiency maximization in UAV-enabled mobile relaying. In Proceedings of the 2017 IEEE International Conference on Communications (ICC), Paris, France, 21–25 May 2017; Volume 5, pp. 1–6.
19. Zhang, S.; Zhang, H.; He, Q.; Bian, K.; Song, L. Joint trajectory and power optimization for UAV relay networks. *IEEE Commun. Lett.* **2017**, *22*, 161–164. [CrossRef]
20. Xue, Z.; Wang, J.; Ding, G.; Wu, Q. Joint 3D location and power optimization for UAV-enabled relaying systems. *IEEE Access* **2018**, *6*, 43113–43124. [CrossRef]
21. Shrestha, K.; Dubey, R.; Singandhupe, A.; Louis, S.; La, H. Multi Objective UAV Network Deployment for Dynamic Fire Coverage. In Proceedings of the 2021 IEEE Congress on Evolutionary Computation (CEC), Kraków, Poland, 28 June–1 July 2021; pp. 1280–1287. [CrossRef]
22. Zhong, X.; Huo, Y.; Dong, X.; Liang, Z. Deep Q-Network Based Dynamic Movement Strategy in a UAV-Assisted Network. In Proceedings of the 2020 IEEE 92nd Vehicular Technology Conference (VTC2020-Fall), Victoria, BC, Canada, 18 November–16 December 2020; pp. 1–6. [CrossRef]
23. Shrestha, K.; La, H.M.; Yoon, H.-J. A Distributed Deep Learning Approach for A Team of Unmanned Aerial Vehicles for Wildfire Tracking and Coverage. In Proceedings of the 2022 Sixth IEEE International Conference on Robotic Computing (IRC), Naples, Italy, 5–7 December 2022; pp. 312–319. [CrossRef]
24. Michailidis, E.T.; Nomikos, N.; Bithas, P.S.; Vouyioukas, D.; Kanatas, A.G. Optimal 3-D aerial relay placement for multi-user MIMO communications. *IEEE Trans. Aerosp. Electron. Syst.* **2019**, *55*, 3218–3229. [CrossRef]
25. Yang, L.; Chen, J.; Hasna, M.O.; Yang, H.-C. Outage performance of UAV-assisted relaying systems with RF energy harvesting. *IEEE Commun. Lett.* **2018**, *22*, 2471–2474. [CrossRef]
26. Wang, F.; Zhang, C.; Tang, F.; Jiang, H.; Wu, Y.; Liu, Y. Lightweight Object-level Topological Semantic Mapping and Long-term Global Localization based on Graph Matching. *arXiv* **2022**, arXiv:2201.05977. [CrossRef]
27. Song, C.; Fan, Y.; Xu, S. New convergence analysis of coverage control on a circle subject to unknown but bounded measurement errors. *Automatica* **2022**, *141*, 110216. [CrossRef]
28. Song, C.; Liu, L.; Feng, G.; Fan, Y.; Xu, S. Coverage control for heterogeneous mobile sensor networks with bounded position measurement errors. *Automatica* **2020**, *120*, 109118. [CrossRef]
29. Qu, Y.; Xu, H.; Song, C.; Fan, Y. Coverage control for mobile sensor networks with time-varying communication delays on a closed curve. *J. Frankl. Inst.* **2020**, *357*, 12109–12124. [CrossRef]
30. Braga, R.G.; da Silva, R.C.; Ramos, A.C.B.; Mora-Camino, F. Collision avoidance based on Reynolds rules: A case study using quadrotors. In *Information Technology-New Generations*; Springer: Berlin/Heidelberg, Germany, 2018; pp. 773–780.

31. Quintero, S.A.; Collins, G.E.; Hespanha, J.P. Flocking with fixed-wing UAVs for distributed sensing: A stochastic optimal control approach. In Proceedings of the 2013 American Control Conference (IEEE, 2013), Washington, DC, USA, 17–19 June 2013; pp. 2025–2031.
32. Zou, Y.-B.; Liu, H.-P. An Efficient NLOS Errors Mitigation Algorithm for TOA-Based Localization. *Sensors* **2020**, *20*, 1403. [CrossRef] [PubMed]
33. Xiong, W.; Schindelhauer, C.; So, H.C.; Wang, Z. Maximum Correntropy Criterion for Robust TOA-Based Localization in NLOS Environments. *Circuits Syst. Signal Process.* **2021**, *40*, 6325–6339. [CrossRef]

sensors MDPI

Article

An H-GrabCut Image Segmentation Algorithm for Indoor Pedestrian Background Removal

Xuchao Huang [1,2], Shigang Wang [1,2,*], Xueshan Gao [1,2,*], Dingji Luo [3], Weiye Xu [1,2], Huiqing Pang [1,2] and Ming Zhou [4]

1 School of Automation, Guangxi University of Science and Technology, Liuzhou 545000, China; 221068359@stdmail.gxust.edu.cn (X.H.)
2 Key Laboratory of Intelligent Sensing and Control, Liuzhou 545000, China
3 Mechanical and Electrical College, Beijing Institute of Technology, Beijing 100190, China
4 Hangke Jinggong Co., Ltd., Beijing 102400, China
* Correspondence: 100001772@gxust.edu.cn (S.W.); xueshan.gao@bit.edu.cn (X.G.); Tel.: +86-185-7726-2003 (S.W.); +86-188-0772-7899 (X.G.)

Abstract: In the context of predicting pedestrian trajectories for indoor mobile robots, it is crucial to accurately measure the distance between indoor pedestrians and robots. This study aims to address this requirement by extracting pedestrians as regions of interest and mitigating issues related to inaccurate depth camera distance measurements and illumination conditions. To tackle these challenges, we focus on an improved version of the H-GrabCut image segmentation algorithm, which involves four steps for segmenting indoor pedestrians. Firstly, we leverage the YOLO-V5 object recognition algorithm to construct detection nodes. Next, we propose an enhanced BIL-MSRCR algorithm to enhance the edge details of pedestrians. Finally, we optimize the clustering features of the GrabCut algorithm by incorporating two-dimensional entropy, UV component distance, and LBP texture feature values. The experimental results demonstrate that our algorithm achieves a segmentation accuracy of 97.13% in both the INRIA dataset and real-world tests, outperforming alternative methods in terms of sensitivity, missegmentation rate, and intersection-over-union metrics. These experiments confirm the feasibility and practicality of our approach. The aforementioned findings will be utilized in the preliminary processing of indoor mobile robot pedestrian trajectory prediction and enable path planning based on the predicted results.

Keywords: indoor pedestrian segmentation; image enhancement; H-GrabCut algorithm

Citation: Huang, X.; Wang, S.; Gao, X.; Luo, D.; Xu, W.; Pang, H.; Zhou, M. An H-GrabCut Image Segmentation Algorithm for Indoor Pedestrian Background Removal. *Sensors* **2023**, *23*, 7937. https://doi.org/10.3390/s23187937

Academic Editors: Zhun Fan and Gang Peng

Received: 12 July 2023
Revised: 8 September 2023
Accepted: 11 September 2023
Published: 16 September 2023

1. Introduction

In recent years, rapid advancements in robot technology have led to its extensive application in various domains such as goods delivery, elderly care, environment cleaning, and blind navigation. In indoor environments, the presence of numerous dynamic objects, predominantly pedestrians, adversely affects self-localization and path planning. Additionally, uneven indoor lighting conditions result in blurred pedestrian images with low brightness, making it difficult to identify distinct feature points and capture the background along with pedestrians. As a consequence, obtaining accurate distance measurements between pedestrians becomes challenging, which hampers obstacle avoidance and precise trajectory prediction for robots. Extracting pedestrians as regions of interest can eliminate irrelevant background information behind them and enable the accurate retrieval of their position coordinates in RGB images, thus facilitating the calculation of distances between pedestrians and depth cameras. This information is vital for pedestrian trajectory recovery or prediction. Therefore, the recognition and segmentation of dynamic obstacles, specifically pedestrians, from images are crucial in enabling robots to respond promptly to such obstacles. The research conducted in this study holds significant importance in the field of robot obstacle avoidance.

To address the issues of unstable noise and indoor illumination in the segmentation algorithm, the implementation of image enhancement can be considered. While traditional histogram equalization technology may enhance image brightness and contrast, it is likely to cause over-enhancement and lacking robustness. To overcome these challenges, scholars have proposed various image enhancement algorithms. For instance, the adaptive histogram equalization method [1–3] and the image difference algorithm [4,5] can effectively avoid unnatural contrast and distortion effects. Nonetheless, current methods are inadequate in improving contrast while retaining the integrity of image boundary features. Various researchers have proposed a Retinex enhancement algorithm based on image fusion and illumination image estimation [6–10]. This algorithm is capable of producing high-quality images, but the contour edges are typically not complete. In conclusion, there is currently no single optimal method to address the issues of contrast enhancement and the preservation of image boundary features simultaneously.

In order to extract distinctive feature points, depth cameras are used to segment pedestrians in the input images as regions of interest. Currently, both deep learning-based and traditional interactive algorithms are commonly employed for image segmentation. Examples of these methods include U-Net segmentation [11], the multi-scale high-order attention algorithm [12], DeepLabv3+ image segmentation [13], the motion object segmentation algorithm [14], and the local alignment network model [15]. While deep learning-based segmentation algorithms can achieve accurate segmentation of target objects, they also increase the complexity of the algorithms and lead to the excessive and redundant use of computational resources when extracting low-level features repeatedly. In addition, traditional interactive image segmentation algorithms have been a focus of research by many scholars, for instance, the GraphCut algorithm [16] and the GrabCut algorithm [17], where the GrabCut algorithm is simple, efficient, and provides high segmentation accuracy. There are also improved algorithms that combine the local sampling of the foreground and background [18], the adaptive k-means clustering algorithm [19], and the utilization of YOLO-V4 for foreground detection [20,21]. These algorithms have demonstrated a certain reference value in practical applications. However, the aforementioned segmentation algorithms are only suitable for the background segmentation of non-target pedestrians and are not applicable for avoiding dynamic obstacles. Indoor environments vary in complexity, lighting conditions change frequently, and images often contain noise. When moving pedestrians are close to the camera, their imaging brightness is low and clothing edge information tends to be blurred and easily fused.

This paper proposes an improved image segmentation algorithm called H-GrabCut to address the aforementioned issues and achieve background removal in complex pedestrian backgrounds. The H-GrabCut (Human-GrabCut) algorithm presented in this paper enhances the GrabCut algorithm's clustering features by incorporating the BIL-MSRCR (Bilateral-MSRCR)image enhancement algorithm and introduces two-dimensional information entropy, UV component distance, and LBP texture features. This optimization effectively resolves the resource limitations associated with nesting multiple deep learning algorithms, while enabling more accurate distance measurement between robots and pedestrians, beyond relying solely on the closest distance for ranging. Furthermore, the algorithm demonstrates excellent performance in experiments, offering a new option for precise indoor pedestrian image segmentation in mobile robotics.

2. Method Construction

In indoor environments, the presence of numerous dynamic objects, primarily pedestrians, has a detrimental impact on the self-localization and path planning of indoor navigation robots. Additionally, uneven indoor lighting conditions result in pedestrian images captured by the visual system being characterized by low brightness, blurred edges, unclear contours, and prominent noise. These factors degrade the robustness of the traditional GrabCut segmentation algorithm, rendering feature points indistinct and possibly including background information, thereby impeding accurate distance estimation between pedestrians and introducing errors that affect the precision of obstacle avoidance and pedestrian trajectory prediction by the robot. To address the aforementioned issues, an improved algorithm called H-GrabCut (Human-GrabCut) is proposed. The algorithm consists of the following steps, as illustrated in Figure 1: (1) the preliminary noise reduction of the input image using range adaptive fast median filtering; (2) initial segmentation region partitioning through pedestrian detection using YOLO-V5; (3) the extraction of cropped images for each pedestrian bounding box, generating an image sequence for further processing; (4) the application of the improved BIL-MSRCR (Bilateral-MSRCR) algorithm to the pedestrian images in the sequence; and (5) the segmentation of the processed images using the H-GrabCut algorithm to obtain the final segmented image with pedestrian RGB coordinates.

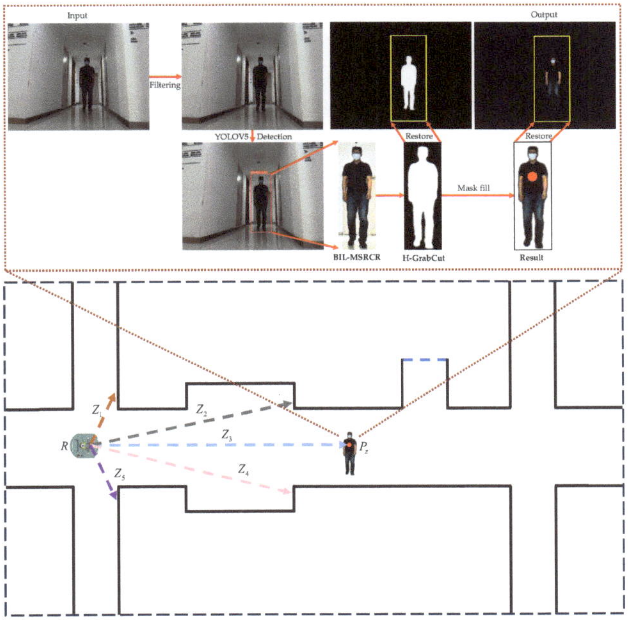

Figure 1. Flowchart of the H-GrabCut algorithm. (**Top**) The input is an RGB original image, and the output of the H-GrabCut algorithm is a background-removed image with pedestrian RGB coordinates. (**Bottom**) This image demonstrates the distance measurement of various obstacles (pedestrians) by the mobile robot R in an indoor environment using a depth camera. The distances between robot R and the walls are represented by Z_1, Z_2, Z_4, and Z_5 (orange, black, pink, and purple), while the Z_3 (blue) indicates the distance between robot R and the pedestrian P_z (red).

2.1. Range Adaptive Fast Median Filtering

As shown in Figure 2, the depth camera based on binocular structured light 3D sensor technology mainly consists of left and right infrared cameras, a laser projection module, and a depth computing processor. The laser projection module is used to project the structured light pattern (speckle pattern) to the target scene, and the left infrared camera and the right infrared camera collect the left infrared structured light image and the right infrared structured light image of the target respectively. After receiving the left infrared structured light image and the right infrared structured light image, the depth calculation processor executes the depth calculation algorithm and outputs the depth image of the target scene.

$$Z = \frac{f \times B}{x_1 - x_2} \quad (1)$$

In Formula (1), f represents the focal length of the camera imaging plane; B denotes the baseline between the left and right infrared cameras; and $x_1 - x_2$ signifies the disparity value of each pixel.

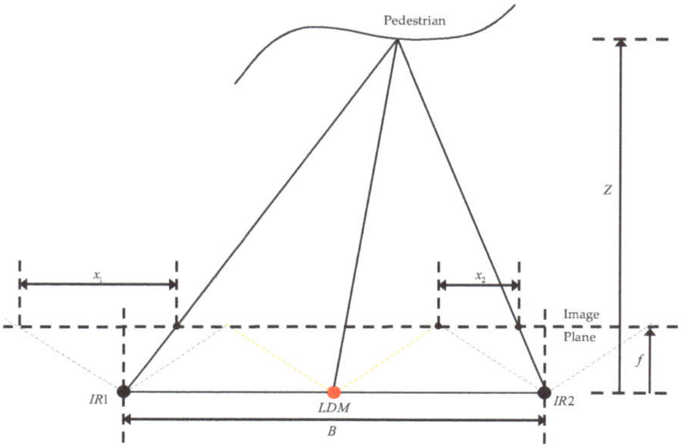

Figure 2. A schematic diagram of a binocular structured light 3D sensor is shown. In this diagram, $IR1$ and $IR2$ denote the infrared modules, and LDM stands for the laser module. The green line signifies the laser module's maximum scanning angle, while the brown line indicates the maximum receiving angle of the infrared modules.

In this paper, the distance between the pedestrian and the robot was measured using a depth camera. When the distance exceeded one meter, a fast median filter was used for background filtering. The fast median filter selects the middle value of pixels in the filter window and counts pixels using a histogram. It then compares the pixels moving out of the window with the pixels moving in. The most effective filter results were achieved using a 3×3 window.

2.2. *YOLO-V5 Pedestrian Target Detection*

Since this algorithm focuses only on optimizing pedestrian segmentation, it detects pedestrians in the filtered images and assigns anchor boxes to them, with these anchor box points representing the foreground regions. The YOLO-V5 object detection network architecture consists of a backbone network and a detection head. The backbone network utilizes the CSP (cross stage partial) architecture, which includes multiple CSP blocks and an SPP (spatial pyramid pooling) module [21]. The detection head comprises three convolutional layers used for predicting the object's category and bounding box. The overall image and coordinate information data transmission structure is illustrated in Figure 3.

As our algorithm focuses on indoor pedestrian segmentation, we utilize the YOLO-V5 framework for pedestrian recognition. However, we modify its detection head to output only one pedestrian label and intercept the image in the anchor frame to generate the required image sequence output. This procedure allows us to obtain the necessary foreground target for the pedestrian segmentation algorithm, thereby minimizing user operations. Additionally, we optimize the algorithm to handle numerous images quickly while maintaining high recognition accuracy and stability.

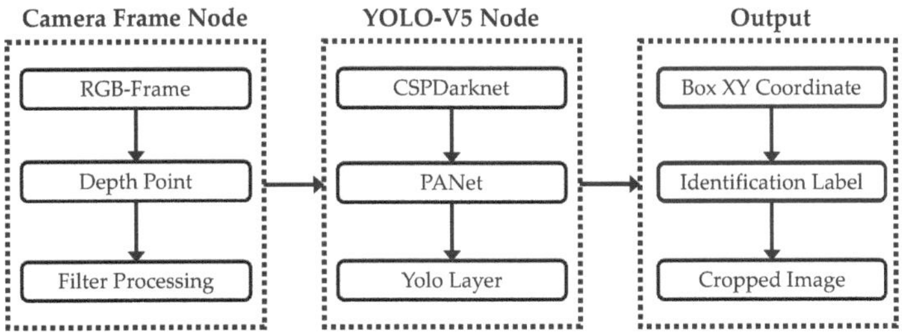

Figure 3. Structure of YOLO-V5 target recognition algorithm. The camera frame node is responsible for capturing the essential camera image data and applying filtering processes to produce an output image. This output image is subsequently fed into the YOLO detection layer for pedestrian annotation. Upon completion of annotation, the XY coordinates of the output anchor boxes can be obtained, along with the RGB images cropped based on these coordinates and their corresponding recognition labels.

2.3. *MSRCR Image Enhancement Algorithm Combined with Bilateral Filtering*

In indoor environments, the image captured by a depth camera is often noisy due to poor lighting, and when pedestrians move in close proximity to the camera, the overall image brightness is low. Moreover, the edge information of pedestrians wearing different clothes is unclear, and similar backgrounds tend to fuse. Thus, image enhancement is necessary. The MSRCR algorithm [22], based on Retinex theory, aims to improve image brightness, contrast, and color balance. The formula is as follows:

$$\log R_i(x,y) = \sum_{k=1}^{N} \omega_k(\log(I_i(x,y)) - \log(I_i(x,y) \times G_k(x,y))) \tag{2}$$

In Formula (2), $k \leq N$ and N represent the number of scales. Generally speaking, when N is 3, Gaussian filters are used to filter the original image on three different scales, respectively. ω_k represents a weight value for each scale, which must satisfy:

$$\sum_{k=1}^{N} \omega_k{=}1 \tag{3}$$

G_k is a Gaussian function with a number of scales k, and its formula is as follows:

$$G_k(x,y){=}\frac{1}{\sqrt{2\pi c_k}} \times \exp\left[-\frac{x^2+y^2}{2c_k^2}\right] \tag{4}$$

In Formula (4), x^2 and y^2 are, respectively, expressed as the distance between the remaining pixels in the neighborhood and the center pixel in the neighborhood, and c_k is the standard deviation of the filter to control the amount of retained spatial details. However, c_k values cannot be modeled and determined theoretically and are basically a balance between enhancing local dynamics (such as showing details in shadows) and color rendering.

The conventional MSRCR algorithm uses a Gaussian filter that employs a Gaussian weighted average of pixels in the vicinity of the center point. Such a filter takes into account only the spatial relationship between pixels but fails to consider pixel similarity. Consequently, this approach blurs the boundaries, including those of anchor points returned by the object detection algorithm, leading to a loss of features. On the other hand, the bilateral filter, a nonlinear filter, can effectively remove noise while maintaining boundary clarity [23]. Bilateral filtering uses both spatial Gaussian weights and color-similarity Gaussian weights simultaneously. The spatial Gaussian function ensures that only pixels in the neighboring area affect the center point, whereas the color-similarity Gaussian function ensures that only pixels with similar values to the center pixels are considered for fuzzy operations. Thus, this study proposes the BIL-MSRCR algorithm, an improved bilateral filtering method applied to the MSRCR algorithm. As a result, the MSRCR algorithm can better preserve the image's edge and texture structure.

The spatial domain of the improved bilateral filtering algorithm is expressed as:

$$h(x_c) = k_d^{-1} \iint f(r)s(r, x_c)\,\mathrm{d}r \tag{5}$$

$$k_d(x_c) = \iint s(r, x_c)\,\mathrm{d}r \tag{6}$$

Formulas (5) and (6), $s(r, x_c)$, represent the spatial distance function between the central pixel x_c and its neighboring pixels r. The color domain in the two-sided filtering algorithm is represented as:

$$h(x_c) = k_r^{-1} \iint f(r)c(f(r), f(x_c))\,\mathrm{d}r \tag{7}$$

$$k_r(x_c) = \iint c(f(r), f(x_c)\,\mathrm{d}r \tag{8}$$

Formulas (7) and (8) represent the similarity between the central pixel value and the pixel value of its neighboring sample. In the above two filtering methods, $f(r)$ represents the pixel value at r and $h(x_c)$ represents the pixel value after filtering at x_c. Combining the above two filtering algorithm ideas, a new bilateral filter is generated, so the bilateral filter can be expressed as:

$$h(x_c) = k^{-1} \iint f(r)s(r, x_c)c(f(r), f(x_c))\,\mathrm{d}r \tag{9}$$

$$k(x_c) = \iint s(r, x_c)c(f(r), f(x_c))\,\mathrm{d}r \tag{10}$$

In this paper, the spatial domain function takes the Gaussian filter as an example and is defined as:

$$s(r, x_c) = \exp\left(-\frac{(r_x - x_c)^2 + (r_y - y_c)^2}{2c_s^2}\right) \tag{11}$$

In Formula (11), (r_x, r_y) represents the coordinate value of the current pixel, (x_c, y_c) represents the coordinate value of the center point of the image, and c_s is the standard deviation parameter of the spatial domain.

The color domain function $c(f(r), f(x_c))$ is defined as:

$$c(f(r), f(x_c)) = \exp\left(-\frac{(f(r_x, r_y) - f(x_c, y_c))^2}{2c_c^2}\right) \tag{12}$$

In Formula (12), $f(r_x, r_y)$ represents the pixel value of the current coordinate, $f(x_c, y_c)$ represents the pixel value of the center point of the image, and c_c is the standard deviation parameter of the color domain.

The global kernel function of bilateral filtering can be written from the above formula:

$$B_k(r_x, r_y) = \exp\left[-\left(\frac{(r_x - x_c)^2 + (r_y - y_c)^2}{2c_s^2} + \frac{(f(r_x, r_y) - f(x_c, y_c))^2}{2c_k^2}\right)\right] \tag{13}$$

Although the kernel function-based bilateral filter in Formula (13) can preserve the edge structure information while denoising, its algorithm operates at a pixel-level. Consequently, the structure preservation effect is not significantly enhanced, and noise will compromise the calculation of the weighting coefficient, resulting in a reduced noise suppression ability. Additionally, the bilateral filter kernel function disregards the image's features and spatial neighborhood. The spatial domain kernel and the color domain kernel in the kernel function receive the same weight, making the filter unsuitable for robust applications. To better preserve the image's edge and detail structure, we assign more weight to the spatial domain kernel. We introduce a coefficient "b", which adjusts between the spatial domain kernel and the color domain kernel.The overall kernel function of improved bilateral filtering is as follows.

$$B_k = \exp\left[-\left(b \times \left(\frac{(r_x - x_c)^2 + (r_y - y_c)^2}{2c_s^2}\right) + (1-b) \times \frac{(f(r_x, r_y) - f(x_c, y_c))^2}{2c_k^2}\right)\right] \tag{14}$$

In Formula (14), b is used to adjust the coefficients of spatial domain kernel and color domain kernel, x_c and y_c are the coordinate values of the image center, $f(x_c, y_c)$ is the pixel value of the image center point, c_s is the standard deviation of Gaussian function in space domain, and c_k is the standard deviation of the Gaussian function in the color domain.

Based on the above improved bilateral filtering, the addition of the weighting coefficient enables the bilateral filtering to better balance the tradeoff between smoothing and preserving details, thus better preserving edge and texture information. Therefore, the overall formula of the BIL-MSRCR algorithm is as follows.

$$\log R_i(x,y) = \sum_{k=1}^{N} \omega_k(\log(I_i(x,y)) - \log(I_i(x,y) \times B_k(x,y)))k \leq N \tag{15}$$

In Formula (15), $R_i(x,y)$ is the pixel output of the ith color channel, $I_i(x,y)$ is the pixel input of the ith color channel of the BIL-MSRCR algorithm, and B_k is the improved bilateral filter kernel function with scale number k.

This paper adopts an improved bilateral filter as the center surround function of Multi-Scale Retinex (MSRCR) for image decomposition and enhancement, allowing for a separation of the light image from its reflection component. The bilateral filter radius, specified in three distinct scales, is set to 5, 10, and 15. The results show that the optimal effect is achieved when b is 0.31 following experimental testing and verification. Subsequent image enhancements significantly enhance the image contrast while prohibiting boundary blurring and maintaining the image's boundary features in its totality.

2.4. GrabCut Algorithm Combining Spatial Information, Chromaticity Information, and Texture Information

Iteratively, the GrabCut algorithm enhances the GMM parameters for modeling the target and background. The Gibbs energy function is then defined, and the algorithm segments the image by solving the energy function's min-cut. The energy function is expressed below:

$$E(\alpha, k, \theta, p) = U(\alpha, k, \theta, p) + V(\alpha, p) \tag{16}$$

$$\theta = \{\pi(\alpha, k), \mu(\alpha, k), \Sigma(\alpha, k), \alpha = (0,1), k = 1 \cdots K\} \tag{17}$$

In Formula (16), α represents the transparency coefficient, 0 is the background, and 1 is the foreground; Formula (17) represents the proportion corresponding to each GMM component; p represents a single pixel; Function U represents a regional data item of the energy function; and the function V represents the smooth term of the energy function.

In order to use the GrabCut algorithm [20], the user is required to mark the area of interest using a rectangular bounding box, whilst the image outside is regarded as the background and the pixels within the possible foreground and background. Employing the k-means algorithm, the pixels located inside the bounding rectangle are clustered into a k class system, which initializes k models of the Gaussian mixture model (GMM). Equation (18) is then used to calculate the RGB value of the pixels within the bounding box to determine its Gaussian component membership.

$$k_n : \arg\min_{k_n} D(\alpha_n, k_n, \theta, p_n) \tag{18}$$

In Equation (18), $\theta = (\mu,\sigma^2)$, in which μ is the mean value and σ^2 is the variance.

The region data item represents the negative logarithm of the probability that a pixel belongs to the background or foreground, which is calculated according to Equation (19):

$$D(\alpha_n, k_n, \theta, p_n) = -\log \pi \times W_n + \frac{1}{2}\log(\det \Sigma W_n) + \frac{1}{2} \times Q_n^T \Sigma W_n^{-1} Q_n \tag{19}$$

In Formula (19), $W_n = (\alpha_n, k_n)$ and $Q_n = [p_n - \mu(\alpha_n, k_n)]$.

After determining the probability that each pixel belongs to the foreground and background, the smooth term is calculated according to Equation (20):

$$V(\alpha, p) = \gamma \times \sum_{(m,n)\in C} dis(m,n)^{-1} \times [\alpha_n \neq \alpha_m] \times \exp\left[-\beta(p_m - p_n)^2\right] \tag{20}$$

In Formula (20), (m,n) represents pixel coordinates; C represents a set of adjacent color pairs; $dis(m,n)^{-1}$ represents the Euclidean distance between adjacent pixels, where 1 is taken; and $\beta = \left(2\left\langle (p_m - p_n)^2 \right\rangle\right)^{-1}$, $\langle \cdot \rangle$ represents the expectation of the image sample.

According to the Gibbs energy term, the s-t graph is constructed, and the image is divided by Equation (21) maximum flow/minimum cut:

$$\min_{\{\alpha_n : n \in T_U\}} \min_{k} E(\alpha, k, \theta, p) \tag{21}$$

In the original GrabCut algorithm, the k-means algorithm only clusters the color information of the image. The following formula is the original pixel feature:

$$z_i = [p_R, p_G, p_B] \tag{22}$$

Formula (22) p_R, p_G, p_B represents the RGB component value of a pixel point z_i. This feature is single, and it is easy to cause the classification error of pixels in the initial classification of images. Considering the overall information of the image, this paper is proposed to integrate spatial information, chroma information, and texture feature information of the image during clustering to optimize the parameters of the GMM model. The following is the optimized features:

$$z_i = [p_R, p_G, p_B, p_H, p_D, p_L] \tag{23}$$

In Formula (23), p_H represents the two-dimensional information entropy of pixel z_i, p_D represents the distance between the UV component of pixel z_i and the clustering center, and p_L represents the LBP texture feature value of pixel z_i.

To improve the accuracy of the clustering algorithm in classifying color information [24], this paper adopts conversion of the image into YUV color space to separate its brightness and chroma components. The Y value in YUV is then used as the gray value, computing its two-dimensional information entropy thereby providing the feature input for clustering. The two-dimensional information entropy of Y component of pixels in clustering features indicates the quantity of information in the image, which is expressed using probability distribution function. To depict the spatial attributes of pixel distribution, this paper adopts the feature quantity (i, j) , where i represents the gray value $(0 \leq i \leq 255)$ of the pixel while j represents the gray level $(0 \leq j \leq 255)$ of neighboring pixels.

$$P_{ij} = \frac{f(i,j)}{N^2} \tag{24}$$

In Formula (24), $f(i, j)$ is the frequency of the occurrence of the feature binary (i, j), and N is the scale of the image; the two-dimensional information entropy of the image is described as:

$$H = -\sum_{i=0}^{255}\sum_{j=0}^{255} p_{ij}\log_2 p_{ij} \tag{25}$$

Image information entropy provides insight into the amount of information contained within an image. To overcome the issue of spatial information depletion caused by using a single feature, adding two-dimensional information entropy to the clustering features accounts for the compositional arrangement of adjacent pixels and the image's spatial distribution characteristics. The traditional k-means clustering algorithm can cause the loss of details in an image as it considers only the gray component whilst ignoring chroma saturation. To address this limitation, this paper introduces a novel distance formula that characterizes the UV component of chroma and the chroma factor of the cluster center point. With this new distance formula, the k-means algorithm can accommodate different image types, thereby improving the algorithm's adaptability.

$$D = \sqrt{(V_C - V_N)^2 + (U_C - U_N)^2} \tag{26}$$

In Formula (26), (U_C, V_C) represents the UV value of the cluster center, and (U_N, V_N) represents the UV value of the current pixel point. Taking the distance between the UV component and the cluster center into the clustering features, the difference between different images can be more accurately reflected, thus improving the robustness of the algorithm.

It is necessary to consider the texture information of an image, in addition to its spatial and color information. The texture feature of each pixel is dependent on its relationship with other pixels within its domain. Distinct objects have unique texture features, which facilitates the better classification of images. This study utilizes the local binary pattern (LBP) operator to describe the image's texture features. LBP possesses rotation and gray invariance. LBP uses a 3×3 template as a unit and adopts the middle pixel as its threshold. A pixel point is marked as 1 if the pixel value of its eight surrounding fields is greater than the middle pixel; otherwise, it is marked as 0. The comparison of these values constructs an 8-bit binary number. The conversion of this binary number to decimal yields the LBP value of the middle pixel and reflects its inherent texture information. The LBP operator is effectively expressed in Equation (27):

$$LBP(p_c) = \sum_{i=0}^{8} 2^i p_i p_i = \begin{cases} 0, p_c > p_i \\ 1, p_c < p_i \end{cases}, i = 1, 2, \ldots, 8. \tag{27}$$

2.5. Algorithms Run Pseudo-Code

Based on the above theory, we can utilize the improved H-GrabCut algorithm to effectively segment images. The specific implementation method of the algorithm can be expressed by the following pseudo-code.

Algorithm 1 introduces our proposed BIL-MSRCR and H-GrabCut algorithms, which adhere to the following principles:

- Apply a range adaptive fast median filter with a window size of 3×3 to the input raw image, resulting in the filtered image I_Q.
- (Lines 4 and 5) Pass the image I_Q to a function that combines the pedestrian bounding box positions obtained from YOLO-V5. Crop the image I_Q based on these bounding box positions, generating a pedestrian image I_H that contains a small background area.

- (Lines 8–14) Apply the BIL-MSRCR algorithm to image I_H, enhancing it to obtain the enhanced pedestrian image I_{BMSR} (Equations (2)–(15)).
- (Line 15) Perform initial k-means clustering on the image to obtain cluster centers. Calculate the distance between the UV components of each pixel and the chromaticity factors of the cluster centers.
- (Line 16) Convert the enhanced I_{BMSR} image from the RGB color space to the YUV color space. Compute the two-dimensional information entropy specifically for the Y channel of the I_{BMSR} image (Equation (25)).
- (Line 17) Incorporate pixel spatial information, chromaticity information, and texture features from the I_{BMSR} image into the k-means clustering process. Initialize the parameters of the Gaussian mixture model (GMM) using LBP components based on defined background pixels, potential foreground pixels, and potential background pixels (Equation (27)).
- (Line 18) Utilize the max-low algorithm to determine the min-cut. Iteratively apply (Equations (18)–(23)) to segment the I_{BMSR} image, resulting in a foreground image mask denoted as I_{MASK}.
- (Line 19) Combine the mask image I_{MASK} with the filtered output image I_Q to generate the segmented image I_{HG}.

Algorithm 1 BIL-MSRCR and H-GrabCut

Input: Camera_Image, b and The $\boldsymbol{xy}$ coordinates returned by YOLO-V5
Output: Mask_Image and Image after segmentation
1: $\boldsymbol{x_i} = [r_x^1, r_y^1, r_x^2, r_y^2]$;
2: $\boldsymbol{w_k} = [w_1, w_2, w_3]$;
3: **while** Camera_Image **and** $\boldsymbol{x_i}$ is not empty **do**
4: $\boldsymbol{I_Q}$ = Range_filtering(Camera_Image);
5: $\boldsymbol{I_H}$ = Cut_Process(I_Q, r_x^1, r_y^1, r_x^2, r_y^2);
6: **for** $i = 0$ **to** scale **step** 1:
7: $\mathbf{log}(\boldsymbol{I_i})$ = Convert_log(I_H);
8: **for** $k = 0$ **to** N **step** 1:
9: $\boldsymbol{B_k}$ = BIL_Process(I_H, w_k, b);
10: $\mathbf{log}(\boldsymbol{I_i} \times \boldsymbol{B_k})$ = Convert_log(I_H, B_k);
11: $\mathbf{log}(\boldsymbol{R_i})$ = Sub(w_k,$\log(I_i \times B_k)$, $\log(I_i)$);
12: $\boldsymbol{I_{BMSR}}$ = Convert_Scale($\log(R_i)$);
13: **end for**
14: **end for**
15: $\boldsymbol{D}$ = Extract_YUV(I_{BMSR});
16: $\boldsymbol{H}$ = Extract_entropy(I_{BMSR});
17: $\boldsymbol{L}$ = Extract_LBP(I_{BMSR});
18: $\boldsymbol{I_{MASK}}$ = HGrabcut_process(I_{BMSR}, D, H, L);
19: $\boldsymbol{I_{HG}}$ = add(I_Q, I_{MASK});
20: **end while**

3. Results

The primary algorithm was implemented using the C++ programming language on the ROS platform. The execution of the algorithm required the OpenCV 4.5.5 library, which was configured in the Clion v2020.3 integrated development environment. The computing device utilized in the experiment had the following specifications: a 64-bit Ubuntu operating system, 3.6 GHz CPU frequency, and 16.0 GB memory. For image recognition and segmentation, an RGB-D camera was necessary. The robot's structure is shown in Figure 4, and the camera's output resolution was set to 640×480 @ 30 fps. To ensure stable and non-shaky images, the robot was configured with a maximum movement speed of 0.1 m/s.

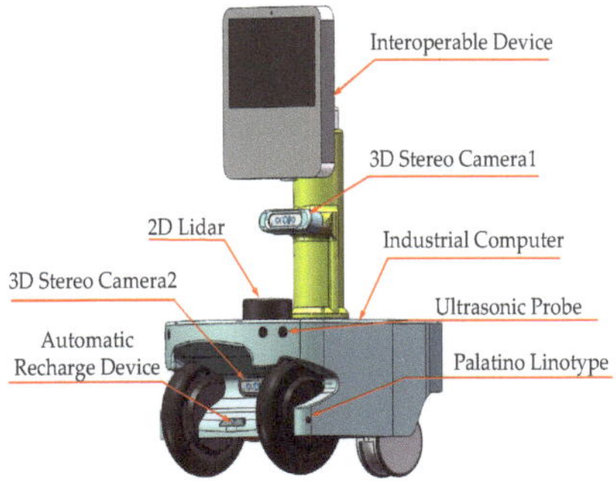

Figure 4. Robot platform sensor layout: interactive device is used to upload sensor data and receive motion control instructions, 3D stereo phase unit is used to return distance information and image information, and the industrial computer is used to run the ROS platform for control.

3.1. *Experimental Analysis of Image Enhancement*

The materials utilized in this experiment are partial images from the INRIA dataset and photographs captured by real machines, as illustrated in Figure 5. To prove the effectiveness and robustness of the proposed algorithm, we use the image enhancement algorithms proposed in the literature, such as CLAHE [3], GSMSR [6], LIME [7], RFVM [8], SRME [9], ALSM [10] as reference, the visual effect and gray-level histogram as indicators. Brightness, contrast, and naturalness in images can be perceived by the human vision system, while the histogram verifies the dynamic range and rationality of the pixel distribution.

Figure 5. Real machine return images (**a**,**b**) and some images of INRIA pedestrian dataset (**c**–**e**).

The enhancement effects of various methods on the image Figure 5c are shown in Figure 6. The original image Figure 6a is dark with low contrast. It is evident that the enhanced image in Figure 6b exhibits significant image distortion and over-enhancement. The brightness enhancement in Figure 6d,e is insufficient as some details in dark areas are still not visible. The overall brightness enhancement in Figure 6c,g is still inadequate, and the distant scenery in dark areas remains unclear. The enhancement results in Figure 6f and our proposed method are better, with suitable overall brightness. However, upon careful observation, it can be noticed that Figure 6f exhibits over-enhancement in distant scenery with minor color distortion, whereas our method Figure 6h achieves appropriate overall enhancement, higher contrast, and better naturalness of the image.

A histogram provides an intuitive evaluation of image quality. An ideal image should meet two criteria. First, it should fully utilize the pixel-level space, ensuring high contrast and clear texture details. Second, pixels should be evenly distributed across the pixel-level space, resulting in appropriate brightness and natural colors. The histograms for the image enhancement algorithms in Figure 6a are displayed in Figure 7. Figure 7a shows that while the original image utilizes the entire pixel-level space, it mainly concentrates pixels at low gray levels, which darkens the image and obscures details. The histograms of the enhancement algorithms all utilize the full gray scale space, yielding good contrast. However, Figure 7b shows over-enhanced pixels with excessive color contrast. Figure 7c exhibits a relatively even pixel distribution, with slight unevenness in high gray scales. In Figure 7d and g, the pixel distribution peaks closer to higher gray scales, displaying uneven areas and obscuring details in darker regions. Figure 7e has a relatively even pixel distribution but leans slightly towards lower levels, resulting in an overall darker appearance. Both Figure 7f and our method offer even pixel distribution, suitable brightness, natural colors, and rich texture and details. However, Figure 7f exhibits over-enhanced pixels at specific levels.

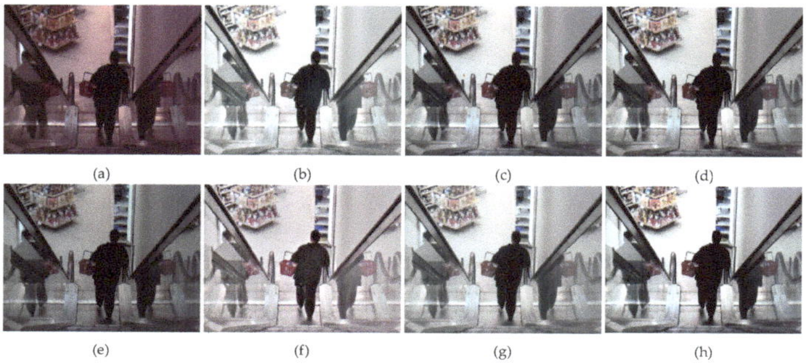

Figure 6. (**a**) is the RGB original of Figure 5c, (**b**) is the CLAHE transformation algorithm used in reference, (**c**) adopts the gray scale adaptive msr algorithm GSMSR, and (**d**) adopts the LIME algorithm. (**e**) adopts the improved RFVM algorithm, (**f**) is the SRME algorithm using the improved regular term, (**g**) is the ALSM algorithm, and (**h**) is the BIL-MSRCR algorithm proposed in this paper.

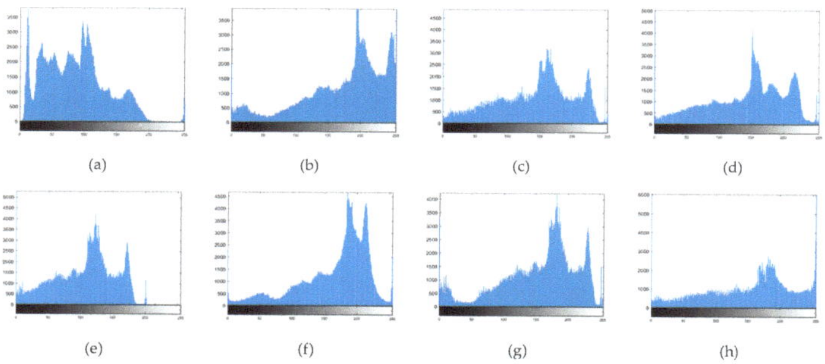

Figure 7. (**a**) is the gray scale histogram of the original Figure 5c, (**b**) is the gray scale histogram of the output image of the CLAHE algorithm, (**c**) is the gray scale histogram of the output of the GSMSR, (**d**) is the histogram of the LIME algorithm, (**e**) is the output histogram of the RFVM algorithm, and (**f**) is the histogram of the output of the SRME algorithm, (**g**) adopts the histogram effect output by ALSM, and (**h**) is the output histogram of the algorithm in this paper.

3.2. Experimental Analysis of Segmentation Algorithm

We utilized the YOLO-V5 object detection algorithm, initially pretrained on the INRIA pedestrian dataset containing 614 images. To enhance algorithm performance in detecting pedestrians under diverse conditions, including occlusions, lighting variations, and various scenes, we expanded the dataset with an additional 100 real-world images. To validate our proposed target detection and segmentation algorithm, we tested it using both real-world images and the INRIA pedestrian dataset. YOLO-V5 was employed to identify foreground targets with a confidence threshold of 0.85 for automatic pedestrian segmentation. You can find the experimental results in Figure 8.

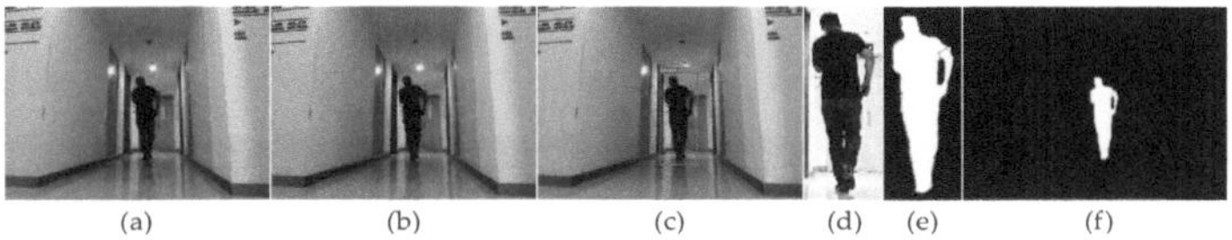

Figure 8. (a) is the RGB original image of Figure 5a, (b) the fast median filter is used to remove the salt and pepper noise caused by poor light conditions; YOLO-V5 target recognition algorithm was further used to identify pedestrians and reduce them according to the anchor frame size, as shown in (c). The BIL-MSRCR algorithm was used to improve the overall brightness of the cropped image, while preserving key details such as edges, as shown in (d). Finally, H-GrabCut algorithm was used to segment the pedestrian image with the background removed, as shown in (e). At last, the mask generated by H-GrabCut was covered back to the original image, as shown in (f).

3.3. Contrast Experiment

We conducted experiments to assess the performance of our proposed algorithm for intricate background removal in comparison to existing algorithms. In our evaluations, we employed a dataset consisting of 300 pedestrian images, comprising 250 images sourced from the INRIA test dataset and an additional 50 real-world images. These images posed various challenges, encompassing diverse and complex backgrounds, occlusions, varying lighting conditions, and scene compositions. We compared the outcomes of our algorithm with those of the conventional GrabCut algorithm [17]. as well as algorithms proposed in prior literature LSGC [18], A-KGC [19]. Figure 9 illustrates a subset of the comparative results.

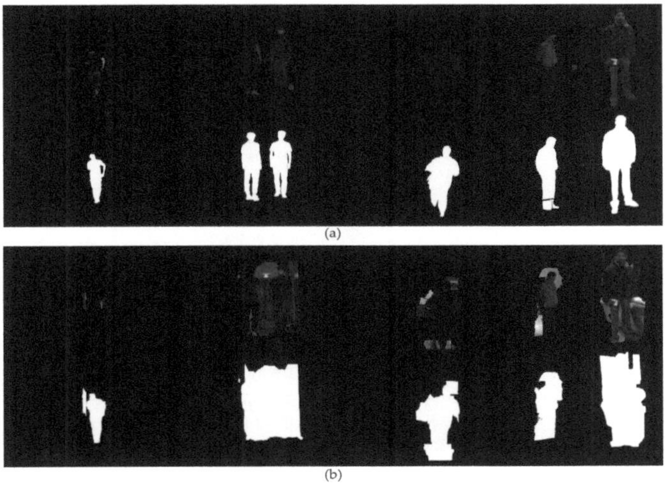

Figure 9. *Cont.*

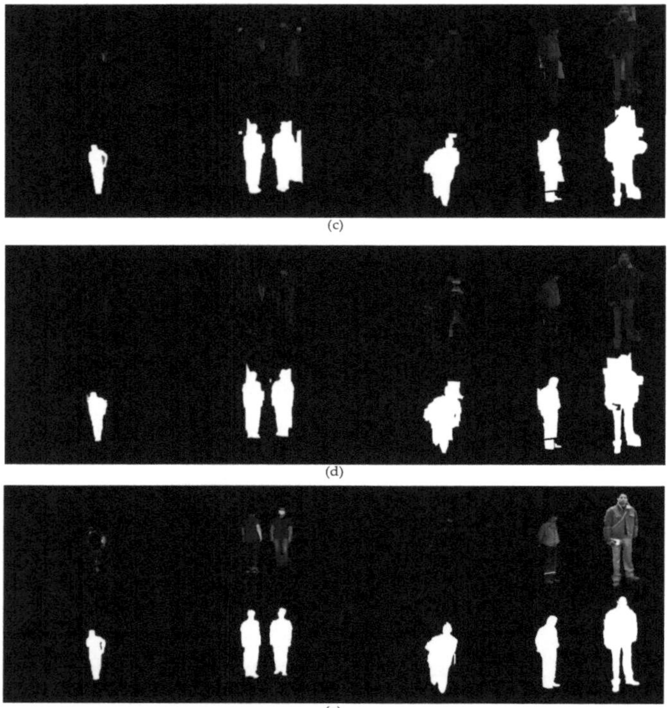

(c)

(d)

(e)

Figure 9. (**a**) portrays the pedestrian standard segmentation in Figure 5, whereas (**b**) shows the pedestrian segmentation acquired via the original GrabCut algorithm. The results of the improved GrabCut algorithm utilizing local sampling LSGC and adaptive *k*-means A-KGC can be seen in (**c**,**d**), respectively. Lastly, in (**e**), the segmentation results of the algorithm presented in this paper are depicted.

Figure 9 illustrates that the original GrabCut algorithm's lack of an object recognition feature necessitates the manual division of foregrounds, resulting in superfluous background elements. Furthermore, it lacks image filtering and enhancement tools, contributing to its sensitivity to noise and low robustness to illumination. Consequently, indistinct pedestrian edges and undersegmentation often occur, as shown in Figure 9b,c. Although the algorithm proposed by reference [18] achieves desirable pedestrian background removal in simple scenes, it may fail to accurately segment the pedestrian area in complex backgrounds with uneven image gray levels, as shown in Figure 9c. In the case of complex indoor pedestrian backgrounds, local sampling becomes more sensitive to noise and outliers leading to over-segmentation. Conversely, the algorithm demonstrated in Figure 9d by reference [19] provides superior results in pedestrian background removal. In comparison to other segmentation algorithms, the proposed algorithm generates clear pedestrian contours with complete details, as shown in Figure 9e.

3.4. Segmentation Result Analysis

In terms of segmentation effect measurement, the following performance indicators were used to further evaluate our method [21]: sensitivity (*Sen*), error segmentation rate

(E), intersection over union (IOU), accuracy (Acc), and relative performance improvement (RPI). All of the above assessment indicators are calculated by the following formula:

$$Acc = \frac{TP + TN}{TP + TN + TN + FP} \tag{28}$$

$$Sen = \frac{TP}{TP + FN} \tag{29}$$

$$IOU = \frac{TP}{TP + FN + FP} \tag{30}$$

$$E = \frac{FP}{TP + FP + FN + TN} \tag{31}$$

$$RPI = \frac{(Original\_Time - Improved\_Time)}{Original\_Time} \times 100\% \tag{32}$$

where TP (true example) represents the number of correctly segmented pixels. FP (false positive example) indicates the number of pixels separated by the error. TN (true negative example) is correctly unsegmented pixels. FN (false negative example) is the number of pixels that are not segmented by errors. The performance indicators of each image are shown in Figure 10, and the specific performance indicators are shown in Table 1:

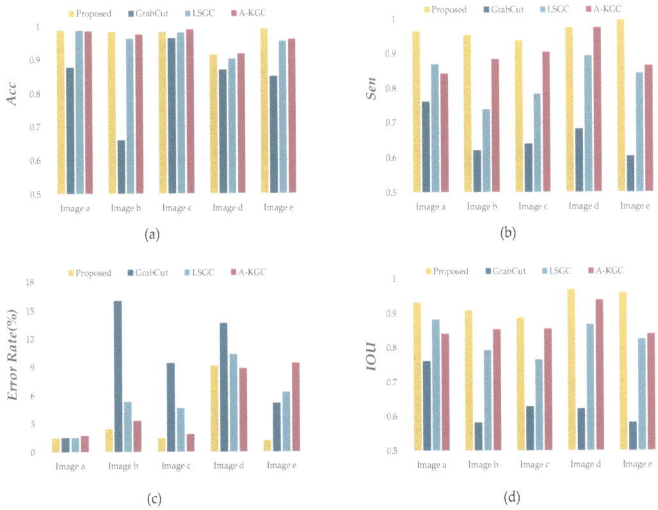

Figure 10. (**a**) is the Acc score of each image in Figure 5 in different segmentation algorithms; (**b**) Figure is the Sen score bar chart of these 5 images; (**c**) is the binary Error segmentation rate, expressed as a percentage; and (**d**) is the IOU score in 4 different segmentation algorithms.

Table 1 shows that the proposed H-GrabCut image segmentation algorithm outperforms other segmentation techniques in terms of sensitivity (Sen), error segmentation rate (E), intersection over union (IOU), accuracy (Acc), relative performance improvement (RPI), and average processing time. This outcome suggests that the proposed algorithm is efficient at eliminating noise. For images with complicated backgrounds, we implemented an enhanced image segmentation approach, which significantly reduced the segmentation error rate. The results of this approach indicate that post-segmentation pedestrian edges are smooth with clear delineation, further affirming the efficacy of the proposed algorithm.

Table 1. Performance index data of different segmentation algorithms.

Method	ACC (%)	SEN (%)	IOU (%)	E (%)	RPI (%)	Time (s)
Grab Cut	84.334	66.102	63.462	8.7650	0.00	6.35
LSGC [18]	95.728	82.563	82.647	5.8062	20.78	5.03
A-KGC [19]	96.482	89.307	86.457	5.0058	50.23	3.16
Proposed	97.133	96.481	93.032	3.1277	69.29	1.95

4. Conclusions

Inspired by the distance between moving obstacles and robots in robot collision avoidance, this study proposes an improved H-GrabCut segmentation algorithm to remove the background of indoor pedestrians in the RGB space. Experimental research is conducted to introduce clustering features that are more effective than those of traditional algorithms. These features preserve the integrity of pedestrian image boundaries and effectively reduce the number of algorithm iterations, providing the optimal measurement point for pedestrian distance and improving the accuracy and reliability of image segmentation.

We conducted tests on the enhanced H-GrabCut algorithm using both real-world scenarios and the INRIA pedestrian dataset. We compared its segmentation results with those of the original GrabCut, local sampling GrabCut (LSGC), and adaptive *k*-Means GrabCut (A-KGC) algorithms. While all three algorithms showed similar performance in segmenting the central regions of pedestrians, a comprehensive comparison involving error segmentation rates, segmentation accuracy, and average intersection over union (IOU) revealed that H-GrabCut outperformed the others. The experiments confirmed the feasibility and practicality of our proposed H-GrabCut algorithm, providing convenience for indoor pedestrian trajectory prediction and obstacle avoidance research in our laboratory. We have developed a system for precise background segmentation on a mobile robot and plan to extend our research to indoor pedestrian trajectory prediction and obstacle avoidance, while considering the stability of background segmentation under varying robot speeds.

Author Contributions: Conceptualization, X.H.; methodology, X.H. and S.W.; software, X.H. and W.X.; validation, X.G., H.P. and W.X.; formal analysis, X.H. and X.G.; investigation, W.X. and H.P.; resources, M.Z.; data curation, D.L.; writing—original draft preparation, X.H. and D.L.; writing—review and editing, X.H. and S.W.; visualization, X.H. and S.W.; supervision, S.W. and X.G.; project administration, M.Z.; and funding acquisition, S.W. and X.G. All authors have read and agreed to the published version of the manuscript.

Funding: This work was supported by National Key R&D Program "Active Health and Aging Science and Technology Response" Special Project (Grant No. 2020YFC2008503) and Guangxi Science and Technology Base and Talent Project (Grant No. AD22080004).

Institutional Review Board Statement: Not applicable.

Informed Consent Statement: Not applicable.

Data Availability Statement: Not applicable.

Conflicts of Interest: The authors declare no conflict of interest.

References

1. Hayati, M.; Muchtar, K.; Maulina, N. Impact of CLAHE-based image enhancement for diabetic retinopathy classification through deep learning. *Procedia Comput. Sci.* **2023**, *216*, 57–66. [CrossRef]
2. Aboshosha, S.; Zahran, O.; Dessouky, M.I. Resolution and quality enhancement of images using interpolation and contrast limited adaptive histogram equalization. *Multimed. Tools Appl.* **2019**, *78*, 18751–18786. [CrossRef]

3. Alwazzan, M.J.; Ismael, M.A.; Ahmed, A.N. A hybrid algorithm to enhance colour retinal fundus images using a Wiener filter and CLAHE. *J. Digit. Imaging* **2021**, *34*, 750–759. [CrossRef] [PubMed]
4. Subramani, B.; Veluchamy, M. Fuzzy gray level difference histogram equalization for medical image enhancement. *J. Med. Syst.* **2020**, *44*, 103. [CrossRef] [PubMed]
5. Bhandari, A.K.; Kandhway, P.; Maurya, S. Salp swarm algorithm-based optimally weighted histogram framework for image enhancement. *IEEE Trans. Instrum. Meas.* **2020**, *69*, 6807–6815. [CrossRef]
6. Chang, J.; Liu, W.; Bai, J. A Retinex image enhancement algorithm based on image fusion technology. *Comput. Eng. Sci.* **2018**, *40*, 1624–1635. [CrossRef]
7. Guo, X.; Li, Y.; Ling, H. LIME: Low-light image enhancement via illumination map estimation. *IEEE Trans. Image Process.* **2016**, *26*, 982–993. [CrossRef] [PubMed]
8. Gu, Z.; Li, F.; Fang, F. A novel retinex-based fractional-order variational model for images with severely low light. *IEEE Trans. Image Process.* **2019**, *29*, 3239–3253. [CrossRef] [PubMed]
9. Li, M.; Liu, J.; Yang, W. Structure-revealing low-light image enhancement via robust retinex model. *IEEE Trans. Image Process.* **2018**, *27*, 2828–2841. [CrossRef] [PubMed]
10. Wang, Y.F.; Liu, H.M.; Fu, Z.W. Low-light image enhancement via the absorption light scattering model. *IEEE Trans. Image Process.* **2019**, *28*, 5679–5690. [CrossRef] [PubMed]
11. Fan, Z.; Liu, K.; Hou, J. JAUNet: A U-Shape Network with Jump Attention for Semantic Segmentation of Road Scenes. *Appl. Sci.* **2023**, *13*, 1493. [CrossRef]
12. Liang, C.; Xiao, B.; Cheng, B. XANet: An Efficient Remote Sensing Image Segmentation Model Using Element-Wise Attention Enhancement and Multi-Scale Attention Fusion. *Remote Sens.* **2022**, *15*, 236. [CrossRef]
13. Zhang, Y.; Zhang, Y.; Zhang, Q. Semantic Segmentation of Traffic Scene Based on DeepLabv3+ and Attention Mechanism. In Proceedings of the 2023 3rd International Conference on Neural Networks, Information and Communication Engineering (NNICE), Guangzhou, China, 24–26 February 2023.
14. Lahmyed, R.E.; Ansari, M.; Kerkaou, Z. A novel visible spectrum images-based pedestrian detection and tracking system for surveillance in non-controlled environments. *Multimed. Tools Appl.* **2022**, *81*, 39275–39309. [CrossRef]
15. Xie, B.; Yang, Z.; Yang, L. Multi-scale fusion with matching attention model: A novel decoding network cooperated with NAS for real-time semantic segmentation. *IEEE Trans. Intell. Transp. Syst.* **2021**, *23*, 12622–12632. [CrossRef]
16. Boykov, Y.Y.; Jolly, M.P. Interactive graph cuts for optimal boundary & region segmentation of objects in ND images. In Proceedings of the Eighth IEEE International Conference on Computer Vision, ICCV 2001, Vancouver, BC, Canada, 7–14 July 2001.
17. Rother, C.; Kolmogorov, V.; Blake, A. "GrabCut" interactive foreground extraction using iterated graph cuts. *ACM Trans. Graph.* **2004**, *23*, 309–314. [CrossRef]
18. Yao, G.; Wu, S.; Yang, H. GrabCut Image Segmentation Based on Local Sampling. In *Business Intelligence and Information Technology, Proceedings of the International Conference on Business Intelligence and Information Technology BIIT 2021, Harbin, China, 18–20 December 2021*; Springer: Cham, Switzerland, 2021.
19. Prabu, S. Object segmentation based on the integration of adaptive K-means and GrabCut algorithm. In Proceedings of the 2022 International Conference on Wireless Communications Signal Processing and Networking (WiSPNET), Chennai, India, 24–26 March 2022.
20. Wang, Q.; He, X.; Wu, X.; Wu, X.; Teng, Q. An improved segmentation algorithm based on GrabCut. *Inf. Technol. Netw. Secur.* **2021**, *40*, 43–52. [CrossRef]
21. Ünver, H.M.; Ayan, E. Skin lesion segmentation in dermoscopic images with combination of YOLO and grabcut algorithm. *Diagnostics* **2019**, *9*, 72. [CrossRef]
22. Jobson, D.J.; Rahman, Z.; Woodell, G.A. A multiscale retinex for bridging the gap between color images and the human observation of scenes. *IEEE Trans. Image Process.* **1997**, *6*, 965–976. [CrossRef]
23. Li, J.; Han, D.; Wang, X. Multi-sensor medical-image fusion technique based on embedding bilateral filter in least squares and salient detection. *Sensors* **2023**, *23*, 3490. [CrossRef]
24. Wang, W.; Yuan, X.; Chen, Z. Weak-light image enhancement method based on adaptive local gamma transform and color compensation. *J. Sens.* **2021**, *2021*, 5563698. [CrossRef]

sensors

MDPI

Article

Trajectory Planner for UAVs Based on Potential Field Obtained by a Kinodynamic Gene Regulation Network

Juncao Hong [1], Diquan Chen [1], Wenji Li [1,2,3,*] and Zhun Fan [1,2,3,*]

[1] College of Engineering, Shantou University, Shantou 515063, China
[2] Key Laboratory of Intelligent Manufacturing Technology, Shantou University, Ministry of Education, Shantou 515063, China
[3] International Cooperation Base of Evolutionary Intelligence and Robotics, Shantou University, Shantou 515063, China
* Correspondence: liwj@stu.edu.cn (W.L.); zfan@stu.edu.cn (Z.F.)

Abstract: Quadrotor unmanned aerial vehicles (UAVs) often encounter intricate environmental and dynamic limitations in real-world applications, underscoring the significance of proficient trajectory planning for ensuring both safety and efficiency during flights. To tackle this challenge, we introduce an innovative approach that harmonizes sophisticated environmental insights with the dynamic state of a UAV within a potential field framework. Our proposition entails a quadrotor trajectory planner grounded in a kinodynamic gene regulation network potential field. The pivotal contribution of this study lies in the amalgamation of environmental perceptions and kinodynamic constraints within a newly devised gene regulation network (GRN) potential field. By enhancing the gene regulation network model, the potential field becomes adaptable to the UAV's dynamic conditions and its surroundings, thereby extending the GRN into a kinodynamic GRN (K-GRN). The trajectory planner excels at charting courses that guide the quadrotor UAV through intricate environments while taking dynamic constraints into account. The amalgamation of environmental insights and kinodynamic constraints within the potential field framework bolsters the adaptability and stability of the generated trajectories. Empirical results substantiate the efficacy of our proposed methodology.

Keywords: trajectory planning; kinodynamic constraints; environmental perception; potential field; gene regulation network

Citation: Hong, J.; Chen, D.; Li, W.; Fan, Z. Trajectory Planner for UAVs Based on Potential Field Obtained by a Kinodynamic Gene Regulation Network. *Sensors* **2023**, 23, 7982. https://doi.org/10.3390/s23187982

Academic Editor: Gregor Klančar

Received: 29 July 2023
Revised: 12 September 2023
Accepted: 12 September 2023
Published: 20 September 2023

1. Introduction

The successful implementation of autonomous drone navigation has immense value in scientific studies and various industries as it liberates their applications from limitations and paves the way for diverse deployments. Despite this progress, the pursuit of safe and efficient navigation in unknown environments poses significant challenges. To achieve secure and efficient navigation, it is of paramount importance for drones to accurately represent and effectively utilize environmental information during trajectory planning.

A widely adopted technique for representing the distance between a quadrotor and objects in its environment is the Euclidean signed distance field (ESDF), which utilizes gradients. The ESDF has proven to be effective in motion planning for single-agent scenarios, providing valuable insights for autonomous navigation.

Nevertheless, it has limitations in certain application scenarios. For instance, its ability to represent all objects in an environment as obstacles may not adequately capture the complexity of cooperative swarm motions or environments with multiple layers of semantic information. This limitation impedes the successful completion of flight tasks in complex scenarios.

To address these challenges, this study proposed a novel gradient potential field, termed the kino-gene regulatory network (K-GRN) potential field, which is based on a GRN model. By considering the kinematic state of a quadrotor, the objective was to enhance trajectory planning and improve the overall navigation performance. Additionally, we introduced the K-GRN planner, which is a local trajectory planning framework that leverages the kinodynamic GRN potential field to generate optimized flight paths.

Inspired by the intricate interactions observed in organisms, a GRN model was used to generate a gradient potential field. This approach has found extensive application in swarm robot morphology generation and serves as a foundation for the development of autonomous systems. Based on the previous work by Li [1], who utilized the GRN potential field for shape generation and cooperative motion of a swarm of drones without communication, this study further enhanced the potential field by incorporating the kinematic state of the robots. This enhancement enabled the design of a local trajectory planning framework using the proposed potential field, thereby ensuring improved navigation capabilities.

The contributions of this study are as follows. First, we propose an improved GRN potential field that demonstrates its applicability in achieving safe and efficient trajectory planning for drones. Second, we introduce the incorporation of the kinematic characteristics of agents, enabling the extension of the potential field from a spatial domain to a state space, thus, enriching navigation capabilities. Finally, based on these contributions, we establish constraint rules for the trajectory search and develop a comprehensive framework for local planning that further enhances the overall navigation system.

The remainder of this paper is organized as follows. In Section 2, we present a comprehensive review of the related work, encompassing GRN, trajectory planning for unmanned aerial vehicles (UAVs) in positional environments, and the application of bio-inspired algorithms in UAV navigation. Section 3 outlines the enhancements made to the GRN potential field and presents the framework of the proposed method. We present a real-time planning algorithm based on the K-GRN potential field in Section 4. In Section 5, we describe the experimental setup and results for validating the effectiveness of our proposed approach. Finally, in Section 6, we present the conclusions by summarizing the key findings and contributions and provide insights into potential avenues for future research and development.

2. Related Work

2.1. GRN Model in UAV Control

In recent years, GRNs have attracted considerable attention in the design of control methods for robot control [2,3]. Taylor et al. [4] introduced a GRN–cell adhesion model (CAM) control method that combines a CAM with a GRN model. This approach enables the spatial and temporal differentiation of protein expression across swarm robots, which are treated as artificial cells with virtual membranes and artificial cell adhesion molecules. Guo et al. [5,6] applied a GRN model to generate adaptive patterns for swarm robots. Each robot possessed two genes that produced proteins to control its movements in the x and y directions via GRN mechanisms. These proteins enabled the robots to move within the boundaries of predefined shapes.

Furthermore, evolutionary algorithms are utilized to optimize robot motion parameters. However, this method relies on global coordinates, which are often challenging for robots to obtain in real-world applications. To address this issue, Guo et al. [7] proposed a GRN-based method for selecting a reference robot and establishing a local coordinate system. Each robot contributed to the formation of an uneven B-spline pattern using a local coordinate system. Building on this local coordinate system, Jin et al. [8] introduced a hierarchical gene regulatory network (H-GRN) for adaptive pattern generation, specifically for trapping targets in dynamic environments. Oh et al. [9] expanded an H-GRN model to achieve area coverage. To address the challenges related to merging and separating patterns during multi-target entrapment and obstacle avoidance, Oh et al. [10] proposed an enhanced H-GRN (EH-GRN) structure that incorporates obstacles and targets as inputs and evolves the GRN to generate a morphogen gradient space. Similarly, Meng and Guo [11] introduced an evolving GRN method to adjust the structure and parameters of the GRN.

Recently, Wu et al. [12] proposed a collaborative GRN (C-GRN) model that enables peer collaboration between agents. This model enables agents to discover and reinforce weak areas in the formed patterns. Yuan et al. [13] combined a tracking-based H-GRN with a leader–follower model and introduced a TH-GRN model that is suitable for dynamic and complex environments. Fan et al. [14] used a genetic programming approach to design a design automation framework for the gene regulatory network.

In summary, the utilization of a GRN is widespread for generating gradient fields that depict environmental information and direct robot movement. However, the current studies either rely solely on gradient information to guide robot behavior or address a GRN method as a distinct upper-level structure, separate from a lower-level planning controller.

The proposed GRN structure is based on VG-Swarm [1] improvement, which expands the dimensions of the GRN potential field and incorporates the state space of drones. We designed a real-time trajectory-planning method specifically tailored to this framework.

2.2. Bio-Inspired Path Planning

Bio-inspired algorithms have recently attracted substantial attention owing to their efficacy in solving complex optimization problems while maintaining balance among their components. In the domain of path planning problems, studies have increasingly focused on harnessing the power of bio-inspired algorithms to optimize these problems [15–18].

Swarm intelligence is derived from the self-organizational behavior observed in biological systems inspired by this swarm intelligence, we have developed a number of swarm intelligence algorithms, among which the most notable are particle swarm optimization (PSO) and ant colony optimization (ACO). For example, Phung et al. [19] proposed an enhanced discrete PSO algorithm tailored to UAV inspection path planning. They replaced the inspection path-planning problem with a solution derived from the extended traveling salesman problem. ACO, originally introduced by Dorigo et al. in 1999 [20], enables ants to cooperatively discover the shortest path between a nest and its food sources using pheromone guidance. Nonetheless, both the ACO and PSO algorithms encounter challenges related to slow convergence speeds when the computational complexity rapidly increases, which is undesirable for real-time planning in complex environments.

Another noteworthy bio-inspired algorithm is the evolutionary algorithm. Sun et al. employed an adaptive multi-objective differential evolution (DE) algorithm in a multi-robot system [21]. Patlea et al. presented an improved genetic algorithm (GA) that employed a binary code matrix for mobile robot navigation in complex environments [22]. However, similar to the ACO and PSO algorithms, the time complexity of evolutionary algorithms is typically high, posing obstacles to achieving real-time performance in complex environments.

One limitation of the bio-inspired algorithms is their lack of effectiveness in real-time path-planning problems. This is because of the requirement for the learning processes, such as the modification of path selection probabilities using an objective function, as observed in ant colony optimization. Among these bio-inspired methods, a GRN model offers a superior option because of its capability to generate gradient fields in real-time. For the first time, we introduced a real-time trajectory planning framework using a potential field generated from a GRN model.

2.3. UAVs Navigation with Gradient Field

Gradient-based motion planning is the dominant approach for generating local trajectories in UAVs, formulating the motion planning as an unconstrained nonlinear optimization problem. The introduction of the Euclidean signed distance field (ESDF) in robotic motion planning by Ratliff et al. [23] has enabled planning frameworks to directly optimize trajectories in a configuration space using its abundant gradient information. However, discrete-time trajectory optimization [23,24] is unsuitable for drones because of its sensitivity to dynamic constraints. To address this limitation, a continuous-time polynomial trajectory optimization method for UAV planning was proposed by [25]. Despite its merits, the method carries a significant computational burden due to the integration of the potential function, and even with random restarts, the success rate of this method approximates only 70%. To address these challenges, Ratliff et al. [23] introduced a B-spline parameterization of trajectories leveraging convex hull properties. In [26], the success rate was significantly improved by finding a collision-free initial path as the front end, and further improvements were made by considering the kinodynamic constraints [27]. Zhou et al. [28] enhanced system robustness by incorporating perception awareness. Among these approaches, the ESDF is crucial in evaluating the distance to nearby obstacles based on the gradient magnitude and direction.

Oleynikova [29] and Han [30] proposed incremental methods for ESDF generation, namely, Voxblox and FIESTA, respectively. Although these methods are highly efficient for dynamic updating, they only provide information on obstacles and neglect other important environmental information in the generated ESDF. Therefore, there is a substantial need to design a gradient field for real-time trajectory planning that can effectively represent environmental information as comprehensively as possible.

3. Kinodynamic GRN Potential Field

Our method improves upon the framework of a GRN model in a VG-Swarm [1], as shown in Figure 1. Within this enhanced framework, the method for computing the GRN concentration in each cell is calculated through Equations (1)–(6).

$$\frac{dT_i}{dt} = \nabla^2 T_i + \gamma_i - T_i \tag{1}$$

$$\frac{dO_j}{dt} = \nabla^2 O_j + \beta_j - O_j \tag{2}$$

$$\frac{dN_m}{dt} = \nabla^2 N_m + \eta_m - N_m \tag{3}$$

$$\begin{cases} \mathcal{N} = \sum_{m=1}^{n_n} N_m \\ \mathcal{T} = \sum_{i=1}^{n_t} T_i \\ \mathcal{O} = \sum_{j=1}^{n_o} O_j \end{cases} \tag{4}$$

$$\begin{aligned} \frac{dM}{dt} = -M + \mathrm{sig}\left(1 - T^2, \theta, k\right) + \mathrm{sig}\left(O^2, \theta, k\right) \\ + \mathrm{sig}\left(N^2, \theta, k\right) \end{aligned} \tag{5}$$

$$\mathrm{sig}(x, \theta, k) = \frac{1}{1 + e^{-k(x-\theta)}} \tag{6}$$

γ_i, β_j, and η_m are positive numbers representing the position of the target, obstacles, and neighboring UAVs, respectively. T_i represents the protein concentration formed by the ith target in the environment (γ_i stands for the position information of the ith target), O_j represents the protein concentration formed by the jth obstacle (β_j stands for the position information of the jth obstacle), and N_m represents the protein concentration formed by the mth neighboring UAV (η_m stands for the position information of the mth neighboring UAV). T, O, and N are the combined concentrations produced by all the targets, obstacles, and neighbors, respectively. ∇ is the Laplace operator, which is defined as the second derivative of T_i, O_j, and N_m. The concentration field obtained by calculating M is used to generate entrapping patterns.

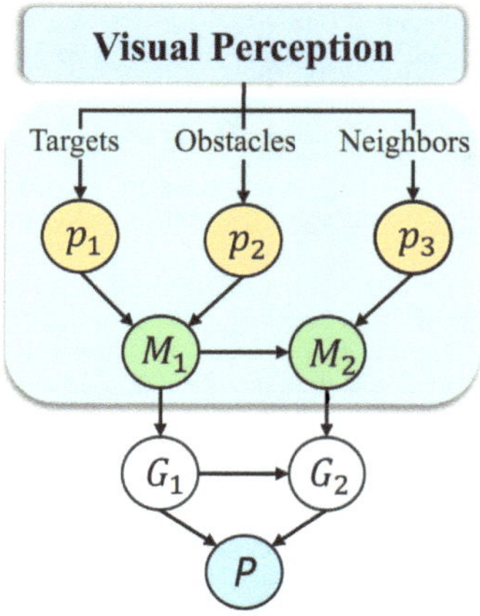

Figure 1. Schematic of the structure of the vision-based GRN. Each cell represents an individual UAV and is composed of an upper and lower layer. In the upper layer, sensory proteins p_1, p_2, and p_3 receive information about the positions of targets, obstacles, and neighboring robots, respectively, forming corresponding concentration fields. Protein M_1 integrates the concentration field from p_1 and p_2, while protein M_2 integrates the concentration field from M_1 and p_3. These proteins influence the production of actuating proteins G_1 and G_2 in the lower layer, which, respectively, represent entrapping patterns and the moving direction of the UAV. G_1 and G_2 also affect the production of protein P, which ultimately determines the dynamic position of the UAV. Moreover, protein P in one cell influences the gene expression of the neighboring cells in the system.

Equations (1) and (2) can be revised to Equations (7) and (8), respectively, which are crucial operations in the GRN model for generating a gradient field.

$$T_i = e^{-td_{tar,i}} \tag{7}$$

$$O_j = e^{-td_{obs,j}} \tag{8}$$

where $d_{tar,i}$ represents the Euclidean distance between the agent and the ith target, and $d_{obs,j}$ represents the distance between the agent and the jth obstacle. Notably, the VG-Swarm considers only the distance between the objects and agents during the guidance control process. To address this limitation, we propose a specialized GRN method for trajectory planning called kinodynamic GRN. Unlike the VG-Swarm, the kinodynamic GRN introduces the concept of the state transition cost. We assume that any point within the grid can serve as an intermediate node for UAV movement. By sampling the states of these intermediate nodes, denoted by $X_{k+1}\ (p,v)$, we can calculate the state transition costs to the obstacle and target nodes, denoted by $\mathcal{L}_{X_{k+1},X_{obs}}$ and $\mathcal{L}_{X_{k+1},X_{tar}}$, respectively. These costs can be expressed using Equations (9) and (10).

$$\mathcal{L}_{x_{k+1},x_{\text{tar}}} = \int_0^{\tau_{k+1,tar}} ||u_{k+1,\text{ tar}}||dt + \rho\tau_{k+1,\text{ tar}} \tag{9}$$

$$\mathcal{L}_{x_{k+1},x_{obs}} = \int_0^{\tau_{k+1,obs}} ||u_{k+1,obs}||dt + \rho\tau_{k+1,obs} \tag{10}$$

where $u_{k+1,tar}$ and $\tau_{k+1,tar}$, respectively, represent the optimal input and optimal time required to transition from the state X_{k+1} to the target state. They are determined using the Pontryagin minimum principle, as expressed in Equation (11).

$$\begin{cases} p_{\mu}^{*}(t) & = \frac{1}{6}\alpha_{\mu}t^{3} + \frac{1}{2}\beta_{\mu}t^{2} + v_{\mu c} + p_{\mu c} \\ \begin{bmatrix} \alpha_{\mu} \\ \beta_{\mu} \end{bmatrix} & = \frac{1}{T^{3}} \begin{bmatrix} -12 & 6T \\ 6T & -2T^{2} \end{bmatrix} \begin{bmatrix} p_{\mu g} - p_{\mu c} - v_{\mu c}T \\ v_{\mu g} - v_{\mu c} \end{bmatrix} \\ \mathcal{J}^{*}(T) & = \sum_{\mu \in \{x,y,z\}} \left(\frac{1}{3}\alpha_{\mu}^{2}T^{3} + \alpha_{\mu}\beta_{\mu}T^{2} + \beta_{\mu}^{2}T \right) \end{cases} \tag{11}$$

Next, the Euclidean distance in Equations (7) and (8) is replaced by the state transition cost, as expressed by Equations (9) and (10), thereby forming the protein concentration expressions (Equations (12) and (13)) for environmental information in the kinodynamic GRN potential field.

$$T_i = e^{-t\mathcal{L}_{x_{k+1},x_{tar}}} \tag{12}$$

$$O_j = e^{-t\mathcal{L}_{x_{k+1},x_{obs}}} \tag{13}$$

In the path-planning process, we also consider the current state of the agent to ensure that the planned path is executable. Hence, the current kinematic state of an individual drone, denoted as X_k, is considered. We define the forward cost, $\mathcal{L}_{X_k,X_{k+1}}$, using Equation (14), which represents the cost of transitioning from state X_k to X_{k+1}.

$$\mathcal{L}_{x_k,x_{k+1}} = \int_{0}^{\tau_{k,k+1}} ||u_{k,k+1}|| dt + \rho\tau_{k,k+1} \tag{14}$$

Finally, the generation of the potential field is expressed using Equations (15)–(19), where M_{k+1} denotes the protein concentration in the sampled state X_{k+1}.

$$\begin{cases} \mathcal{T} = \sum_{i=1}^{n_t} T_i, \\ \mathcal{O} = \sum_{j=1}^{n_o} O_j \end{cases} \tag{15}$$

$$J_0 = e^{-t\mathcal{L}_{x_k,x_{k+1}}} \tag{16}$$

$$M_{k+1} = \text{sig}\left(1 - T^2, \theta, k\right) + \text{sig}\left(O^2, \theta, k\right) + \text{sig}\left(1 - J_0^2, \theta, k\right) \tag{17}$$

$$\text{sig}(\text{x}, \theta, k) = \frac{1}{1 + e^{-k(x-\theta)}} \tag{18}$$

As illustrated in Algorithm 1, during the construction of the K-GRN potential field, the grid map needs to be initialized first. Following that, the grids within M are iterated over, during which the state sampling occurs to acquire candidate states. Subsequently, these candidate states within each grid are revisited to compute their associated costs and integrate them into the GRN model. This process ultimately yields concentration information for the specific candidate states at that particular position.

In this context, M constitutes an assortment of state samples within specific ranges across each grid of the map, indicating the concentration of K-GRN at that position under that particular state. $\mathcal{E}$ encompasses environmental data, which include obstacles, objectives, and optionally, neighboring drones. X represents the state of the UAV, wherein X_k signifies the present state, X_{k+1} denotes the sampled candidate state, and X_{obs} and X_{tar}, respectively, refer to the states when the drone encounters an obstacle and when it reaches the target destination. $\mathcal{L}$ embodies the cost associated with state transitions, wherein the subscripts of $\mathcal{L}$ align with the originating state and the terminal state.

Algorithm 1 Generation of K-GRN Field

Notation: Concentration M, Environment $\mathcal{E}$ cost $\mathcal{L}$ State X_e;
Initialize: $M \leftarrow InitMap(M, X)$
for M_i in M: **do**
  for X_i in X_{k+1}: **do**
    $(\mathcal{L}_{X_{k+1},X_{tar}}) \leftarrow$ EvaluateCostToTarget $(X_j, \mathcal{E})$
    $(\mathcal{L}_{X_{k+1},X_{obs}}) \leftarrow$ EvaluateCostToObstacle $(X_j, \mathcal{E})$
    $(\mathcal{L}_{X_k,X_{k+1}}) \leftarrow$ EvaluateCostToInstance $(X_j, \mathcal{E})$
    if $\neg$ isCollsionFree() **then**
      $M(i,j) \leftarrow$ CulculateKGRNConcentration $(\mathcal{L}_{X_{k+1},X_{tar}}, \mathcal{L}_{X_{k+1},X_{obs}}, \mathcal{L}_{X_k,X_{k+1}})$
    else
      SetInfConcentration $(M(i,j))$
    end if
  end for
end for
return M

In summary, the K-GRN potential field is a specialized set of potentials crafted for the purpose of trajectory planning. In comparison to the alternative approaches, the potential field molded by K-GRN can embrace a wider array of environmental data, incorporating elements such as obstacles, target points, and neighboring information. It formulates distinct potential fields surrounding these elements to effectively steer the trajectory planning procedure. Furthermore, K-GRN factors in more intricate state information across multiple dimensions, encompassing both the UAV's current state and the projected future states. This holistic approach ensures the traceability of trajectories and significantly elevates the safety standards for the drone.

4. Trajectory Planner Based on K-GRN Potential Field

As mentioned in Section 3, the concentration gradient field of the K-GRN incorporates both environmental information and kinematic considerations to facilitate trajectory planning. However, maintaining a continuous gradient field throughout the planning process can be computationally demanding and impractical, especially for real-time planning in dynamic environments. To address this challenge and improve planner efficiency, we propose a sampling-based approach. Instead of computing the concentration values for all the grid cells, we selectively sample the grid points of interest within the drone's feasible input range. These sampled points represent potential locations that can be navigated by the drone. By focusing on these points, we can significantly reduce the computational burden and calculate the concentration information efficiently. This approach enables us to strike a balance between computational efficiency and the accurate representation of the environmental information necessary for effective trajectory planning.

A flowchart of the trajectory planner is shown in Figure 2. Initially, a grid map is created in which each grid cell is affected by the protein concentration produced by objects in the environment, including the targets and obstacles. The current input u_τ and its corresponding time τ for the drone are sampled. The cost of the drone executing a trajectory segment is computed using the sampled final state. The path with the lowest cost is selected, and the search process is continued until the vicinity of the target is reached.

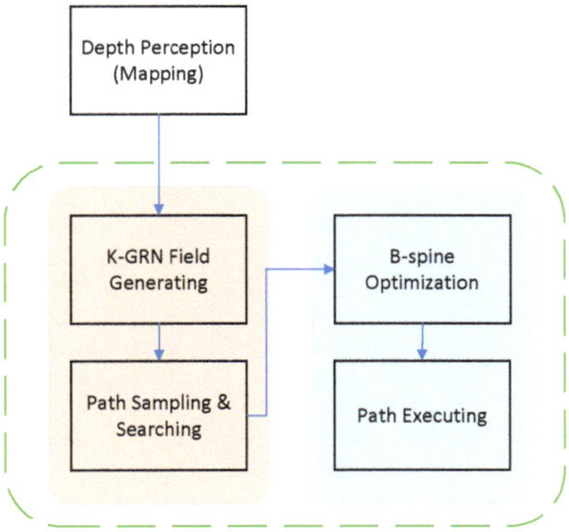

Figure 2. Flow diagram of the trajectory planner.

5. Experiments

The simulation experiments were conducted using UE4 to evaluate the performance of the proposed approach, which can incorporate a wide range of realistic physical models and introduce the various challenges to drone flight. We designed six different scenes, each with two levels of obstacle density, as shown in Figure 3. In each experiment, the obstacles were randomly generated to ensure a diverse and dynamic environment. The navigation task was repeated 100 times for each scene to obtain statistically significant results. In each trial, a random target point within a distance range of 15–20 m was set for the drone. The success or failure of the task was determined based on whether the drone successfully reached the target without colliding with obstacles or experiencing a crash. Through these experiments, we assessed the robustness and effectiveness of the proposed trajectory planner in various challenging scenarios.

(**a**) (**b**)

Figure 3. Illustrations of the two different scenes with varying obstacle densities. (**a**) Dense jungle scene. (**b**) Sparse jungle scene.

To evaluate the effect of different gradient fields on trajectory generation, we conducted experiments to compare various approaches. Specifically, our proposed method was compared against two other approaches: one that utilizes the ESDF-based method, which derives its potential field formation from [30], and another that does not incorporate a potential field at all, as described in [31]. The experimental results are presented in Table 1.

In the table, EF refers to the planning method using ESDF, while NF refers to the planning method that does not use potential fields. "Success" is defined as the drone's capability to traverse from the starting point to the endpoint (within a proximity of roughly 0.3 m) in a jungle setting without colliding with obstacles.

Notably, the proposed method, which integrated both environmental and state information, required a slightly longer computation time than the compared methods. However, the paths generated by the proposed approach strike a balance between safety and efficiency. We observed significant improvements in both the success rate of the trajectory execution and the average flight speed of the drone compared with the benchmark methods. These findings demonstrate the effectiveness of the proposed approach in generating trajectories that consider both safety and efficiency. Despite the slightly longer computation time, the benefits gained in terms of the improved success rate and flight speed render the proposed method a favorable choice for trajectory planning in dynamic environments.

Table 1. Effect of Different Gradient Fields on Trajectory Generation.

Policy	Success Rate	Max Velocity (m/s)	Average Velocity (m/s)	Slove Time (ms)
EF	0.69	**4.13**	2.65	6.55
NF	0.64	3.81	2.47	**1.49**
GRN	**0.72**	3.93	**2.71**	6.88

The numbers in bold represent the optimal data for this column.

We have also conducted experiments to showcase the efficacy of integrating the dynamics information of drones and their candidate points into the potential field. This underscores the essentiality of evolving from GRN to K-GRN, as presented in Table 2. In the table, KF refers to the planning method that integrates kinematic information into the potential field, while NKF refers to other potential field planning methods that do not include dynamic information. NF is a planning method that does not use potential fields. In the identical scenario and utilizing the same experimental approach, we conducted a comparative analysis of the proposed approach, which involves the K-GRN potential field, against the performance of the GRN potential field without the integration of motion dynamics. Additionally, we contrasted these with navigation task planning methods that do not rely on potential fields. The experimental results underscore a distinct advantage in navigation success rates for the trajectory planning method guided by K-GRN. As previously discussed in this paper, one of the key advantages of K-GRN over other approaches lies in its ability to ensure trajectory traceability, thus bolstering drone safety. Concerning the time of generating potential field t_{field}, our method shows a minor drawback in terms of the duration needed for potential field formation. However, when evaluating the overall planning time, which encompasses both t_{field} and t_{plan}, we still maintain a significant edge over alternative planning methods. Despite the slightly longer computational time required for potential field generation, the proposed method demonstrated superior performance in terms of the navigation task success rate and planning time compared with the other two methods. This highlights the effectiveness of the proposed approach in dynamic environments and its potential as an ideal choice for trajectory planning.

Table 2. Performance Comparison of Different Planning Methods After Incorporating UAV Kinematics Information into GRN Potential Field.

Policy	Success Rate	t_{field} (ms)	t_{plan} (ms)
KF	**0.72**	5.22	**0.91**
NKF	0.69	4.88	4.47
NF	0.55	**0**	2.90

The numbers in bold represent the optimal data for this column.

To evaluate the effectiveness of our trajectory planning framework, we compared it with two state-of-the-art methods: the Ego-planner [31] and the Fast-planner [28]. The Fast-Planner utilizes the ESDF for path search and incorporates the kinematic information in the search process, whereas the Ego-planner proposes a real-time planning method that does not rely on the ESDF. We conducted a series of experiments on six scenes with two obstacle-density levels. Both benchmark methods were open-source, and we used their default parameters for the experiments.

Table 3 presents a comparison between K-GRN and two state-of-the-art (SOTA) methods across two distinct scenarios. In this context, employing the same experimental methodology, we conducted separate comparisons of the drone's navigation success rates, average flight speeds, and planned trajectory lengths under each condition.

Table 3. Comparison of UAV Navigation Performance in Scenes with Different Obstacle Densities.

Policy	Dense Scene			Sparse Scene		
	Success Rate	Average Velocity (m/s)	Trajectory Length (m)	Success Rate	Average Velocity (m/s)	Trajectory Length (m)
GRN-planner	**0.56**	**2.01**	71.29	**0.72**	**2.71**	58.13
Fast-planner	0.47	1.95	**65.76**	0.69	2.65	55.43
Ego-planner	0.53	1.88	67.18	0.64	2.47	**49.90**

The numbers in bold represent the optimal data for this column.

Across all scenes, our proposed method significantly outperformed the two benchmark methods in terms of both success rates and average speeds. This demonstrates the effectiveness of the proposed approach in achieving successful navigation while maintaining rapid flight speed. However, our paths tended to be slightly longer than those generated by the other methods, due to our emphasis on flight safety. This trade-off ensures a higher level of safety during the flight.

Furthermore, Figure 4 displays the real-time speed variations among the three algorithms in dense scenes. It is worth highlighting that the presented algorithm showcases notably seamless speed transitions across the entire navigation journey. The pronounced speed reduction captured in the graph aligns with the drone executing obstacle avoidance maneuvers. While navigating through an identical course, our approach strategically adopts a safer path-planning strategy, prioritizing the avoidance of densely wooded narrow passages. Consequently, despite generating a lengthier route, this is a paramount benefit of our method.

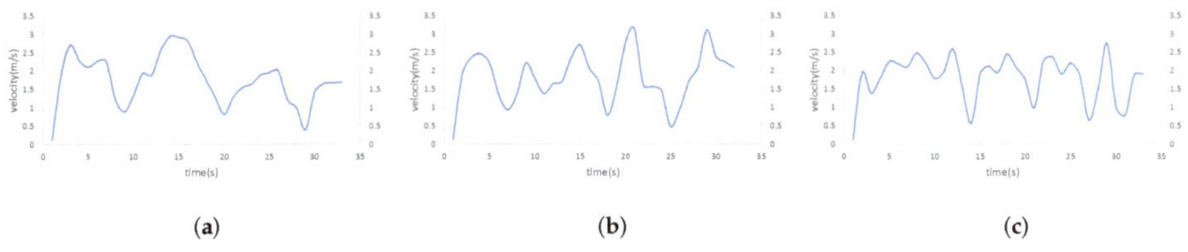

Figure 4. Real-time Velocity Variation of UAV during the Experiment: (**a**) Velocity Curve of GRN-planner. (**b**) Velocity Curve of Fast-planner. (**c**) Velocity Curve of Ego-planner.

To evaluate the practical efficacy of the K-GRN planner, we conducted relevant simulation experiments. The simulation environment was designed to replicate a real jungle scenario. In this scenario, we had the drone shuttles through the jungle. As shown in Figure 5, the UAV's obstacle avoidance process, guided by the K-GRN planner, is depicted. It is clear that the drone could detect obstacles in advance, perform evasive maneuvers, and navigate successfully while maintaining a safe distance from the obstacles. Figure 6 displays the trajectory of the drone as it shuttles through the jungle environment.

(**a**) (**b**) (**c**)

Figure 5. (**a**–**c**): Obstacle avoidance trajectories of a drone guided by the K-GRN planner in a jungle scenario.

Figure 6. Trajectories of the drone as it shuttles through the jungle environment.

In summary, our proposed method outperformed the benchmark methods in terms of success rate and average speed across various obstacle-density scenarios. The experimental results validated the superiority of the proposed approach, highlighting its potential to advance the field of UAV trajectory planning.

6. Conclusions

In this study, we have introduced an improved GRN potential field, termed the K-GRN potential field, specifically designed for trajectory planning. This potential field incorporates the costs of state transitions for intelligent agents to assist planners in generating smooth and efficient trajectories. Additionally, we have designed a planner that searches for the optimal trajectory using input sampling, requiring only a search for the path with the lowest cost within the established potential field. Through benchmark comparisons, we have validated the advantages of incorporating the state space in the GRN potential fields, the effectiveness of the K-GRN potential field in trajectory planning, and the performance of the K-GRN-guided planner compared with state-of-the-art methods. The results demonstrated that the proposed method offered superior paths in terms of both safety and effectiveness for drone flight.

Our approach also displays specific limitations: Driven by safety considerations, we embraced a cautious planning strategy, which results in a difficulty in devising optimal paths within complex scenarios. Similarly, due to this rationale, despite attaining advantages in average speed, our method encountered difficulties in outperforming the SOTA approach with regards to the highest planned speed. In forthcoming endeavors, we aim to enhance the K-GRN methodology by integrating adaptive planning strategies, thereby enabling it to demonstrate diverse obstacle avoidance performances across a spectrum of scenario complexities. Furthermore, we aim to extend the application of the K-GRN potential fields to drone swarm missions and explore their application in achieving cooperative flight in environments without communication capabilities.

Author Contributions: Conceptualization, J.H., W.L. and Z.F.; methodology, J.H., D.C. and Z.F.; software, J.H. and D.C.; validation, J.H., D.C., Z.F. and W.L.; formal analysis, J.H. and D.C.; investigation, J.H. and D.C.; resources, J.H.; data curation, D.C.; writing—original draft preparation, J.H., D.C., Z.F. and W.L.; writing—review and editing, J.H., D.C., Z.F. and W.L.; visualization, J.H. and D.C.; supervision, Z.F. and W.L.; project administration, J.H., D.C., Z.F. and W.L.; funding acquisition, J.H. All authors have read and agreed to the published version of the manuscript.

Funding: This research was supported in part by the National Key R&D Program of China (grant number 2021ZD0111501, 2021ZD0111502), the Key Laboratory of Digital Signal and Image Processing of Guangdong Province, the Key Laboratory of Intelligent Manufacturing Technology (Shantou University), Ministry of Education, the Science and Technology Planning Project of Guangdong Province of China (grant number 180917144960530), the Project of Educational Commission of Guangdong Province of China (grant number 2017KZDXM032), the State Key Lab of Digital Manufacturing Equipment & Technology (grant number DMETKF2019020), National Natural Science Foundation of China (grant number 62176147), STU Scientific Research Foundation for Talents (grant number NTF21001), Science and Technology Planning Project of Guangdong Province of China (grant number 2019A050520001, 2021A0505030072, 2022A1515110660), Science and Technology Special Funds Project of Guangdong Province of China (grant number STKJ2021176, STKJ2021019), and Guangdong Special Support Program for Outstanding Talents (2021JC06X549)

Institutional Review Board Statement: Not applicable.

Informed Consent Statement: Not applicable.

Data Availability Statement: Not applicable.

Conflicts of Interest: The authors declare no conflict of interest.

Abbreviations

The following abbreviations are used in this manuscript:

UAV	Unmanned aerial vehicles
GRN	Gene regulation network
K-GRN	Kinodynamic GRN
ESDF	Euclidean signed distance field
CAM	Cell adhesion model
GRN-CAM	Gene regulation network-cell adhesion model
H-GRN	Hierarchical GRN
EH-GRN	Enhanced H-GRN
C-GRN	Collaborative GRN
TH-GRN	Tracking-based H-GRN
VG-Swarm	A vision-based gene regulation network For UAVs swarm behavior emergence
PSO	Particle swarm optimization
ACO	Ant colony optimization
DE	Differential evolution
GA	Genetical algorithm
UE4	Unreal engine 4
KF	Kinodynamic GRN field
NKF	GRN Field Without Kinodynamic

References

1. Li, H.; Cai, Y.; Hong, J.; Xu, P.; Cheng, H.; Zhu, X.; Hu, B.; Hao, Z.; Fan, Z. VG-Swarm: A Vision-Based Gene Regulation Network for UAVs Swarm Behavior Emergence. *IEEE Robot. Autom. Lett.* **2023**, *8*, 1175–1182. [CrossRef]
2. Guo, H.; Meng, Y.; Jin, Y. Self-adaptive multi-robot construction using gene regulatory networks. In Proceedings of the 2009 IEEE Symposium on Artificial Life, Nashville, TN, USA, 3 March–2 April 2009; pp. 53–60.

3. Guo, H.; Jin, Y.; Meng, Y. A morphogenetic framework for self-organized multirobot pattern formation and boundary coverage. *ACM Trans. Auton. Adapt. Syst. (TAAS)* **2012**, *7*, 15. [CrossRef]
4. Taylor, T.; Ottery, P.; Hallam, J. *Pattern Formation for Multi-Robot Applications: Robust, Self-Repairing Systems Inspired by Genetic Regulatory Networks and Cellular Self-Organisation*; Tech. Rep. EDI-INF-RR-0971; University of Edinburgh: Edinburgh, UK, 2007.
5. Guo, H.; Meng, Y.; Jin, Y. A cellular mechanism for multi-robot construction via evolutionary multi-objective optimization of a gene regulatory network. *BioSystems* **2009**, *98*, 193–203. [CrossRef] [PubMed]
6. Jin, Y.; Guo, H.; Meng, Y. Robustness analysis and failure recovery of a bio-inspired self-organizing multi-robot system. In Proceedings of the 2009 Third IEEE International Conference on Self-Adaptive and Self-Organizing Systems, San Francisco, CA, USA, 14–18 September 2009; pp. 154–164.
7. Guo, H.; Meng, Y.; Jin, Y. Swarm robot pattern formation using a morphogenetic multi-cellular based self-organizing algorithm. In Proceedings of the 2011 IEEE International Conference on Robotics and Automation, Shanghai, China, 9–13 May 2011; pp. 3205–3210.
8. Jin, Y.; Guo, H.; Meng, Y. A hierarchical gene regulatory network for adaptive multirobot pattern formation. *IEEE Trans. Syst. Man, Cybern. Part B* **2012**, *42*, 805–816.
9. Oh, H.; Jin, Y. Adaptive swarm robot region coverage using gene regulatory networks. In Proceedings of the Advances in Autonomous Robotics Systems: 15th Annual Conference, TAROS 2014, Birmingham, UK, 1–3 September 2014; Proceedings 15; Springer: Berlin/Heidelberg, Germany, 2014; pp. 197–208.
10. Oh, H.; Jin, Y. Evolving hierarchical gene regulatory networks for morphogenetic pattern formation of swarm robots. In Proceedings of the 2014 IEEE Congress on Evolutionary Computation (CEC), Beijing, China, 6–11 July 2014; pp. 776–783.
11. Meng, Y.; Guo, H. Evolving network motifs based morphogenetic approach for self-organizing robotic swarms. In Proceedings of the 14th Annual Conference on Genetic and Evolutionary Computation, Philadelphia, PA, USA, 7–11 July 2012; pp. 137–144.
12. Wu, M.; Zhou, Y.; Zhu, X.; Ma, L.; Yuan, Y.; Fang, T.; Wang, J.; Bao, W.; Fan, Z. Cooperation-based gene regulatory network for target entrapment. In Proceedings of the Advances in Swarm Intelligence: 10th International Conference, ICSI 2019, Chiang Mai, Thailand, 26–30 July 2019; Proceedings, Part I 10; Springer: Berlin/Heidelberg, Germany, 2019; pp. 60–69.
13. Yuan, Y.; Fan, Z.; Zhu, X.; Wu, M.; Ma, L.; Fang, T.; Wang, Z.; Bao, W.; Zhou, Y.; Chen, H.; et al. Th-grn model based collective tracking in confined environment. In Proceedings of the Advances in Swarm Intelligence: 10th International Conference, ICSI 2019, Chiang Mai, Thailand, 26–30 July 2019; Proceedings, Part II 10; Springer: Berlin/Heidelberg, Germany, 2019; pp. 33–43.
14. Fan, Z.; Wang, Z.; Li, W.; Zhu, X.; Hu, B.; Zou, A.M.; Bao, W.; Gu, M.; Hao, Z.; Jin, Y. Automated pattern generation for swarm robots using constrained multi-objective genetic programming. *Swarm Evol. Comput.* **2023**, *81*, 101337. [CrossRef]
15. Borges, C.D.B.; Almeida, A.M.A.; Júnior, I.C.P.; Junior, J.J.d.M.S. A strategy and evaluation method for ground global path planning based on aerial images. *Expert Syst. Appl.* **2019**, *137*, 232–252. [CrossRef]
16. Peng, Y.; Li, S.W.; Hu, Z.Z. A self-learning dynamic path planning method for evacuation in large public buildings based on neural networks. *Neurocomputing* **2019**, *365*, 71–85. [CrossRef]
17. Huang, Y.; Chen, J.; Wang, H.; Su, G. A method of 3D path planning for solar-powered UAV with fixed target and solar tracking. *Aerosp. Sci. Technol.* **2019**, *92*, 831–838. [CrossRef]
18. Saha, O.; Dasgupta, P.; Woosley, B. Real-time robot path planning from simple to complex obstacle patterns via transfer learning of options. *Auton. Robot.* **2019**, *43*, 2071–2093. [CrossRef]
19. Phung, M.D.; Quach, C.H.; Dinh, T.H.; Ha, Q. Enhanced discrete particle swarm optimization path planning for UAV vision-based surface inspection. *Autom. Constr.* **2017**, *81*, 25–33. [CrossRef]
20. Dorigo, M.; Maniezzo, V.; Colorni, A. Ant system: Optimization by a colony of cooperating agents. *IEEE Trans. Syst. Man Cybern. Part B* **1996**, *26*, 29–41. [CrossRef] [PubMed]
21. Sun, Z.; Wu, J.; Yang, J.; Huang, Y.; Li, C.; Li, D. Path planning for GEO-UAV bistatic SAR using constrained adaptive multiobjective differential evolution. *IEEE Trans. Geosci. Remote Sens.* **2016**, *54*, 6444–6457. [CrossRef]
22. Patle, B.; Parhi, D.; Jagadeesh, A.; Kashyap, S.K. Matrix-Binary Codes based Genetic Algorithm for path planning of mobile robot. *Comput. Electr. Eng.* **2018**, *67*, 708–728. [CrossRef]
23. Ratliff, N.; Zucker, M.; Bagnell, J.A.; Srinivasa, S. CHOMP: Gradient optimization techniques for efficient motion planning. In Proceedings of the 2009 IEEE International Conference on Robotics and Automation, Kobe, Japan, 12–17 May 2009; pp. 489–494.
24. Kalakrishnan, M.; Chitta, S.; Theodorou, E.; Pastor, P.; Schaal, S. STOMP: Stochastic trajectory optimization for motion planning. In Proceedings of the 2011 IEEE International Conference on Robotics and Automation, Shanghai, China, 9–13 May 2011, pp. 4569–4574.
25. Oleynikova, H.; Burri, M.; Taylor, Z.; Nieto, J.; Siegwart, R.; Galceran, E. Continuous-time trajectory optimization for online uav replanning. In Proceedings of the 2016 IEEE/RSJ International Conference on Intelligent Robots and Systems (IROS), Daejeon, Republic of Korea, 9–14 October 2016; pp. 5332–5339.
26. Gao, F.; Lin, Y.; Shen, S. Gradient-based online safe trajectory generation for quadrotor flight in complex environments. In Proceedings of the 2017 IEEE/RSJ International Conference on Intelligent Robots and Systems (IROS), Vancouver, BC, Canada, 24–28 September 2017; pp. 3681–3688.
27. Ding, W.; Gao, W.; Wang, K.; Shen, S. An efficient b-spline-based kinodynamic replanning framework for quadrotors. *IEEE Trans. Robot.* **2019**, *35*, 1287–1306. [CrossRef]

28. Zhou, B.; Pan, J.; Gao, F.; Shen, S. Raptor: Robust and perception-aware trajectory replanning for quadrotor fast flight. *IEEE Trans. Robot.* **2021**, *37*, 1992–2009. [CrossRef]
29. Oleynikova, H.; Taylor, Z.; Fehr, M.; Siegwart, R.; Nieto, J. Voxblox: Incremental 3d euclidean signed distance fields for on-board mav planning. In Proceedings of the 2017 IEEE/RSJ International Conference on Intelligent Robots and Systems (IROS), Vancouver, BC, Canada, 24–28 September 2017; pp. 1366–1373.
30. Han, L.; Gao, F.; Zhou, B.; Shen, S. Fiesta: Fast incremental euclidean distance fields for online motion planning of aerial robots. In Proceedings of the 2019 IEEE/RSJ International Conference on Intelligent Robots and Systems (IROS), Macau, China, 3–8 November 2019; pp. 4423–4430.
31. Zhou, X.; Wang, Z.; Ye, H.; Xu, C.; Gao, F. Ego-planner: An esdf-free gradient-based local planner for quadrotors. *IEEE Robot. Autom. Lett.* **2020**, *6*, 478–485. [CrossRef]

sensors

MDPI

Article

An Improved Genetic Algorithm for the Recovery System of USVs Based on Stern Ramp Considering the Influence of Currents

Lulu Zhou [1], Xiaoming Ye [1,*], Zehao Huang [2], Pengzhan Xie [1], Zhenguo Song [2] and Yanjia Tong [1]

1 School of Energy and Power Engineering, Huazhong University of Science and Technology, Wuhan 430074, China; zhoululu1225@163.com (L.Z.); xpz564941004@163.com (P.X.); u202010679@hust.edu.cn (Y.T.)
2 China Ship Development and Design Center, Wuhan 430064, China; howardze@163.com (Z.H.); song850624@126.com (Z.S.)
* Correspondence: xmye@hust.edu.cn

Abstract: With the progression of marine exploration and exploitation, as well as the advancements in mechanical intelligence, the utilization of the unmanned surface vehicle (USV) and the design of their guidance system have become prominent areas of focus. However, the stern ramp recovery of the USV is still in its infancy due to its unique attitude requirements and automation design. Furthermore, few studies have addressed the impact of maritime disturbances, with most research limited to simulations. To enhance the efficiency and accuracy of stern ramp recovery, this paper presents the development and construction of a novel recovery system. By incorporating physical modeling of disturbance forces acting on USVs at sea, the practicality of the system is improved. Additionally, an optimized genetic algorithm is introduced in the navigation module to improve convergence rates and subsequently enhance recovery efficiency. A line-of-sight (LOS) algorithm based on average velocity is proposed in this paper to ensure the attainment of unique attitude requirements and to improve the effectiveness of stern chute recovery. This paper provides a detailed description of the independently designed USV hardware system. Moreover, simulations and practical experiments conducted using this experimental platform are presented, offering a new solution for the USV's stern ramp recovery.

Keywords: the unmanned surface vehicle; path planning; genetic algorithm; line of sight method; autonomous guidance and recovery; stern ramp recovery

Citation: Zhou, L.; Ye, X.; Huang, Z.; Xie, P.; Song, Z.; Tong, Y. An Improved Genetic Algorithm for the Recovery System of USVs Based on Stern Ramp Considering the Influence of Currents. *Sensors* **2023**, *23*, 8075. https://doi.org/10.3390/s23198075

Academic Editor: Enrico Meli

Received: 9 August 2023
Revised: 15 September 2023
Accepted: 21 September 2023
Published: 25 September 2023

1. Introduction

If you ask chatGPT how much of the ocean has yet to be explored by mankind, it will tell you that over 80% of the ocean remains unmapped, unobserved, and unexplored. To be honest, we have no idea exactly how much of the ocean is uncharted by humanity. But we cannot deny that there has been an increasing enthusiasm and confidence among researchers in the development of ocean exploration technologies in recent years. Driven by oceanographic research and other marine equipment requirements, the unmanned surface vehicle (USV) has gained worldwide attention for its remarkable autonomy and mission assistance capabilities.

Compared to other vessels, the USV has the advantage of being smaller and operating more agilely, requiring only sufficient space for the relevant sensors and auxiliary navigation devices to make sure the vehicle functions properly. Moreover, the USV can be applied in extreme conditions independently, such as strong waves, tides, and radiation leaks, by means of a predefined system or remote control by professionals; thus, the safety of the operators can be ensured. Due to a variety of these aforementioned advantages, the USV has been well-accepted in both civilian and military fields in recent years [1–4], as shown in Figure 1.

(a) (b)

Figure 1. (**a**) Unmanned Surface Vehicle (USV) for civilian use; (**b**) USV for military use [5–12].

Unlike maritime navigation, the recovery situation of the USV does not need to consider the International Regulations for Preventing Collisions at Sea (COLREGS) for its immediate environment, close to the stern ramp of its large target ship, without the interference of other vessels. The intelligence of the stowage and release process of the USV will have a direct impact on the efficiency of its mission performance. It is mainly divided into hoist-based recovery, well-deck recovery, and stern ramp recovery. Examples are shown in Figure 2. The hoist-based recovery method shows advanced development with minimal modifications to the mother ship's hull, but recovery speed is highly affected by water surface conditions, leading to challenging bracket alignment and extended retrieval times. In contrast, the well-deck recovery technique accommodates complex water surface conditions and achieves rapid retrieval speeds but requires greater structural demands on the mother ship and the inclusion of a spacious docking bay.

(a) (b) (c)

Figure 2. (**a**) Hoist-based recovery requires the installation of davits, parking racks, and towing devices on board the mother ship. To complete the unhooking operation, manual assistance is needed [13]. (**b**) The recovery enables small boat stowage on racks in a well-deck, and once the water reaches a certain depth, the release mechanism activates for deployment [14]. (**c**) The stern ramp of a ship [15].

The stern ramp recovery technology exhibits superior overall performance in terms of the mother ship's hull structure requirements, adaptability to water surface conditions, and stowage efficiency, making it highly suitable for the rapid autonomous recovery of USVs. This method not only allows the mother ship to retract and release small boats at higher speeds, but also adapts to sea conditions of level 6. This advantage allows the mother ship to choose the appropriate timing for flushing. The primary operating principle of the stern ramp recovery system is that during recovery, the USV navigates a predetermined trajectory toward the mother ship's aft section. During this approach, the USV continuously adjusts its heading angle to ensure alignment with the centerline of the mother ship's stern ramp. In addition, it maintains a specified distance from the aft section of the mother ship while staying on course. However, achieving precise centering and control near the stern ramp remains challenging. The low dependency on manual assistance from the mothership during the recycling process results in high autonomous navigation accuracy and strong resistance to wake interference from the USV. This places higher demands on the convergence rate of the USV path planning algorithm and the accuracy of the recovery system. The development of autonomous recovery technology for USVs is still in its nascent stages.

Due to the special requirements of stern ramp recovery for the path and USV's attitude, the question of which path planning algorithm to choose is particularly important for the establishment of the recovery system. We hope that factors such as resistance to ocean current disturbance, smooth motion, accurate heading angle, and reasonable dynamic parameters of the USV are comprehensively considered in the proposed recycling system. Therefore, the algorithm used should have a multidimensional evaluation system rather than a single optimization method. In comparison with other intelligent algorithms, the genetic algorithm (GA) has significant advantages due to its adaptive evaluation indicators. Also, the genetic algorithm does not fall into the local optimum problem as easily as gradient-based optimization algorithms. For this reason, we chose this algorithm and optimized it to be combined with the recycling system. Furthermore, a suitable control system for the dynamics of the USV is imperative from a hardware perspective. To enhance the realism of the simulation in a maritime environment, we also conducted modeling of the interference environment and incorporated it into the tracking algorithm. Our main contributions are summarized as follows.

- We conducted physical modeling of currents and waves to analyze the impact of USV movement in a water environment from a force perspective. Subsequently, we applied these findings to our tracking algorithm in order to enhance the accuracy and effectiveness of the tracking process.
- To meet the specific requirements of stern ramp recovery, we enhanced the genetic algorithm by incorporating three-dimensional modeling and genetic manipulation techniques. This resulted in an improved convergence rate of the algorithm. Additionally, we designed a multi-dimensional evaluation index fitness function to streamline the screening mechanism.
- We improved the tracking algorithm and introduced a new tracking algorithm specifically designed for average speed based on line of sight (LOS). This exclusive algorithm ensures the effectiveness of waypoint tracking.
- We independently constructed the experimental platform and devised a hardware connection system within the USV. This system serves as a solution for implementing the algorithm in the actual operation of the USV.

2. Related Works

Based on the detailed introduction of the recycling method provided in the previous section, this section will primarily focus on presenting the relevant methods employed by scholars for unmanned boat path planning. Furthermore, we will elucidate the rationale behind our selection of the genetic algorithm and LOS as the research methodology.

2.1. The Planning Algorithm

Currently, path planning algorithms used for USVs in domestic and international research can be classified into four main categories: graph search algorithms, virtual potential field methods, random sampling algorithms, and intelligent algorithms.

Graph search algorithms are considered classical path planning algorithms; however, their usage has declined in recent years due to the slow computational speed of individual underlying logic. Among these algorithms, the A* algorithm has garnered attention for its remarkable scalability. Singh et al. [16] considered moving ships as quasi-static entities and other ships as static obstacles during the current planning time for map modeling. They used the A* algorithm to search for collision avoidance paths by incorporating the notion of ship safety. Simulation results demonstrated the effectiveness of the algorithm in effectively avoiding moving ships while achieving commendable real-time performance.

The basic concept of the virtual potential field (VPF) method is to construct a virtual potential field within the map using certain techniques. The path is then generated using the gradient descent algorithm. However, this method suffers from inherent limitations such as local minima and oscillations. Kim et al. [17] integrated the method with the velocity obstacle method by incorporating a repulsive force field associated with the encountered velocity obstacle between two ships into the classical potential field. This fusion approach enabled dynamic collision avoidance for USVs. Similarly, Sang et al. [18] used a hierarchical programming approach. They first used the A* algorithm to generate an initial path and then used an improved artificial potential field (APF) method to generate the desired USV formation.

Random sampling methods exhibit varying convergence rates, making them extensively employed in the context of Multi-Agent Path Finding (MAPF) due to their inherent efficacy. Lee et al. [19] introduced a novel approach called the grafting RRT algorithm to achieve dynamic collision avoidance in USVs which uses the speed obstacle method to detect potential collisions and calculates the grafting angle. By generating and inserting grafting points based on this angle, successful dynamic collision avoidance is achieved.

In terms of obstacle avoidance, Khan et al. introduce a penalty term in the objective function, so that the tracking control and obstacle avoidance problems are unified into a constrained optimization problem, and active rewards are given to obstacle evaders [20]. As objectives become more complex, intelligent algorithms, such as the genetic algorithm, excel in path planning. It utilizes genetic manipulations like mutation and crossover to enhance diversity in the sample population, enabling efficient multi-target search. However, the traditional genetic algorithm exhibits low convergence rates in USV path planning and heavily relies on well-designed genetic operators. To address these limitations, Wang et al. [21] propose a combination of GA and fuzzy APF for hierarchical path planning, effectively adapting to unpredictable environments. Nonetheless, challenges persist in handling time-varying dynamic obstacles. Xin et al. [22] suggest mitigating these issues by increasing the number of superior offspring through multi-domain inversion and second fitness evaluation. Nevertheless, this approach introduces additional complexities to the algorithm and encoding process.

2.2. The Tracking Algorithm

The purpose of tracking control is to manipulate the propulsion system, such as the propeller and rudder, based on a rational tracking control law. This enables the USV to navigate along the intended trajectory determined by the path planning module. Depending on whether the trajectory includes a time dimension, the problem can be divided into two categories: path tracking and trajectory tracking. The stern ramp recovery involves dynamic obstacle avoidance and requires a higher speed than the mother ship during the slope flushing phase. Therefore, it is a trajectory tracking problem.

Pettersen [23] used the LOS algorithm to compute the desired heading angle and combined it with a cascaded feedback controller to control the yaw torque. This approach achieved linear tracking of the USV at a constant velocity. Fossen [24], using kinematic models for both USV and Unmanned Aerial Vehicle (UAV), rigorously proved the unified semi-global exponential stability of path tracking under LOS control. This contribution enriched the theoretical foundation of the LOS guidance law.

To improve the tracking stability and accuracy of the USV under environmental disturbances such as wind, waves, and currents, Caharija [25] introduced an integral term into the classical LOS guidance. This term compensated for the lateral drift of the USV and mitigated tracking biases caused by environmental disturbances. Building on this work, Fossen [26] used a nonlinear adaptive controller to achieve two-dimensional Dubins curve path tracking. Simulation results demonstrated the successful application of this tracking control method in accurately tracking the USV even under significant drift angles induced by wind, waves, and power disturbances.

Liu [27] proposed an improved LOS tracking algorithm based on prediction. This algorithm was designed for tracking underactuated USVs, effectively compensating the uncontrollable sideslip caused by marine environmental disturbances through adaptive terms. In addition, this study cascaded the tracking error system with the prediction error system and rigorously demonstrated the consistent global stability of the entire system.

3. Recovery System of USVs Based on Stern Ramp

Due to the particular environment of the autonomous recovery using a stern ramp, most of the recovery strategies have less concern about interference caused by currents as well as the effects of malposition on recovery effects. For a successful ramp approach, it is essential to have sufficient power and to approach the ramp at an optimal angle. This combination guarantees a seamless and proficient transition onto the ramp surface, facilitating a smooth and efficient docking process. To ensure the safety and efficiency of the recovery process, we have developed an innovative stern ramp recovery system that prioritizes three key factors: precise visibility, seamless guidance, and stable navigation. The recovery system structure is shown in Figure 3.

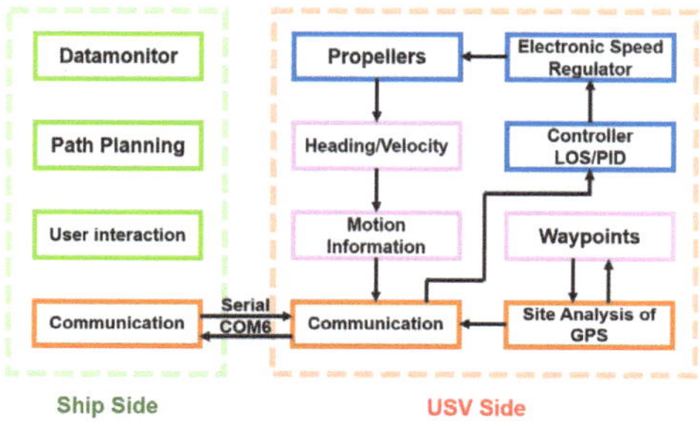

Figure 3. The recovery system structure of our USV.

Enter the initial parameters for the planning algorithm into the upper computer, wait for it to complete the calculations, and then package and transmit the set of waypoint information generated by the genetic algorithm to the USV end. This set of information includes the coordinates, real-time velocities, and headings of each waypoint. Upon receiving the information, the PID controller actuates the electronic speed regulators to regulate the propellers, thereby adjusting the speed and heading of the USV. Meanwhile, the LOS controller accurately tracks the waypoint to prevent interference such as currents from causing the USV to deviate from the planned route. In the meantime, the actual position information of the USV obtained by the Global Positioning System (GPS) will also be sent to the upper computer together with the current speed and the heading.

3.1. *Precise Visibility Establishment of a Current Interference Model for Unmanned Surface Vehicles*

The recovery process of the USV is affected by external environmental disturbances, which have a significant impact on the speed and stability of the centering procedure. Therefore, it is essential to develop a suitable mathematical model to account for these external disturbances. Factors such as sea currents and waves particularly affect the autonomous guided recovery motion.

The motion of the USV can be summarized into 6° of freedom, and in order to accurately describe the motion state of the USV, a combination of a geodetic coordinate system and follower coordinate system is used to describe the position and attitude information, as shown in Figure 4.

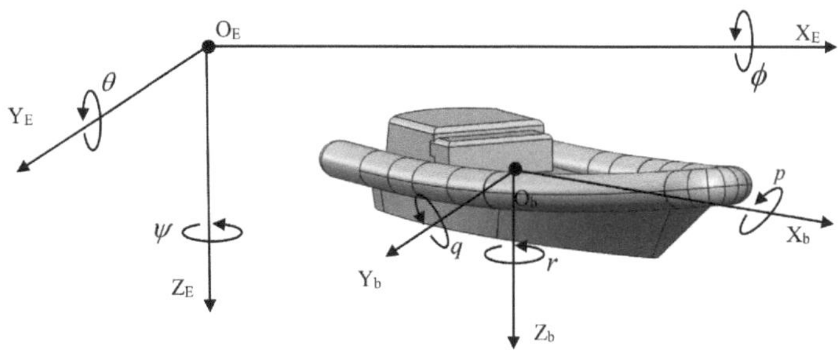

Figure 4. The geodetic coordinate system and the follower coordinate system.

Assuming that the center of gravity of the USV aligns with the center of the follower coordinate system, we can deduce the equation of motion governing the forces exerted on the USV during its movement. This equation is derived from the Newton–Euler equations for rigid body motion in a fluid medium and is given as Equation (1).

$$\begin{cases} m\dot{u} - mvr = F_x \\ m\dot{v} - mvr = F_y \\ I_G\dot{r} = F_n \end{cases} \tag{1}$$

where m denotes the quality of the USV, v and u present the transverse and longitudinal velocities, r is the bow-swinging angular velocity, I_G is the moment of inertia of the USV on the z-axis, and F_x, F_y, and F_n are the force in the x-direction and y-direction and the moment.

3.1.1. Current Interference Model

The disturbance force and disturbance moment experienced by the USV due to sea currents can be attributed to two main components: (1) viscous resistance, arising from the friction and pressure difference between the USV and the surrounding fluid; (2) inertial resistance, resulting from the circulation patterns near the USV and the influence of the free liquid surface. The equations are given by Equation (1).

$$\begin{cases} F_x^{current} = \frac{1}{2}\rho S_f V_c^2 C_x(\beta) \\ F_y^{current} = \frac{1}{2}\rho S_r V_c^2 C_y(\beta) \\ F_n^{current} = \frac{1}{2}\rho S_r V_c^2 C_n(\beta) \end{cases} \tag{2}$$

where V_c is the flow velocity; S_f is the forward projected area of the USV above the waterline; S_r is the side projected area of the USV above the waterline; β is the encounter angle; $C_x(\beta)$, $C_y(\beta)$, $C_n(\beta)$ are the test coefficients.

3.1.2. Wave Interference Model

The wave force can be categorized into two components: the first-order wave force and the second-order wave force. The former induces vertical oscillation of the USV with a fixed period, without altering its total mechanical energy. On the other hand, the latter causes the USV to experience time-varying horizontal thrust, resulting in horizontal displacement [28], which influences the autonomous guided recovery process. Hence, we disregard the impact of the first-order wave force and focus our analysis on the effect of the second-order wave force on the USV during its autonomous guided recovery.

During the recovery process, the wave force primarily originates from the wind and the wake of the mother ship. The drift force model proposed by Daidola [29] enables us to derive the wave force from the wave model, as depicted in Equation (3):

$$\begin{cases} F_x^{wave} = \frac{1}{2}\rho L\xi^2 \cos\chi W_x(\lambda_w) \\ F_y^{wave} = \frac{1}{2}\rho L\xi^2 \sin\chi W_y(\lambda_w) \\ F_n^{wave} = \frac{1}{2}\rho L\xi^2 \sin\chi W_n(\lambda_w) \end{cases} \quad (3)$$

where ξ is the average wave amplitude; χ is the encounter angle; λ_w is the wave wavelength; $W_x(\lambda_w)$ and $W_y(\lambda_w)$ are the wave drift coefficients; $W_n(\lambda_w)$ is the drift moment coefficient around the z-axis. The specific calculations are shown in Equation (4).

$$\begin{cases} \xi = h_w/2 \\ W_x(\lambda_w) = 0.05 - 0.2(\frac{\lambda_w}{L}) + 0.75(\frac{\lambda_w}{L})^2 - 0.51(\frac{\lambda_w}{L})^3 \\ W_y(\lambda_w) = 0.46 + 6.83(\frac{\lambda_w}{L}) - 15.65(\frac{\lambda_w}{L})^2 + 8.44(\frac{\lambda_w}{L})^3 \\ W_n(\lambda_w) = -0.11 + 0.68(\frac{\lambda_w}{L}) - 0.79(\frac{\lambda_w}{L})^2 + 0.21(\frac{\lambda_w}{L})^3 \end{cases} \quad (4)$$

where h_w is the wave height. The relationship between the height of the wave h_w, the wave period T_{wave}, the wavelength λ_w, and the wind speed V_R is shown in Equation (5).

$$\begin{cases} h_w = 0.015V_R^2 + 1.5 \\ T_{wave} = -0.001V_R^3 + 0.042V_R^2 + 5.6 \\ \lambda_w = 0.54T_{wave}^2 \end{cases} \quad (5)$$

3.2. *Seamless Guidance Path Planning Algorithm for Stern Ramp Recovery*

The quality, efficiency, and convergence of path planning algorithms are crucial factors to consider. The genetic algorithm, known for its intelligent nature and multi-evaluation index mechanism, offers significant advantages in global path planning. However, traditional genetic algorithms rely on two-dimensional raster graphs, which suffer from time-consuming computations, unstable algorithmic architectures, and non-uniqueness in convergence rate performance. To address these limitations, we propose a novel approach that combines the characteristics and requirements of stern chute recovery, focusing on optimizing waypoint distribution, improving population initialization, and utilizing advanced genetic methods to establish a more suitable path planning algorithm.

3.2.1. Novel Collision Avoidance Based on Three-Dimensional Modeling

Considering the wide range of actual distances in latitude and longitude recorded by GPS, it becomes necessary to perform additional processing on the coordinates. In our simulation process and program design, we use the following coordinate information transformation formula for actuation:

$$\begin{cases} X_{USV} = (X_{GPS} - X_0)/100{,}000 \\ Y_{USV} = (Y_{GPS} - Y_0)/100{,}000 \end{cases} \quad (6)$$

where X_{USV} and Y_{USV} denote the current location of the USV, X_{GPS} and Y_{GPS} present the information GPS received, and X_0 as well as Y_0 indicate the position receiver placed.

It should be noted that there is a trade-off between the computational time and the resolution [30]; if the resolution is higher, a longer computational time will be required. However, while streamlining environment modeling can improve algorithm efficiency and convergence rate, as well as reduce spatiotemporal complexity, it also introduces potential drawbacks. These include sharper corners and excessively long sections in the planned path, resulting in rougher paths that hinder smooth evolution. Therefore, it is crucial to establish a robust algorithmic operation mechanism based on a well-designed pre-processing method for the modeling environment. This ensures the practical feasibility of the planning path.

Considering the sparsely distributed obstacles on the water surface, we have opted to use circular bounding boxes to represent these obstacles, as illustrated in Figure 5a. The outer circle of the actual obstacle boundary is used as the obstacle in the planning space, aligning with the cornering motion dynamics of the USV in a water environment. This approach significantly reduces memory consumption associated with environment modeling, as it only requires storing a few essential data points such as center coordinates, radius, and movement speed for each obstacle. Taking into account the shifting caused by interference and the error of being labeled as a particle, we add the compensation radius to the boxes. The equation is given by Equation (7).

$$R_l = R_{obs} + D_{USV} + R_c \tag{7}$$

where R_l denotes the minimum radius of collision avoidance distance from the USV to the center of the circular bounding box, R_{obs} represents the initial radius of the box, and the whole compensation radius is the length of the hull D_{USV} plus the possible displacement R_c caused by environmental interference, which is usually taken as 1 m.

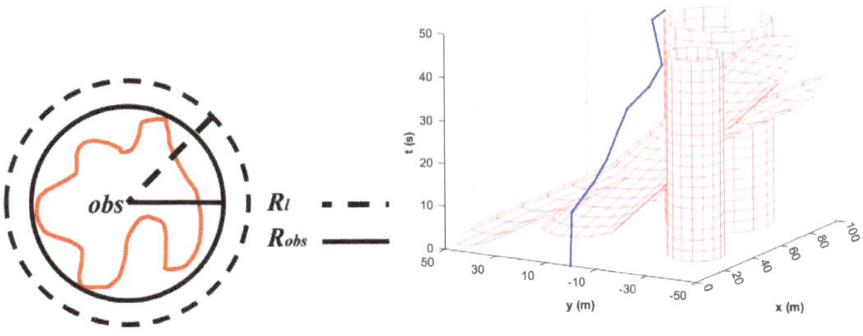

(**a**) (**b**)

Figure 5. (**a**) The circular bounding boxes. (**b**) The planning path in 3D coordinate system. The blue line presents the planning path.

Upon introducing the time dimension into the modeling, the bounding boxes of the obstacles are represented as oblique or straight cylinders in the x-y-t coordinate system. The expression of the three-dimensional geometry is depicted in Equation (8).

$$\begin{cases} (X - X_{obs} + V_x t)^2 + (Y - Y_{obs} + V_y t)^2 = R_{obs}^2 \\ V_x = 0\ \mathrm{m/s}\ \textit{when static obstacle} \\ V_y = 0\ \mathrm{m/s}\ \textit{when static obstacle} \end{cases} \tag{8}$$

where X_{obs} and Y_{obs} denote the abscissa and ordinate of the center of the circular bounding boxes, R_{obs} represents its radius, and V_x and V_y denote the velocity in the x-direction and velocity in the y-direction.

In order to simplify the mathematical modeling of obstacle avoidance, we made the following assumptions.

- Dynamic obstacles move in uniform linear motion.
- The USV performs segmented uniform acceleration linear motion, with each consecutive pair of waypoints forming a segment.

By solving for temporal intersections between the motion trajectory of the USV and the movement trajectory of the obstacles, we can determine whether the proposed planning will result in a collision. The relevant equation is given by Equation (9).

$$
\begin{aligned}
&\left[\left(\overline{V_{Uy}}+V_{Oy}\right)^2+\left(\overline{V_{Ux}}+V_{Ox}\right)^2\right]T^2 \\
&+2\left[\left(\overline{V_{Uy}}+V_{Oy}\right)\left(Y_{usv}-Y_{obs}\right)+\left(\overline{V_{Ux}}+V_{Ox}\right)\left(X_{usv}-X_{obs}\right)\right]T \\
&+\left[\left(X_{usv}-X_{obs}\right)^2+\left(Y_{usv}-Y_{obs}\right)^2-R_{obs}^2\right]=0
\end{aligned} \tag{9}
$$

where $\overline{V_{Ux}}$ and $\overline{V_{Uy}}$ are the average velocities of the certain segment, and V_{Ox} and V_{Oy} are the velocities of the obstacle bounding boxes.

Given the particularity of stern ramp recovery, we adopt the horizontal line representing the stern ramp as the x-axis in a 2D plane. This division separates the water into two symmetrical parts, one above and one below the stern ramp. The distribution of obstacles and the performance of the planned path are shown in Figure 5b.

3.2.2. Improvements in Genetic Manipulation

In this paper, the genetic algorithm selects the population as the solution set for the recovery paths. Each path is represented by a distinct genetic individual. These individuals encompass three-dimensional coordinate information of multiple waypoints, storing the X, Y, and T coordinates along with real-time speed. The enhanced global planning algorithm autonomously determines the number of gene loci based on the input start and end points. This effectively enhances the computational capability of the genetic algorithm. Rather than generating the waypoints randomly, we designate the X-axis direction as the step direction and establish the range of intervals for each waypoint based on the step size D_{SL}. The length of D_{SL} varies according to different scenarios and accuracy requirements.

After that, the waypoints will then be generated in each interval in turn. The code conversion between binary and decimal is shown below.

$$
\begin{cases}
X_{tr}=X_i-D_{SL}\cdot(i-1)-1 \\
Y_{tr}=Y_i+64 \\
V_{tr}=V_i\cdot 5
\end{cases} \tag{10}
$$

where X_{tr} and X_i denote the abscissa of binary system and decimal system. Ordinate and velocity are presented in a similar way.

$$
\begin{cases}
X_i=\sum\limits_{n=1}^{3} m\cdot 2^{(n-1)}+1 \\
Y_i=\sum\limits_{n=4}^{10} m\cdot 2^{(n-1)}-64 \\
V_i=\sum\limits_{n=11}^{14} \frac{m\cdot 2^{(n-1)}}{5} \\
T_i=\sum\limits_{n=1}^{N-1} \frac{\sqrt{(X_{n+1}-X_n)^2+(Y_{n+1}-Y_n)^2}\cdot 2}{V_i+V_{i+1}}
\end{cases} \tag{11}
$$

where n represents the number of bits in the binary sequence and N represents the total number of waypoints. An example of the encoding is shown in Figure 6.

Decimal		Binary
Point1: (85,10,0)		Point1: 10010010100000
Point2: (76, 6,2.4)		Point2: 00010000000001
Point3: (63, 0,2.6)		Point3: 10010000000011
Point4: (56, 7,1.4)		Point4: 01101101010001
Point5: (44, −3,1.2)	⇨	Point5: 00110000101111
Point6: (35, −6,0.6)		Point6: 00101111111010
Point7: (24, −4,2.6)		Point7: 01110000010011
Point8: (13, 9,1)		Point8: 10001110110010
Point9: (6, −5,1.8)		Point9: 11110000011011
Point10:(0,0,0)		Point10: 00010000000000

Figure 6. The translation process of the encoding.

The binary sequence serves as the fundamental basis for genetic algorithm mutation. However, a common issue arises with value overflow subsequent to decoding. Consequently, a mapping procedure is introduced to the binary sequence, involving binary conversion, decoding, and subsequent restoration of the original decimal sequence through inverse mapping. This measure effectively addresses the challenge of overflow encountered in the process.

The collision avoidance operations of the traditional genetic algorithm, based on 2D raster maps, consume significant memory resources and require additional time. To address this issue, as mentioned in the previous section, we employed a memory-efficient obstacle design during the 3D modeling process. This approach enables collision avoidance operations during population initialization. As illustrated in Figure 7, this preheating process facilitates the generation of high-quality initial individuals and enhances the convergence rate of the algorithm.

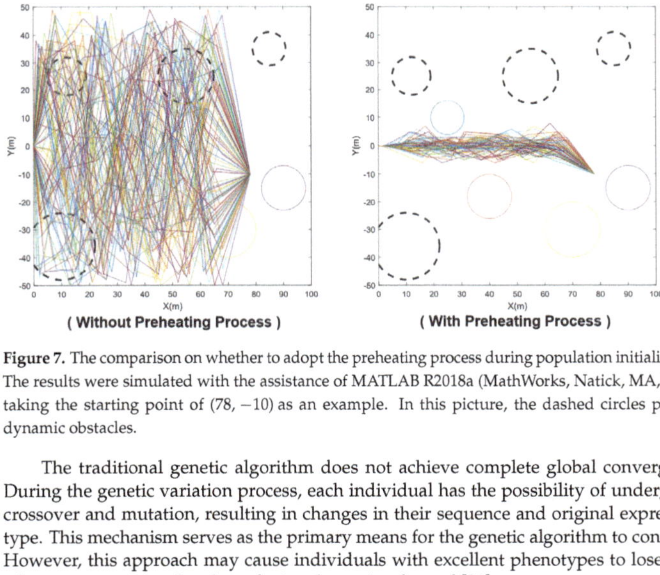

Figure 7. The comparison on whether to adopt the preheating process during population initialization. The results were simulated with the assistance of MATLAB R2018a (MathWorks, Natick, MA, USA), taking the starting point of (78, −10) as an example. In this picture, the dashed circles present dynamic obstacles.

The traditional genetic algorithm does not achieve complete global convergence. During the genetic variation process, each individual has the possibility of undergoing crossover and mutation, resulting in changes in their sequence and original expression type. This mechanism serves as the primary means for the genetic algorithm to converge. However, this approach may cause individuals with excellent phenotypes to lose their advantageous genes, thereby reducing the retrieval speed [31].

Many scholars have demonstrated that the genetic algorithm with Elitism Strategy achieves global convergence [32]. After population initialization, the fitness function of genetic individuals is evaluated, and the superior individuals are preserved in the elite group. This prevents the alteration of advantageous phenotypes during the selection process. In each generation, the individuals in the elite group will be replicated to the next generation, continuously updating the set [33]. Typically, around 10% of the elite sequences are retained. Incorporating this method helps improve the convergence speed of the genetic algorithm and reduces the number of iterations.

Due to the requirements of rollover prevention caused by inappropriate heading and large turns in the surface environment, we add dynamic metrics to the fitness function to evaluate the performance of USVs. In cases where the generated path does not satisfy the constraints of obstacle avoidance and dynamic metrics, the fitness function is assigned a value of 0. By implementing the Elitism Strategy, individuals with superior genetic traits are preserved, thus minimizing the impact of discarding individuals with undesirable traits on the overall population. In addition, the presence of defective individuals during the evolutionary process consumes computational memory resources, resulting in reduced computational efficiency and increased traversal time of the algorithm [34]. To address this issue, this study proposes a novel strategy called the Strategy of Sacrifice and Intraspecific Hybridization (SSIH), as shown in Figure 8. This strategy emulates natural processes observed in biological populations, ensuring both the feasibility of the generated planning path and improving the convergence rate.

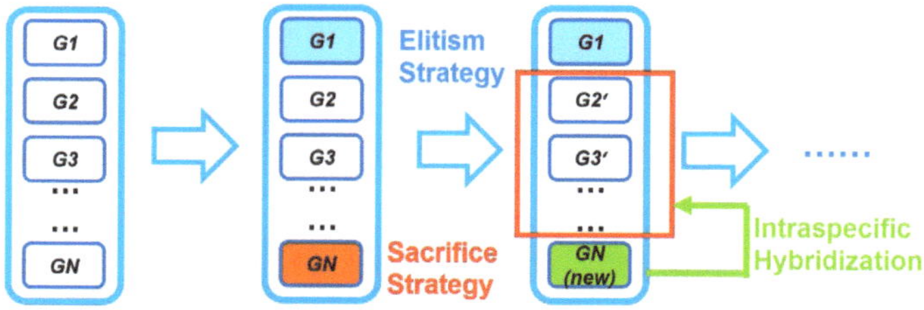

Figure 8. The novel genetic manipulation.

Individuals with a fitness function of 0 are treated as individuals that have died after the population iteration, and their memory is released when the algorithm is implemented. Subsequently, a local initialization scheme is employed to allocate a new path planning to the discarded genetic individuals, considering them as newly generated offspring within the population. In the subsequent genetic mutation operation, these regenerated offspring are hybridized with other genes present in the population. Simulation experiments demonstrate that this operation reduces the convergence time by approximately 10 s and minimizes the number of iterations required, thereby endowing the genetic algorithm with a notable real-time advantage in dynamic obstacle avoidance.

The genetic algorithm is designed to seek optimal solution outcomes. In the event that the quality of the generated path fails to meet the threshold established by the program, the iteration process will persist. In order to uphold real-time planning efficacy, we have imposed a maximum iteration limit of 500. Through extensive simulations and empirical observations, we have determined that this value significantly surpasses the number of iterations necessary to yield results that align with the desired objectives. The pseudo code is shown in Algorithm 1.

Using an ex ante method to analyze, the computational complexity of the proposed genetic algorithm is $O(N^2)$. But the embedded do-while loop is used for population initialization, which has a very small computational footprint. In practical recovery applications, the execution time of the algorithm is about 2 s and the relevant results will be discussed in Section 2.

3.2.3. New Evaluation Mechanism for the Fitness Function

The implementation of a fitness function aims to enhance the initially subtle distinctions among individuals in the population through calibration and dynamic mapping. This process facilitates the clear manifestation of the variations in phenotypic excellence between different individuals and enables a more effective selection of genetically qualified individuals.

However, while penalty functions can be applied to constraints of any type (linear or nonlinear), their performance is not always satisfactory. Achieving constraint optimality through penalty functions relies on the discovery of appropriate penalty parameters. Even so, determining these parameters often necessitates numerous experiments and has posed a challenge for researchers [35]. Khan et al. proposes an algorithm that uses nature-inspired optimization methods in 2020 to directly solve nonlinear optimization problems without the need for any transformations in robot tracking methods [36].

Algorithm 1 Optimized Genetic Algorithm

Input: (X_0, Y_0): start point; $width, length$:the size of map; $(X_{USV}, Y_{USV}, V_{USV})$: the current information of the USV;
Output: getPath(): the planning path which consists of waypoints;

```
set D_L = D_best, set popsize = Num, Fitness_value = F_Max, Nmax;
Foundation(); //generate initial population
m = 0, k = 0;
while m < popsize; do
    i = 1, n = 0;
    while i < genePoint; do
        if isCollision(Point_i,Point_{i-1}) == False then
            put Point_i into the geneList;
        else
            restart point generating;
        end if
        i + +, n + +;
        if n > Nmax then
            break; //setting a threshold to prevent program crashes
        end if
    end while
    m + +;
end while
Fitness_calculating(); //get the Fitness_max;
while Fitness_max < Fitness_value do
    Elit_Strategy(); //implement the Elit Strategy
    Tournament(); //select a portion of individuals with strong adaptability
    TransferBinary(); //binary encoding conversion
    Variation_binary(); //genetic variation
    TransferDecimal(); //decimal encoding conversion
    TimeReset();
    FitnessReset(); //recalculate fitness function
    Intraspecific();
    k++;
    if k >= 500 then
        break; //set maximum convergence generation to prevent too much time cost
    end if
end while
getPath();
```

We propose a set of adaptation functions that incorporate a penalty weight allocation mechanism. A novel nonlinear fitness relation is established by penalizing the total distance, time spent, degree of alignment, and amount of cornering in recovery planning. In addition, this approach combines the constraints related to collision avoidance with those concerning the dynamics of the USV. While we currently cannot overcome this limitation, our approach simplifies this process by introducing various indicators.

Quantitative indicators and fitness can be calculated by the following equation:

$$\begin{cases} Fitness(x) = G_{acc}G_{vel}G_{col}\frac{A}{a\cdot Fit_d + b\cdot Fit_T + c\cdot Fit_a + d\cdot Fit_m} \\ Fit_d = \frac{\sum D_{segment,i}}{\sqrt{(x_{end}-x_{start})^2+(y_{end}-y_{start})^2}} \\ Fit_T = \frac{\sum \Delta T_{segment,i}}{N-1} \\ Fit_{mal} = \frac{\sum_{i=1}^{N} |y_i|}{N} \\ Fit_{ang} = \frac{\sum_{i=1}^{N} (\pi-\theta)}{\pi} \end{cases} \tag{12}$$

where Fit_d, Fit_T, Fit_{mal}, and Fit_{ang} denote the evaluation functions for distance, time cost, malposition, and turning angle, respectively. And G_{acc}, G_{vel}, and G_{col} present the hard kinetic evaluation index of acceleration, velocity, and collision avoidance. a,b,c,d are parameters set by the researcher.

The fitness function mechanism is the key to ensuring good convergence and screening mechanism in the genetic algorithm. We restricted many variables relating to dynamics in the fitness function. To control the motion attitude and smoothness of the vehicle, we optimized the acceleration of each route segment and screened the randomly generated waypoint velocities.

Due to the fact that all waypoint information is packaged through the communication system to the USV end, which is controlled and processed by controllers to optimize path details and tracking effects, optimizing variables such as turns and malposition in path planning algorithms can not only reduce the computational burden of controllers such as LOS, but also facilitate the effective control of the mechanical motion of the propeller by controllers such as PID. The USV's motion is based on controlling the left and right propellers, and smooth propeller motion is conducive to dealing with ocean current disturbances.

3.3. Stable Navigation Trajectory Tracking Control for Stern Ramp Recovery

The trajectory tracking algorithm calculates the heading angle and velocity of the USV based on its current position, velocity state, and desired trajectory curve. This algorithm guides the USV to continuously approach the target trajectory, similar to missile guidance principles. Notable guidance laws for trajectory tracking include LOS, Follow the Carrot (FTC), Constant Bearing (CB), and Proportional Navigation (PN). Among these, the LOS algorithm stands out for its ease of use, reliability, and robustness, which have been extensively validated through research investigations [37]. Therefore, this study focuses on investigating the trajectory tracking algorithm using the LOS method.

As shown in Figure 9a, the classical LOS algorithm operates by determining the target heading for guiding the USV based on the forward viewpoint along the target path.. Initially, the forward way distance R is established. This is accomplished by considering the current position of the USV as the circle's center and setting the radius of the forward way distance R. The intersection point between the circle and the desired path represents the forward viewpoint. Subsequently, the target heading angle is derived from the line connecting the current point to the forward way point. This approach transforms the problem of tracking the desired path into a heading angle control problem. By utilizing the line between the current point and the forward way point, the target heading angle is obtained, thereby resolving the challenge of path tracking through heading angle control. The selection of the forward way distance R significantly impacts the tracking performance. If R is excessively large, rapid correction of track deviations becomes challenging, while an excessively small R may induce track oscillation [38].

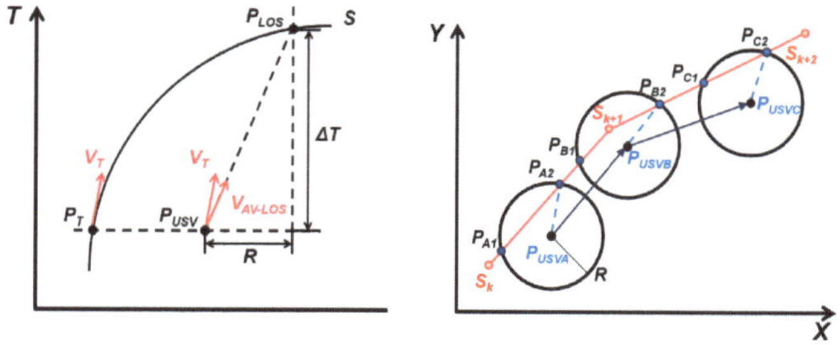

(a) (b)

Figure 9. (**a**) The theory of the traditional los. (**b**) The theory of Average Velocity Line of Sight (AV-LOS).

The current LOS and its improved algorithms are limited to guidance in the spatial domain only, which means that it can only derive a target heading from the forward viewpoint and has a very limited ability to handle speed. This means that it can only derive a target heading from the forward viewpoint, and has a very limited ability to handle speed. For this purpose, we introduce an enhanced algorithm called Average Velocity Line of Sight (AV-LOS).

$$V_{AV-LOS} = \frac{R}{\Delta T} \tag{13}$$

where ΔT denotes the time lag between the current position and the forward viewpoint.

As shown in Figure 9b, the planned trajectory S is projected on the two-dimensional plane (X, Y, Z) and contains the path points S_K, S_{K+1}, S_{K+2}, and the actual trajectory of the USV passes through the point P_{USVA}, P_{USVB}, P_{USVC} in turn, and the forward viewpoints determined by the forward view distance R are P_{A1}, P_{A2}, P_{B1}, P_{B2}, P_{C1}, P_{C2}, and at the beginning moment, two forward viewpoints P_{A1} and P_{A2} are solved by the forward view circle and the segment $S_K - S_{K+1}$. At the beginning moment, two forward viewpoints P_{A1} and P_{A2} are solved from the forward view circle and the line segment $S_K - S_{K+1}$, and at this time, P_{A2}, which is closer to the point S_{K+1}, is selected as the solution forward viewpoint. When the USV moves to P_{USVB}, there is only one forward viewpoint on the track segment $S_K - S_{K+1}$, and at this time, P_{B2} is obtained by incorporating $S_{K+1} - S_{K+2}$ into the solution and is used as the solution forward viewpoint. When the USV is located in P_{USVC}, S_K is out of the team and P_{C2} is selected as the solution forward viewpoint. When the unmanned boat is in P_{USVC}, S_K is out of the group and P_{C2} is chosen as the forward viewpoint of the solution.

In order to enhance the anti-interference capability of the tracking algorithm, we conducted data tests on simulated ocean waves using the aforementioned methods. Wind and wave data for various sea conditions, such as wave height, were obtained from the International Sea State Scale. Since it is not feasible to measure the level of interference caused by waves during actual operation, we could only apply pre-set compensation measures against such interference. Multiple experiments were conducted in the ship pool laboratory, revealing a maximum deviation of 4.94° for the heading angle and 0.16 m for alignment.

After performing calculations based on the formula presented in Section 3.1, the compensated acceleration is incorporated into the AV-LOS algorithm to enhance the anti-interference capability of the recovery system.

4. Experiments

In this section, we will present the hardware model of the stern ramp recovery system along with the underlying framework logic. Additionally, we will showcase the outcomes of simulation and field experiments conducted using the platform. The results will be analyzed using quantitative indicators and graphical representations for a comprehensive evaluation.

4.1. Experimental Conditions and Hardware System

At the mother ship's end, a fixed steel transom chute is installed at the shore to replicate the mother ship for simplicity. The steel transom chute features an upward-sloping U-shaped opening with cylindrical rolling sheaves on the inside, as depicted in Figure 10. The hardware system is constructed on a sturdy fixed steel transom chute structure, which offers ample space and structural stability. Similarly, the hardware system at the mother ship end can be categorized into three sub-modules: computation and control module, sensor module, and power module.

As shown in Figure 11, the boat-side controller described in this paper uses the STM32F103ZET6, and the controller has a wealth of interface resources. We assembled the hardware system, where the hardware system was structured as shown in Figure 12a. The USV used in the experiment with its information is shown in Figure 12b.

To streamline the development process, we opted for Visual Studio 2022 (Microsoft, SEA, Redmond, WA, USA) and utilized the C# language to create a Windows application. The upper computer was developed using the C# language, and its software interface is illustrated in Figure 13.

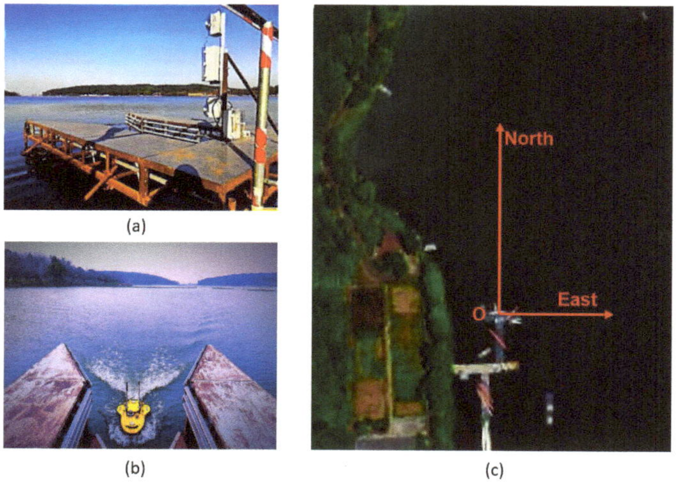

Figure 10. The actual situation of the experimental platform. (**a**) The photo of our platform, which acts as a model of stern ramp. (**b**) The picture records the moment when our USV finishes the recovery process. (**c**) The satellite map of our platform and the directions given are relevant to the coordinate system in the following pictures that we show to describe our experimental results.

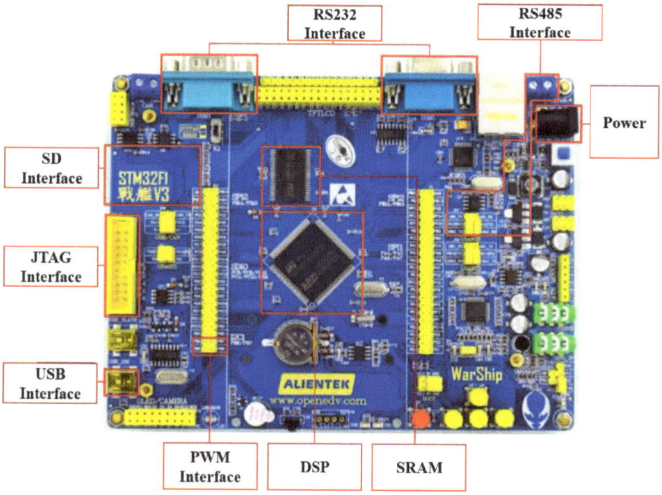

Figure 11. The layout of MCU. Equipment (Guangzhou YuanziDianzi Electronic Technology Co., Ltd, Guangzhou, China) has been modified.

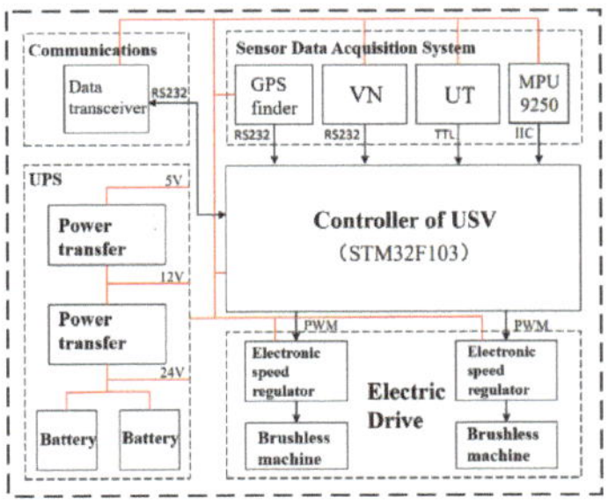

Specific parameters of the USV	Parameter
Volumn	1000 ×650 × 300 mm
Weight	7 kg
Maximum Payload	25 kg
Point-positioning precision	1.5 m in plane, 2.5 m in vertical
Directional accuracy	0.2° (1 m baseline)
Maximum speed	8 m/s
Update period of GPS	150 ms

(a) (b)

Figure 12. (**a**) The framework of the hardware system of our USV. (**b**) The main parameters of our USV. (Shell from Shanghai Huace Navigation Technology Ltd., Shanghai, China).

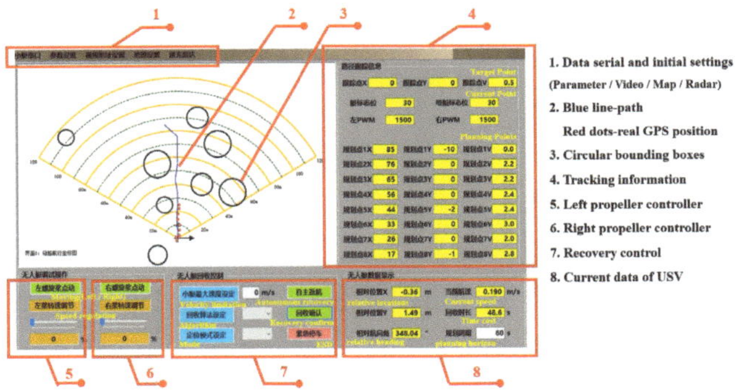

Figure 13. Operation interface of the upper computer.

After constructing the hardware/software system for the recovery system as outlined in the preceding section, a series of tests were conducted in lake waters. The primary objective was to validate the fundamental functionalities of the recovery system. Upon successfully debugging the system's hardware and software components, a dedicated recovery scenario was devised to facilitate a series of recovery tests. These tests aimed to evaluate the effectiveness of the improved genetic algorithm (IGA) path planning algorithm and the AV-LOS tracking algorithm.

4.2. The Simulation of IGA and AV-LOS

By distinguishing between the length of the route plan, the size of the corners, and the number of surrounding obstacles, we designed four arithmetic examples with different starting points, as shown in Table 1.

By incorporating control variables, we have formulated these computational scenarios. The simulation outcomes will be depicted in Figure 14, where EG1 denotes the absence of population initialization enhancement, EG2 signifies the absence of SSIH, EG3 employs a simplistic linear penalty function, and EG4 indicates the lack of an Elitism Strategy.

Table 1. Specific information of planning examples for recovery .

Case	Start Point	End Point
1	(85, 10)	(0, 0)
2	(55, −15)	(0, 0)
3	(43, 6)	(0, 0)
4	(92, 20)	(0, 0)

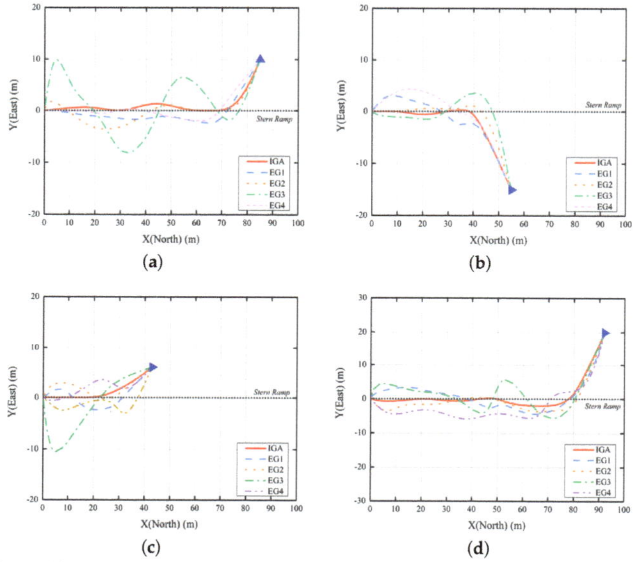

Figure 14. The comparative results of simulation outcomes of four examples. The blue triangle represents the approximate orientation of the starting point towards the USV.(**a**) Case 1; (**b**) Case 2; (**c**) Case 3; (**d**) Case 4.

After generating each set of operating algorithms, we applied curve fitting using β-spline curves to all pathways and conducted a comparative analysis. It is evident that the IGA item displays a relatively strong performance of malposition across all four sets of examples, particularly in terms of proximity to the end of the central axis of the stern ramp, denoted by (0, 0).

Moreover, the deficiency of a multidimensional evaluation system has contributed to the notably poor performance of the EG3 item across all four algorithms. It is apparent that there are numerous challenging corners in practical implementation, resulting in compromised path smoothness. In conjunction with the previously mentioned evaluation metrics formulas, we conducted a comprehensive comparison of the evaluation metric values for malposition and large angles in the IGA and EG3 samples across the four examples. The results have been carefully tabulated and are presented in Table 2. Based on the table, it is evident that the smoothing metrics of the IGA and EG3 in all cases exhibit a significant difference of approximately two-fold. This observation suggests that the unimproved fitness function lacks efficacy in terms of path smoothing selection.

Table 2. Scalar analysis of smoothing indicators for modified fitness functions.

Case	$Fit_{ang,IGA}$	$Fit_{ang,EG3}$	$Fit_{mal,IGA}$	$Fit_{mal,EG3}$
1	0.48	1.22	1.30	2.40
2	0.42	0.86	2.57	3.86
3	0.37	0.69	2.17	3.00
4	1.44	2.75	4.55	5.82

This is because the convergence and screening mechanism of the genetic algorithm is closely related to the design of the fitness function. The fitness function of IGA strictly limits the recovery attitude requirements for stern ramp recovery to moderate and heading. In addition, EG3 also lacks constraints on the path smoothness parameter indicators, resulting in uneven speed and frequent sharp turns.

To validate the effectiveness of the preheating process in improving the genotype quality of the initial population and consequently reducing the number of iterative generations required, we conducted 100 experiments utilizing an arithmetic case sample with a starting point of (85, 10). These experiments were carried out both with and without the preheating operation. We set the number of elite groups for the experiments to 10. Moreover, to ensure better visual interpretation of the results, we enforced a fixed number of iterations per dataset at 500, which significantly exceeds the number of iterations typically needed for path planning. By recording the average fitness function values for the elite group in each generation, we analyzed and presented the results in Figure 15.

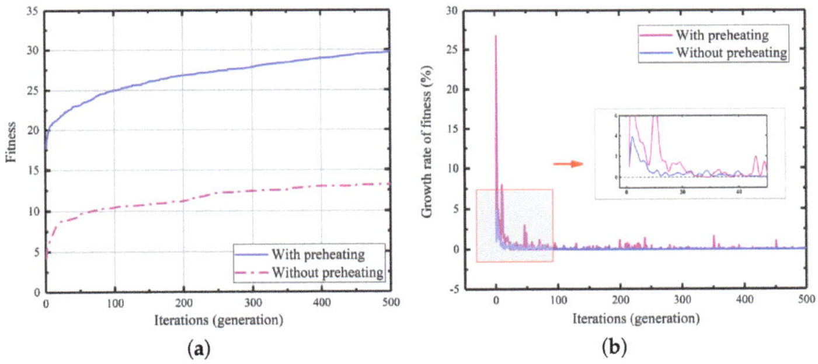

Figure 15. (**a**) The actual fitness function values. (**b**) The growth rate of fitness.

Based on the experimental results, it is evident that the example utilizing the preheating process exhibits a higher fitness function value in the initial stage of the algorithm, indicating its effectiveness in improving the genotype quality. As depicted in Figure 15, although the algorithm without the preheating process demonstrates a higher growth rate of fitness values, the fitness function value of the offspring progeny is still constrained by the parent population. This indicates that the parent genes that underwent population initialization optimization have good phenotypes. Conversely, the algorithm incorporating the preheating process tends to reach a saturation point in its growth rate within approximately 20 generations. In our extensive simulations, we have observed that this is indeed the number of iterations required for the IGA to successfully accomplish path planning.

The Elite Strategy and SSIH contribute to an accelerated convergence rate primarily by enhancing genetic manipulation techniques, as evidenced by the time taken for computational planning displayed in Table 3. Additionally, they exhibit some positive impact on path smoothness. Table 4 presents the comparative metrics illustrating these effects.

Table 3. Comparison of the time consumption for iterations.

Case	$Time_{IGA}$	$Time_{EG2}$	$Time_{EG4}$
1	2.067	11.392	12.769
2	2.683	8.430	8.319
3	1.556	7.436	6.366
4	3.792	15.985	62.00

Table 4. Comparison of evaluation metrics for the planning results of Improved Genetic Algorithm (IGA), EG2, and EG4 algorithms.

Case	$Fit_{ang,IGA}$	$Fit_{ang,EG2}$	$Fit_{ang,EG4}$	$Fit_{mal,IGA}$	$Fit_{mal,EG2}$	$Fit_{mal,EG4}$	$Fit_{d,IGA}$	$Fit_{d,EG2}$	$Fit_{d,EG4}$	$Fit_{T,IGA}$	$Fit_{T,EG2}$	$Fit_{T,EG4}$
1	0.48	0.59	0.60	1.30	2.00	2.00	1.05	1.14	1.15	5.18	5.40	5.34
2	0.42	0.83	0.63	2.57	3.14	2.71	1.08	1.13	1.15	5.17	5.68	4.80
3	0.37	0.76	0.75	2.17	2.17	2.33	1.02	1.11	1.09	4.95	5.60	5.60
4	1.44	1.44	3.44	4.55	4.55	3.73	1.19	1.19	1.17	5.09	5.00	5.74

It is evident that the incorporation of the Elite Strategy or SSIH yields advantages in enhancing the convergence rate. As the planning difficulty increases, there is a notable increase in computation time. However, these methods operate through distinct mechanisms. The former safeguards the retention of high-quality individuals during the genetic process by preserving their genotypes in each generation. On the other hand, the latter facilitates genetic diversity by releasing redundant memory and generating novel samples.

Based on the comparison of dimensionless evaluation metrics, it is observed that the planning results achieved by IGA exhibit superior performance in terms of smoothness, path distance, and time consumption compared to the other two algorithms. However, it should be noted that there are instances where the differences in metrics for some data are not particularly significant. To improve the overall quality of the planned path, our fitness function adopts a weighting strategy. Even if the fitness function values of the final screening are significantly different, some qualitative indicators may be similar.

The AV-LOS algorithm simulation is implemented using MATLAB. As an illustrative example, the desired trajectory S in (X, Y, T) space is composed of three motion segments concatenated together. The segments consist of $P_1(0, 0, 0)$, $P_2(0, 20, 8)$, $P_3(30, 20, 14)$, and $P_4(30, 0, 22)$. These segments collectively form an inverted 'U' shape in the X/Y plane, effectively simulating tracking scenarios involving acceleration, deceleration, and right-angle turns typical for unmanned boats. The simulation control period is denoted as T = 0.2, and the algorithm utilizes a forward-looking distance of $R = 3$, and the following assumptions are made:

- The USV utilizes a fixed control period T for speed and heading control. At the start of each period, the desired heading and desired speed are determined by solving the guidance law.
- The hysteresis effect in heading control is disregarded, and changes in speed are treated as linear transformations that either increase or decrease over time.

The simulation results are shown in Figure 16.

The formula for trajectory deviation $B(t)$ can be expressed as follows:

$$B(t) = \sqrt{(X_t(t) - X_r(t))^2 + (Y_t(t) - Y_r(t))^2} \tag{14}$$

where $X_t(t)$ and $Y_t(t)$ denote the coordinates of the desired trajectory and $X_r(t)$ and $Y_r(t)$ represent the coordinates of the real trajectory.

It is evident that the deviation of the 2D trajectory is minimal. Nonetheless, owing to the inclusion of the forward view distance, the actual trajectory undergoes smooth transitions with certain curvature during right-angle turns. Consequently, a deviation emerges at the termination point of the 3D trajectory. The LOS trajectory deviation takes the form of a three-step staircase since it directly adopts the isochronous target speed as the desired speed. Employing a random function, generate 50 sets of initial coordinates and repeat the simulation 10 times for each set. Calculating the average trajectory deviation for each algorithm, the resulting average deviation for the LOS algorithm is determined to be 2.0402 m, whereas, for the AV-LOS algorithm, it is found to be 0.8161 m.

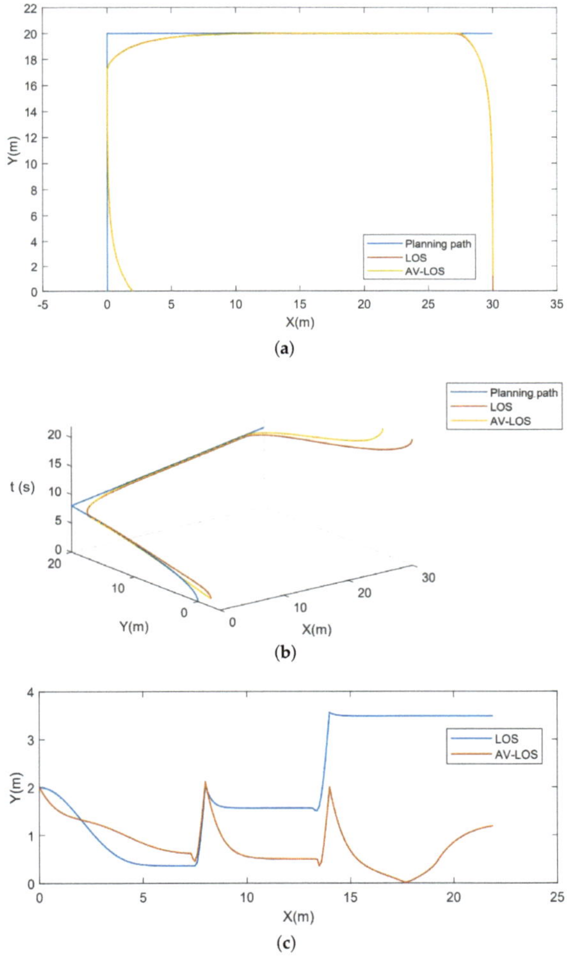

Figure 16. The simulation result of AV-LOS. (**a**) Two-dimensional projection for trajectory tracking. (**b**) Three-dimensional projection for trajectory tracking. (**c**) Tracking simulation with starting point (2, 0, 0).

4.3. The Field Tests of IGA Incorporating AV-LOS

We selected the initial three samples, which demonstrate significant variations in their planning characteristics, for conducting field experiments. In order to investigate the trajectory tracking performance of AV-LOS under various wind and wave interference scenarios, we recorded the first set of field-tested waypoint planning data and replicated the experiment twice at different times. The resulting actual paths of each experiment are denoted as Real path1, Real path2, and Real path3, respectively. The real GPS coordinates obtained from these experiments are visualized using a β-spline plot, as depicted in Figure 17.

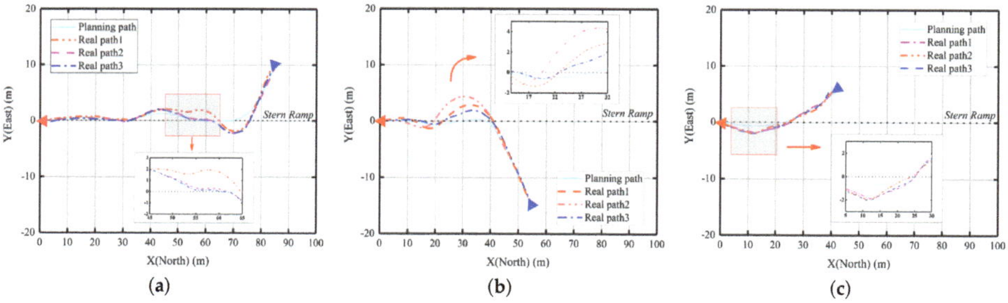

Figure 17. The result of field tests using IGA incorporating AV-LOS. (**a**) Case 1; (**b**) Case 2; (**c**) Case 3.

It is evident that the actual planning results closely resemble the simulation results, with a slight deviation observed during larger corners. This discrepancy can be attributed to varying disturbance conditions, which result in different magnitudes of translation for the USV during the cornering process. This phenomenon is closely associated with the tracking radius R set in the AV-LOS algorithm. As depicted in Figure 18, setting a tracking radius R that is too large causes an arc offset in the waypoints during tracking, while setting it too small hinders the upper computer's ability to assist the USV in tracking the next waypoint, leading to the USV making a turn with the current waypoint as the center.

Figure 18. (**a**) The example when the USV is unable to meet the demand for malposition because the tracking radius R is too big. After the position in the third picture, the USV will hit the steel plate when heading to the next waypoint. (**b**) When the tracking radius R is too small, the USV made a turn at the current waypoint.

Additionally, it is observed that the actual trajectory tends to have larger corners compared to the planning path. This is due to the necessity for the USV to undergo a dynamics adjustment process while making turns in a water environment. However, it is crucial for path planning to avoid sharp corners, as they can render the waypoint tracking ineffective during the dynamics adjustment process and result in a significant deviation from the intended course. Although the simulation results demonstrate that AV-LOS exhibits a small tracking error, a strategy solely focused on speed-oriented tracking is not an optimal solution for this issue.

To provide a more comprehensive analysis of the system's actual planning effectiveness, we conducted a thorough examination of the collected heading and speed data, as shown in Figures 19 and 20.

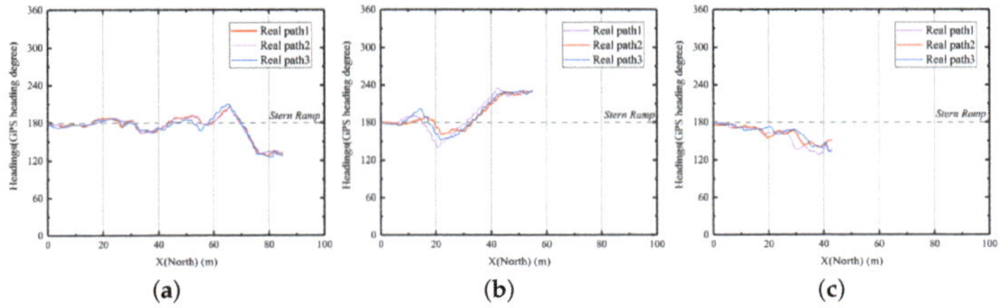

Figure 19. The result of field tests using IGA incorporating AV-LOS. (**a**) Case 1; (**b**) Case 2; (**c**) Case 3.

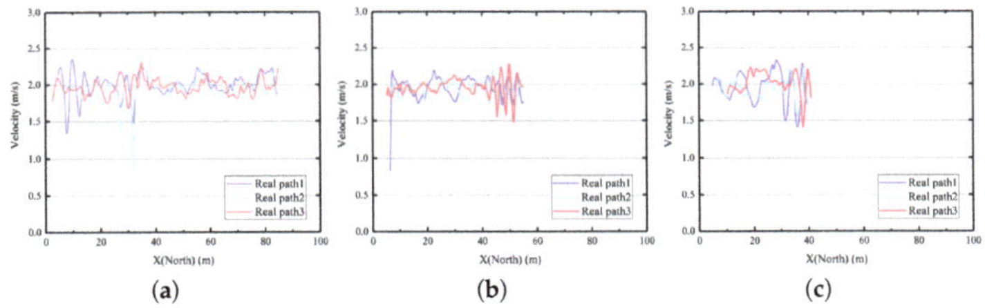

Figure 20. The result of field tests using IGA incorporating AV-LOS. (**a**) Case 1; (**b**) Case 2; (**c**) Case 3.

From Figure 19, we can see that the heading of the USV in water tends to be basically stable without significant fluctuations. Due to the influence of water flow, there are slight differences in the headings of the three path instances. It can be observed that as the x coordinate decreases, the heading gradually converges to 180°. This is due to the alignment of the USV to the centerline of the stern ramp during the recovery process and preparation for pitch correction. The difference between the heading angle and 180° does not exceed 60° in any of the examples, indicating that there were few sharp turns.

We removed the process of rapid speed changes during the start and end phases and analyzed the speed changes of the USV during the driving phase. It can be observed from Figure 20 that when the planning start point is far from the target, the planning path of the USV is longer and the speed uniformity is better. However, when the planning path is short and the longitudinal distance between the start and end points is large, the speed stability is not as good as the former. However, the difference between the maximum speed and the minimum speed usually does not exceed 1 m/s, which indicates that the anti-interference ability of the AV-LOS algorithm has a certain effect. This also indicates that there are still many areas for improvement in our work in the future.

5. Discussion

This paper presents an improved genetic algorithm for a stern ramp recovery system of the USV. The genetic algorithm, known for its global planning capabilities and retrieval efficiency, has proven to be highly practical. However, traditional two-dimensional modeling approaches and single fitness evaluation systems significantly increase the time required for path planning.

In the realm of algorithm optimization, we validated the viability of a genetic algorithm utilizing 3D modeling alongside a novel coding mechanism. By enhancing genetic operations to boost progeny diversity, we achieved a substantial improvement in the convergence rate. Furthermore, the multifactor screening of the fitness function enhances efficiency, albeit placing higher demands on the formulation of linear coefficients. By incorporating the forces exerted on the USV in real water environments through accurate modeling, we have integrated them into the enhanced AV-LOS algorithm. Furthermore, a method diagram for the AV-LOS algorithm is provided. Through simulation experiments, it has been demonstrated that the AV-LOS algorithm outperforms the traditional LOS algorithm in terms of tracking performance.

We have built a USV experimental platform and designed a hardware system. In practice, the optimized genetic algorithm demonstrates adaptability when dealing with longer distances that involve larger detours. However, due to surface disturbances, the actual path generated using the IGA might experience partial shifts, potentially resulting in less favorable detours. Nonetheless, it possesses an advantage in terms of malposition and being able to return to the original route promptly with the assistance of AV-LOS.

However, our system still has several shortcomings. One is that the movement of the obstacles set up in the experiment follows a regular pattern, which may not fully reflect real-world scenarios. The second is that the attitude adjustment of the USV during navigation may occasionally widen the turning path segment, even though the disturbance of the water surface environment is taken into account. And finally, when planning paths with larger turns and smaller total lengths, minor speed fluctuations may occur. In future efforts, we aim to further optimize the hardware system and the structure of the algorithm, while proposing reasonable solutions to deal with these limitations.

Author Contributions: Methodology, L.Z., X.Y. and P.X.; writing, L.Z.; reviewing, X.Y. and Z.H.; software, L.Z. and X.Y.; validation, L.Z. and P.X.; data curation, L.Z., Y.T. and P.X.; project administration, X.Y., Z.H. and Z.S. All authors have read and agreed to the published version of the manuscript

Funding: This research received no external funding.

Institutional Review Board Statement: Not applicable.

Informed Consent Statement: Not applicable.

Data Availability Statement: Data sharing not applicable.

Acknowledgments: Thank you to Shaojie You and Xiang Liu, who have graduated after designing the prototype for this project. Shaojie You laid a lot of foundation for the modeling and control of ocean currents. Xiang Liu has put a lot of effort into the design of software and the concept of path tracking.

Conflicts of Interest: The authors declare no conflicts of interest.

References

1. Arzamendia, M.; Espartza, I.; Reina, D.G.; Toral, S.; Gregor, D. Comparison of eulerian and hamiltonian circuits for evolutionary-based path planning of an autonomous surface vehicle for monitoring ypacarai lake. *J. Ambient. Intell. Humaniz. Comput.* **2019**, *10*, 1495–1507. [CrossRef]
2. Peng, Z.; Wang, J.; Wang, D.; Han, Q.L. An overview of recent advances in coordinated control of multiple autonomous surface vehicles. *IEEE Trans. Ind. Inform.* **2020**, *17*, 732–745. [CrossRef]
3. Song, R.; Liu, Y.; Bucknall, R. Smoothed A* algorithm for practical unmanned surface vehicle path planning. *Appl. Ocean. Res.* **2019**, *83*, 9–20. [CrossRef]
4. Liu, Z.; Zhang, Y.; Yu, X.; Yuan, C. Unmanned surface vehicles: An overview of developments and challenges. *Annu. Rev. Control.* **2016**, *41*, 71–93. [CrossRef]
5. Weihai Skysailing Technology. SKYSAILING. 2021. Available online: http://tianfanzhineng.com/product/info/1/ (accessed on 3 August 2023).

6. SUITOR. Unmanned System. 2023. Available online: https://www.4008075595.com/content_2_12_122.html (accessed on 3 August 2023).
7. Ku, Y.N.Z. Analysis of the Development of USV's Technology and Operational Styles Abroad. 2022. Available online: https://www.eet-china.com/mp/a158815.html (accessed on 3 August 2023).
8. Carboniferous. USVs Launched to Assist in Water Environmental Protection. Available online: https://stjep.cn/article (accessed on 3 August 2023).
9. China Economic Net. Israel Unveils Unmanned Combat Marine System. 2014. Available online: http://en.ce.cn/World/Middleeast (accessed on 4 August 2023).
10. Observer, S. Türkiye Has Developed the "Marlin" Unmanned Boat, and the Era of Unmanned Combat at Sea is Coming? 2022. Available online: https://export.shobserver.com/baijiahao/html/538455.html (accessed on 4 August 2023).
11. Xie, Y. China's First Comprehensive Geological Survey of Coastal Zones Using Unmanned Boats. 2017. Available online: https://www.shu.edu.cn/info/1055/58977.htm (accessed on 4 August 2023).
12. Yunzhou Intelligence. Alumni Enterprise Yunzhou Intelligent Will Participate in the Tongji University Alumni Industry Expo. 2019. Available online: https://www.163.com/dy/article/ESNP6V960518SJIU.html (accessed on 4 August 2023).
13. Eads, M.K.E. Well Deck Entry. 2017. Available online: https://www.defense.gov/Multimedia/Photos/igphoto/2002038005/ (accessed on 25 July 2023).
14. Chuang, G.Z. The Invention Patent for the Stern Slide of Guangzhou Shipyard Won the first Chinese Patent Silver Award. 2019. Available online: https://www.sohu.com/picture/286333857 (accessed on 25 July 2023).
15. Sohu. What Kind of Wonderful Sights Can You See If You Dive into the 7000-Meter-Deep Sea? 2022. Available online: http://it.sohu.com/a/576939130_121118997 (accessed on 25 July 2023).
16. Singh, Y.; Sharma, S.; Sutton, R.; Hatton, D.; Khan, A. A constrained A* approach towards optimal path planning for an unmanned surface vehicle in a maritime environment containing dynamic obstacles and ocean currents. *Ocean. Eng.* **2018**, *169*, 187–201. [CrossRef]
17. Kim, Y.H.; Son, W.S.; Park, J.B.; Yoon, T.S. Smooth path planning by fusion of artificial potential field method and collision cone approach. *MATEC Web Conf.* **2016**, *75*, 05004. [CrossRef]
18. Sang, H.; You, Y.; Sun, X.; Zhou, Y.; Liu, F. The hybrid path planning algorithm based on improved A* and artificial potential field for unmanned surface vehicle formations. *Ocean. Eng.* **2021**, *223*, 108709. [CrossRef]
19. Lee, H.C.; Yaniss, T.; Lee, B.H. Grafting: A path replanning technique for rapidly-exploring random trees in dynamic environments. *Adv. Robot.* **2012**, *26*, 2145–2168. [CrossRef]
20. Khan, A.H.; Li, S.; Luo, X. Obstacle avoidance and tracking control of redundant robotic manipulator: An RNN-based metaheuristic approach. *IEEE Trans. Ind. Inform.* **2019**, *16*, 4670–4680. [CrossRef]
21. Wang, N.; Xu, H.; Li, C.; Yin, J. Hierarchical path planning of unmanned surface vehicles: A fuzzy artificial potential field approach. *Int. J. Fuzzy Syst.* **2021**, *23*, 1797–1808. [CrossRef]
22. Xin, J.; Zhong, J.; Yang, F.; Cui, Y.; Sheng, J. An improved genetic algorithm for path-planning of unmanned surface vehicle. *Sensors* **2019**, *19*, 2640. [CrossRef] [PubMed]
23. Pettersen, K.Y.; Lefeber, E. Way-point tracking control of ships. In Proceedings of the 40th IEEE Conference on Decision and Control (Cat. No. 01CH37228), Orlando, FL, USA, 4–7 December 2001; Volume 1, pp. 940–945.
24. Fossen, T.I.; Pettersen, K.Y. On uniform semiglobal exponential stability (USGES) of proportional line-of-sight guidance laws. *Automatica* **2014**, *50*, 2912–2917. [CrossRef]
25. Caharija, W.; Pettersen, K.Y.; Sørensen, A.J.; Candeloro, M.; Gravdahl, J.T. Relative velocity control and integral line of sight for path following of autonomous surface vessels: Merging intuition with theory. *Proc. Inst. Mech. Eng. Part M J. Eng. Marit. Environ.* **2014**, *228*, 180–191. [CrossRef]
26. Fossen, T.I.; Pettersen, K.Y.; Galeazzi, R. Line-of-sight path following for dubins paths with adaptive sideslip compensation of drift forces. *IEEE Trans. Control. Syst. Technol.* **2014**, *23*, 820–827. [CrossRef]
27. Liu, L.; Wang, D.; Peng, Z. Predictor-based line-of-sight guidance law for path following of underactuated marine surface vessels. In Proceedings of the 2015 Sixth International Conference on Intelligent Control and Information Processing (ICICIP), Wuhan, China, 26–28 November 2015; pp. 284–288.
28. Wu, N.; Chen, H. Simulation design on dynamic positioning system of vessels. *Ship Sci. Technol.* **2013**, 103–106.
29. Song, J.; Du, J.; Li, W.; Sun, Y.; Chen, H. Simulation of ship dynamic positioning system disturbance due to wave. *J. Dalian Marit. Univ.* **2011**, *37*, 6–8.
30. Zhuang, J.; Su, Y.; Liao, Y.; Sun, H. Unmanned Surface Vehicle Local Path Planning Based on Marine Radar. *J. Shanghai Jiao Tong Univ.* **2012**, *46*, 1371–1375.
31. Zhu, C.; Liang, X. Novel genetic algorithm with multi-elitist preservation method. *J. Comput. Appl.* **2008**, *28*, 939–941. [CrossRef]
32. Ma, D. Unmanned Surface Vehicle Path Planning Based on Immune Genetic Algorithm. Master's Thesis, Huazhong University of Science and Technology, Wuhan, China, 2018.
33. de Jong Kenneth, A. *Analysis of the Behavior of a Class of Genetic Adaptive Systems*; Technical Report No. 185; Department of Computer and Communication Sciences, University of Michigan: Ann Arbor, MI, USA, 1975.
34. Zarándy, Á.; Nemeth, M.; Nagy, Z.; Kiss, A.; Santha, L.; Zsedrovits, T. A real-time multi-camera vision system for UAV collision warning and navigation. *J. Real-Time Image Process.* **2016**, *12*, 709–724. [CrossRef]

35. Deb, K. An efficient constraint handling method for genetic algorithms. *Comput. Methods Appl. Mech. Eng.* **2000**, *186*, 311–338. [CrossRef]
36. Khan, A.H.; Li, S.; Chen, D.; Liao, L. Tracking control of redundant mobile manipulator: An RNN based metaheuristic approach. *Neurocomputing* **2020**, *400*, 272–284. [CrossRef]
37. Breivik, M.; Fossen, T.I. Applying missile guidance concepts to motion control of marine craft. *IFAC Proc. Vol.* **2007**, *40*, 349–354. [CrossRef]
38. You, S. A Dissertation Submitted in Partial Fulfillment of the Requirements for the Master Degree in Engineering. Master's Thesis, Huazhong University of Science and Technology, Wuhan, China, 2022.

sensors

Article

An Infrared Small Target Detection Method Based on Attention Mechanism

Xiaotian Wang [1,2], Ruitao Lu [3,*], Haixia Bi [1] and Yuhai Li [4]

1 Unmanned System Research Institute, Northwestern Polytechnical University, Xi'an 710072, China; xtwang0111@xjtu.edu.cn or 18710993786@163.com (X.W.)
2 School of Information and Communication Engineering, Xi'an Jiaotong University, Xi'an 710049, China
3 Department of Missile Engineering, Rocket Force University of Engineering, Xi'an 710025, China
4 National Key Laboratory of Electromagnetic Space Security, Tianjin 300308, China; liyuhai.cn@gmail.com
* Correspondence: luruitao10@nudt.edu.cn

Abstract: The human visual attention system plays an important role in infrared target recognition because it can quickly and accurately recognize infrared small targets and has good scene adaptability. This paper proposes an infrared small target detection method based on an attention mechanism, which consists of three modules: a bottom-up passive attention module, a top-down active attention module, and decision feedback equalization. In the top-down active attention module, given the Gaussian characteristics of infrared small targets, the idea of combining knowledge-experience Gaussian shape features is applied to implement feature extraction, and quaternion cosine transform is performed to achieve multi-dimensional fusion of Gaussian shape features, thereby achieving complementary fusion of multi-dimensional feature information. In the bottom-up passive attention module, considering that the difference in contrast and motion between the target and the background can attract attention easily, an optimal fast local contrast algorithm and improved circular pipeline filtering are adopted to find candidate target regions. Meanwhile, the multi-scale Laplacian of the Gaussian filter is adopted to estimate the optimal size of the infrared small target. The fast local contrast algorithm based on box filter acceleration and structure optimization is employed to extract local contrast features, and candidate target regions can be obtained by using an adaptive threshold. Besides, the mean gray, target size, Gaussian consistency, and circular region constraint are used in pipeline filtering to extract motion regions, and the false-alarm rate is reduced effectively. Finally, decision feedback equalization is adopted to obtain real targets. Experiments are conducted on some real infrared images involving complex backgrounds with sea, sky, and ground clutters, and the experimental results indicate that the proposed method can achieve better detection performance than conventional baseline methods, such as RLCM, ILCM, PQFT, MPCM, and ADMD. Also, mathematical proofs are provided to validate the proposed method.

Keywords: infrared small target detection; feature frequency domain fusion; attention mechanism; feature weighting adjustment

Citation: Wang, X.; Lu, R.; Bi, H.; Li, Y. An Infrared Small Target Detection Method Based on Attention Mechanism. *Sensors* **2023**, *23*, 8608. https://doi.org/10.3390/s23208608

Academic Editors: Gang Peng and Zhun Fan

Received: 17 July 2023
Revised: 22 August 2023
Accepted: 25 August 2023
Published: 20 October 2023

1. Introduction

Detecting and tracking infrared small targets has been a challenging and extensively studied topic in the military field. This technique is commonly used in infrared search and tracking systems, infrared weapon guidance, and missile early warning systems [1–3]. In infrared scenes, small targets are often surrounded by strong background clutter and noise, while the limited imaging area and lack of distinct features such as shape and texture make detection and tracking infrared small targets difficult at a long imaging distance. Especially for backgrounds with sea, sky, and ground clutters, the detection difficulty further increases due to dynamic changes, low signal-to-noise ratio, fluctuations in target energy, and even submergence. The imaging characteristics and complex and diverse backgrounds of the aforementioned small infrared targets pose great challenges to the current detection task. Therefore, it is urgent to study how to detect infrared small targets accurately, robustly, and in real-time [4–7].

The primary methods of small infrared target detection can be categorized as filtering-based methods, low-rank sparse restoration-based methods, and feature continuity-based methods. The filtering-based method exploits the difference between the target and background clutter in the spatial and frequency domains to detect the target. Various filters have been introduced, including the adaptive toggle operator [8], nonnegativity-constrained variational mode decomposition [9], and frequency-tuned salient region detection [10]. These methods have a small amount of calculation, but the detection effect is poor. They can only suppress the uniform background to a certain extent, and the suppression effect is poor in complex backgrounds with sea, sky, and ground clutters. The low-rank sparse restoration-based method utilizes the difference in frequency characteristics of the target and the background clutter to detect the target [11–13]. However, the large number of calculations limit the engineering application of this type of method. The feature continuity-based method fully utilizes the temporal continuity of the shape and gray level of the target, as well as prior information such as the continuity of the motion trajectory to distinguish the target from the background clutter [14–16]. This type of method has a good detection effect for fast-moving targets, but it will fail for stationary targets. Therefore, the existing methods cannot be used in engineering practice to realize accurate, robust, and real-time detection in complex backgrounds.

In recent years, the human visual attention system has been introduced into the research of infrared small target detection. The human visual attention system can be divided into two types: the data-driven bottom-up passive attention mechanism and the task-dependent top-down active attention mechanism. Although infrared small targets are very weak, there are certain differences between the targets and the local background, and the human visual attention system can capture these differences and locate these areas quickly. Inspired by active attention and passive attention mechanisms, researchers have innovatively proposed various algorithms and made great breakthroughs [17–20].

By imitating the human visual attention mechanism, local contrast is adopted to effectively enhance the target area and suppress background clutter and noise, such as LCM (Local Contrast Measure) [21], ILCM (Improved Local Contrast Measure) [22], LDAM (Local Difference Adaptive Measure) [23], RLCM (Relative LCM) [24], DLCM (Double-Layer Local Contrast Measure) [25], MLHM (Multiscale Local Homogeneity Measure) [26], LIG (Local Intensity and Gradient) [27], HB-MLCM (High-Boost-based Multiscale Local Contrast Measure) [28], MPCM (Multiscale Patch-based Contrast Measure) [29], WSLCM (Weighted Strengthened Local Contrast Measure) [30] and ADMD (Absolute Directional Mean Difference) [31], etc. Based on strong representational capabilities, many deep learning methods have also begun to be widely used [32,33].

The local contrast-based methods mentioned above only use passive attention information to detect small infrared targets, but ignore active attention information, which can introduce many false alarms. Meanwhile, it is necessary to select features that can accurately distinguish between real targets and false-alarm targets. To overcome the aforementioned limitations, this paper proposes a small target detection method based on an attention mechanism. The proposed method employs passive and active attention information to improve detection performance. The main contributions of this paper are summarized as follows:

(1) A top-down active attention module is proposed to obtain target knowledge-experience Gaussian shape features, and the quaternion cosine transform is used to achieve multi-dimensional fusion of Gaussian shape features, thereby significantly improving the signal-to-clutter ratio gain (SCRG) and background suppression factor (BSF).

(2) The difference in contrast and motion between the target and the background is exploited to design a bottom-up passive attention module; meanwhile, an optimal fast local contrast algorithm and improved circular pipeline filtering are adopted to

find candidate target regions, using passive attention features with a discriminative ability to accurately detect small infrared targets.

(3) The decision feedback equalization considers the results of active and passive attention mechanisms to find real targets, which can better adapt to environmental changes in different scenarios.

(4) The remainder of this paper is organized as follows. In Section 2, the related work—such as the attention method, quaternion method, and ILCM—is reviewed. Section 3 presents the general framework of the proposed method and provides an in-depth explanation of its mathematical proofs. Section 4 outlines the experiments to validate the proposed method. Finally, Section 5 concludes this paper.

2. Related Work

In this section, the attention method, the quaternion method, and LCM are reviewed in brief, all of which have a strong connection to our research.

2.1. Attention Method

The human visual attention system can be divided into two types: the data-driven bottom-up passive attention mechanism and the task-dependent top-down active attention mechanism, as shown in Figure 1. In the bottom-up passive attention mechanism, the prefrontal cortex (PFC) and posterior parietal cortex (PPC) integrate the original physical characteristics of external stimuli that are transmitted through the visual pathway to form a complete saliency map in the brain. This causes eye movement controlled by the superior colliculus (SC) and makes it easy for us to pay attention to the yellow rectangular target in the visual scene (the task here is to search the yellow rectangle). The physical features here include color, intensities, orientations, etc. The top-down active attention mechanism actively searches for targets with priority map characteristics in the prefrontal cortex (PFC) of the brain according to known task information and knowledge experience; it causes eye movement so that we can easily pay attention to the desired target.

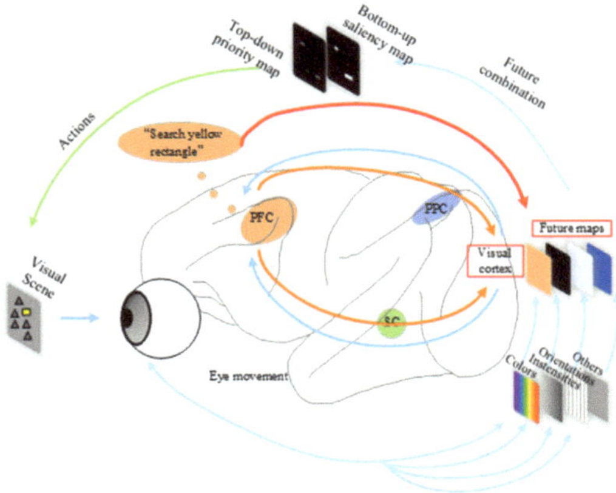

Figure 1. The human visual attention system.

Inspired by the human visual attention system, the simulation and simplification of each attention module can realize the accurate detection of infrared small targets.

2.2. Quaternion Method

The quaternion discrete cosine transform (QDCT) algorithm combines discrete cosine transform (DCT) with quaternion algebra, and it is widely used in image processing.

Object detection has been realized through input variables by applying QDCT in recent times. The stronger the feature generalization ability, the higher the detection performance of the algorithm.

The quaternion representation is:

$$I_q = x_1 + x_2 i + x_3 j + x_4 k \tag{1}$$

where x_1, x_2, x_3 and x_4 are four real numbers, and they satisfy the operation rule $i^2 = j^2 = k^2 = ijk = -1, ij = -ji = k, jk = -kj = i, ki = -ik = j$. Each quaternion I_q is a linear combination of 1, i, j, and k.

Assume that the resolution of I_q is $M \times N$, where M and N represent the length and width of the image, respectively. The QDCT of I_q is given by

$$Q(u,v) = QDCT(I_q) = \alpha_u^M \alpha_v^N \sum_{m=0}^{M-1} \sum_{n=0}^{N-1} \mu_Q I_q N(u,v,m,n) \tag{2}$$

where α_u^M and α_v^N are the coefficients. $N(u,v,m,n)$ and μ_Q are given by

$$\begin{cases} N(u,v,m,n) = \cos\left[\frac{\pi}{M}(m+\frac{1}{2})u\right] \cos\left[\frac{\pi}{N}(n+\frac{1}{2})v\right] \\ \mu_Q = -\sqrt{\frac{1}{3}}i - \sqrt{\frac{1}{3}}j - \sqrt{\frac{1}{3}}k \end{cases} \tag{3}$$

The corresponding inverse quaternion cosine transform (IQDCT) is shown below:

$$I_q' = IODCT(Q(u,v)) = \sum_{u=0}^{M-1} \sum_{v=0}^{N-1} \alpha_u^M \alpha_v^N \mu_q Q(u,v) N(u,v,m,n) \tag{4}$$

The quaternary cosine transform can be employed to fuse the input feature variables to realize the complementary fusion of feature information of different dimensions and achieve target detection.

2.3. LCM Algorithm

The traditional LCM algorithm uses a sliding window to traverse the entire image pixel by pixel from left to right and top to bottom to calculate the local contrast of the image, consequently achieving accurate detection of small targets. The sliding window contains 3×3 sub-windows in total. Among the sub-windows, the one in the center denoted by "0" represents the target area, and the ones in the surrounding area denoted by "1", "2", "3", "4", "5", "6", "7", and "8" represent the background area.

The average gray value of each sub-window can be expressed as

$$m = \frac{1}{N} \sum_{i,j} I_d(i,j) \tag{5}$$

where N is the number of pixels in each sub-window, and $I_d(i,j)$ is the gray value in the sub-window whose center is located at the coordinates (i,j). When the sliding window passes through a certain pixel of the image, the LCM of the pixel can use the central sub-window and the eight adjacent background sub-windows, and its expression is:

$$C = \min(L_0 \times m_0 / m_k)(k = 1, 2, \ldots, 8) \tag{6}$$

where L_0 and m_0 are the maximum gray value and mean gray value of the sub-window "0", respectively, while $m_k (k = 1, 2, \ldots, 8)$ represents the mean gray value of the eight surrounding sub-windows. The eight surrounding sub-windows serve as the local background (see Figure 2).

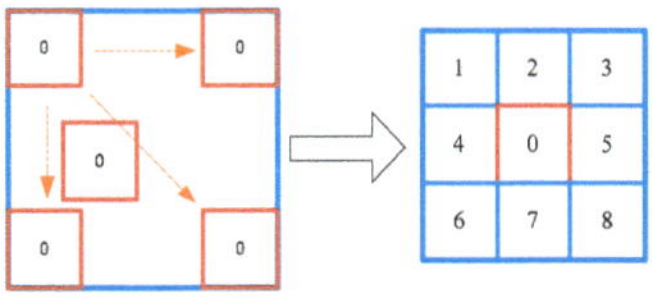

Figure 2. The calculation principle of LCM.

The fast local contrast algorithm based on box filter acceleration and structure optimization is an improved version of the LCM, and it has a faster calculation speed while ensuring accuracy.

3. Methodology

In the top-down active attention module, to fully utilize the knowledge-experience information of infrared small targets, this paper extracts Gaussian shape features and uses QDCT to achieve complementary fusion of multi-dimensional Gaussian shape features. In the bottom-up passive attention module, the difference in contrast and the motion between the target and the background can attract attention easily, and based on this, this paper proposes an optimal fast local contrast algorithm and uses improved circular pipeline filtering to find candidate target regions, which can reduce the false alarm rate and improve detection precision effectively. Additionally, decision feedback equalization is adopted to detect real targets, and it is effective in different scenarios to better adapt to environmental changes. Figure 3 shows the flowchart of the proposed method, followed by a detailed description of the specific execution steps. The introduction of the top-down active attention module is outlined in Section 3.1, followed by the explanation of the bottom-up passive attention module in Section 3.2. Section 3.3 discusses decision feedback equalization, and Section 3.4 presents the optimal fast local contrast algorithm along with its mathematical proofs.

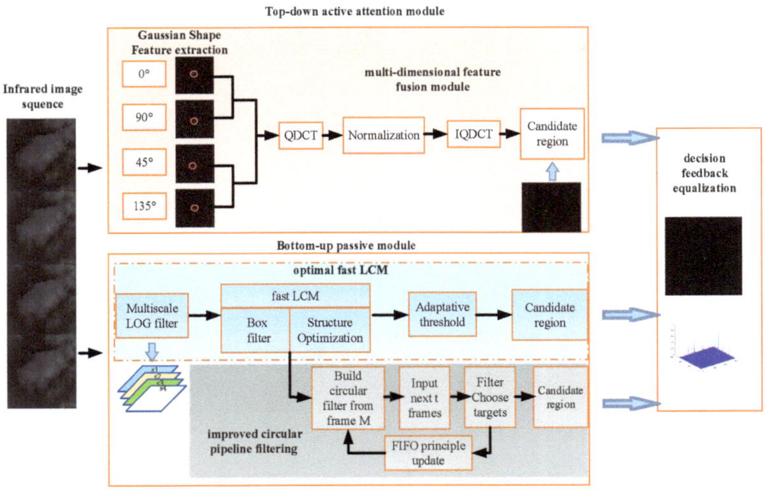

Figure 3. The flowchart of the proposed method.

3.1. Top-Down Active Attention Module

The top-down active attention module exploits the prior knowledge of the infrared small target to design a discriminator and effectively detect the infrared small target. Due to the use of an optical imaging system in Gaussian shape feature extraction, infrared small targets appear as isotropic Gaussian-like spots while the background typically exhibits a uniform direction. To distinguish the Gaussian-like spots and strip-like textures that have a single direction, the second-order directional derivative filter is adopted to construct a continuous function within a specific area using discrete points, and then, the directional derivative of the function is determined. As shown in Figure 4, this method can effectively extract the desired features.

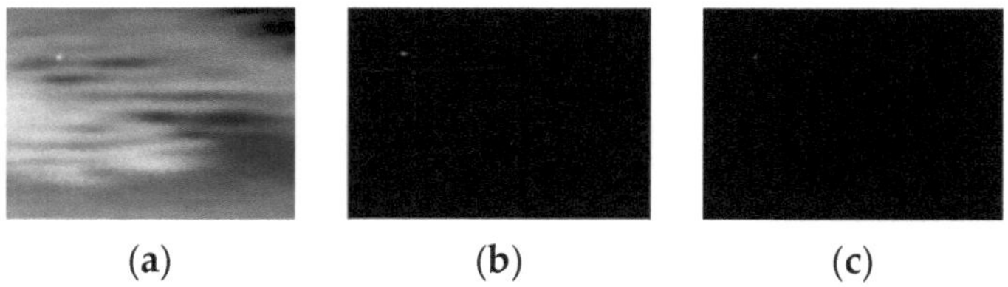

Figure 4. The effect of the second-order directional derivative filter. (**a**) The original image; (**b**) the processing effect in the horizontal direction; (**c**) the processing effect in the vertical direction.

Denote the gray value of the image at the coordinate point (x, y) as $f(x, y)$, and denote the directional derivative D_i of the direction vector l as:

$$\begin{aligned} D_i &= \left.\frac{\partial^2 f(x,y)}{\partial l^2}\right|_{(x_0,y_0)} \\ &= \left.\left[f_{xx}(x,y)\cos^2\alpha + 2f_{xy}(x,y)\cos\alpha\cos\beta + f_{yy}(x,y)\cos^2\beta\right]\right|_{(x_0,y_0)} \end{aligned} \tag{7}$$

where α is the angle between the direction vector l and the X-axis (row of image), and β is the angle between the direction vector l and the Y-axis (column of image).

Because the target has the same Gaussian characteristics, the Gaussian-like target is retained in the detection process. Meanwhile, to better suppress the background clutter, four directions $\alpha = 0°, 45°, 90°, 135°$ are selected for calculation. So, the Gaussian shape features of the target can be obtained

$$[D_1(x,y), D_2(x,y), D_3(x,y), D_4(x,y)] \tag{8}$$

In the multi-dimensional feature fusion module, the quaternion cosine transform is used to achieve complementary fusion of multi-dimensional Gaussian shape features [34]. The Gaussian shape features are represented by $D_1(x,y)$, $D_2(x,y)$, $D_3(x,y)$, and $D_4(x,y)$ respectively. They are taken as four data channels to construct quaternion representation:

$$q(x,y) = \lambda_1 * D_1(x,y) + \lambda_2 * D_2(x,y)\vec{i} + \lambda_3 * D_3(x,y)\vec{j} + \lambda_4 * D_4(x,y)\vec{k} \tag{9}$$

where $\vec{i}, \vec{j}$, and $\vec{k}$ represent three imaginary axes, and $\lambda_1, \lambda_2, \lambda_3$, and λ_4 correspond to different data channel weights; here, they are set to 0.15, 0.15, 0.35, and 0.35, respectively.

Assume that the resolution of $q(x,y)$ is $M \times N$, where M and N represent the length and width of the image, respectively. The QDCT of $q(x,y)$ is given by

$$Q(u,v) = QDCT(q(x,y)) = \alpha_u^M \alpha_v^N \sum_{m=0}^{M-1} \sum_{n=0}^{N-1} \mu_Q q(x,y) N(u,v,m,n) \tag{10}$$

where α_u^M and α_v^N are the coefficients. $N(u,v,m,n)$ and μ_Q are given by Equation (3).

The normalization function is used for $Q(u,v)$ to suppress the low-frequency information of the background and enhance the high-frequency information of the target. The frequency-domain normalization function is shown below:

$$Q' = \mathrm{sgn}(Q) = \begin{cases} \frac{x_0}{|Q|} + \frac{x_1}{|Q|}i + \frac{x_2}{|Q|}j + \frac{x_3}{|Q|}k & |Q| \neq 0 \\ 0 & |Q| = 0 \end{cases} \tag{11}$$

where x_0, x_1, x_2, and x_3 are four components of Q, respectively. $|Q|$ is the magnitude of the quaternion. Finally, the corresponding IQDCT is used to obtain the spatial domain saliency map as follows:

$$\begin{aligned} q'(x,y) &= IODCT(\mathrm{sgn}(QDCT(q(x,y)))) \\ &= IQDCT(Q') \\ &= \sum_{u=0}^{M-1}\sum_{v=0}^{N-1} \alpha_u^M \alpha_v^N \mu_q Q'(u,v) N(u,v,m,n) \end{aligned} \tag{12}$$

To make the saliency map smoother, the smoothing process is performed by using the Gauss smooth filter to obtain the final map. The formula is as follows:

$$S(x,y) = g(x,y,\sigma = 1.5) * \left[q'(x,y) \odot \bar{q}'(x,y)\right] \tag{13}$$

Then, the candidate region of the top-down active attention module can be obtained and denoted as R_a.

3.2. Bottom-Up Passive Attention Module

Based on the contrast and motion characteristics of infrared small targets, the bottom-up passive attention module designs a discriminator to effectively detect targets. An optimal fast local contrast algorithm is proposed in this paper to find the candidate target regions by using contrast characteristics of infrared small targets, and improved circular pipeline filtering is adopted to find candidate target regions using motion characteristics of infrared small targets. The difference in contrast and motion between the target and the background can attract passive attention easily, which is a discriminative characteristic for accurately detecting small infrared targets.

Since the scale of infrared small targets has a certain range, it is necessary to cover all targets with scales ranging from 2×2 to 9×9. When contrast feature extraction is performed by calculating the local contrast at multiple scales, the target and clutter background will be enhanced at the same time, and the detection effect is not ideal. To resolve this problem, this paper proposes an optimal fast local contrast algorithm to extract candidate regions. In this algorithm, the multiscale Laplacian of Gaussian (LOG) filter is used to estimate the optimal scale of the infrared small target. Meanwhile, the fast local contrast algorithm based on box filter acceleration and structure optimization is employed to extract local contrast features, and candidate target regions can be obtained by an adaptive threshold. The flowchart of the proposed optimal fast local contrast algorithm is shown in Figure 5. To obtain the sub-window scale of the corresponding image pixel $f(x,y)$, the LOG filter is used to estimate the sub-window scale in advance and obtain the optimal scale of the infrared small target. The formula of the multiscale LOG filter with a scale-space constant $s_i(i = 1,2,3,4)$ is shown as follows

$$LoG(x,y,s_i) = \frac{1}{2\pi {s_i}^2}\left[\frac{x^2 + y^2 - 2{s_i}^2}{{s_i}^4}\right] e^{\frac{-x^2-y^2}{2{s_i}^2}} \tag{14}$$

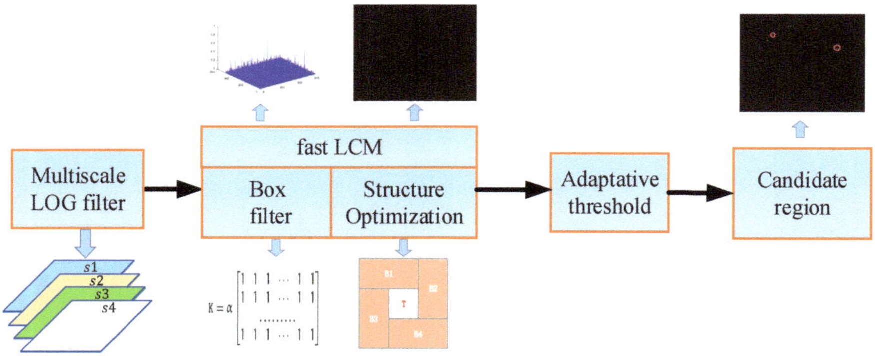

Figure 5. The flowchart of the proposed optimal fast local contrast algorithm.

The scale-space constant for detecting small infrared targets should satisfy the criterion that the size of the target is slightly smaller than $2\sqrt{2}s_i$, i.e., the scale-space constant corresponding to small infrared targets (around 3 × 3 pixels) should be 1.1. In this study, the scale of a small target is smaller than 9 × 9 pixels. To guarantee the real-time performance of the algorithm, four typical scales are selected. The scale-space constant s_i corresponding to small infrared targets (3 × 3 pixels, 5 × 5 pixels, 7 × 7 pixels, and 9 × 9 pixels) are [1.1 1.77 2.48 3.19]. The optimal sub-window scale is defined as

$$F = \max(F_i) = LoG(x,y,s_i) \otimes f(x,y)(i = 1,2,3,4) \tag{15}$$

At each position, the optimal sub-window scale can be obtained to design the local contrast algorithm. Based on box filter acceleration and structure optimization, a fast local contrast algorithm is proposed in this paper to extract the local contrast features. As shown in Figure 6, the sub-window in the center denoted by "0" represents the target area, and sub-windows in the surrounding area are four rectangular sub-windows denoted by "11", "22", "32", "44", respectively [3]. The original sub-windows of the LCM algorithm denoted by "1" and "2" are combined into "11", those denoted by "3" and "5" are combined into "22", those denoted by "7" and "8" are combined into "33", and those denoted by "4" and "6" are combined into "44". The difference d_i between the central sub-window "0" and its four neighboring sub-windows with the optimal sub-window scale σ_i is defined as

$$d_i(x,y,\sigma_i) = \frac{(m_0^{\sigma_i} - m_{Bi i}^{\sigma_i})m_0^{\sigma_i}}{m_{Bi i}^{\sigma_i}}(i = 1,2,\ldots,4) \tag{16}$$

where $m_0^{\sigma_i}$ and $m_{Bi i}^{\sigma_i}$ are the mean gray value of the central sub-window "0" and its four neighbor sub-windows with the optimal sub-window scale σ_i, respectively.

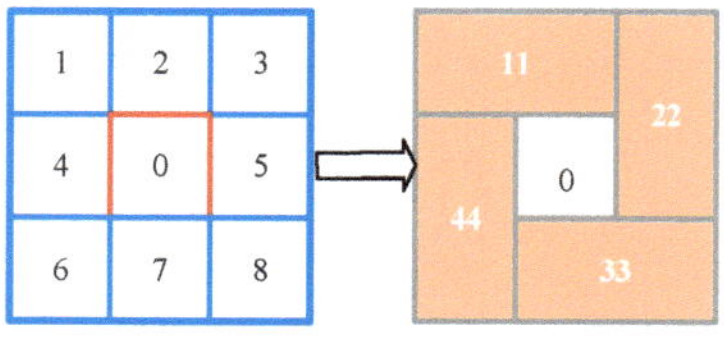

Figure 6. The structure optimization of the fast local contrast algorithm.

The fast local contrast map with the optimal sub-window scale σ_i can be defined as

$$S(x, y, \sigma_i) = \min\{d_1 * d_3, d_2 * d_4\} \tag{17}$$

To accelerate the mean gray calculation of four rectangular neighbor sub-windows, a box filter is introduced. As shown in Figure 7, the box filter mainly converts the gray of the image pixel $f(x', y')$ into the sum of the gray value of the corresponding diagonal region from the upper left corner to any point (x, y). Its mathematical formula is

$$I(x, y) = \sum_{x' \leq x, y' \leq y} f(x', y') \tag{18}$$

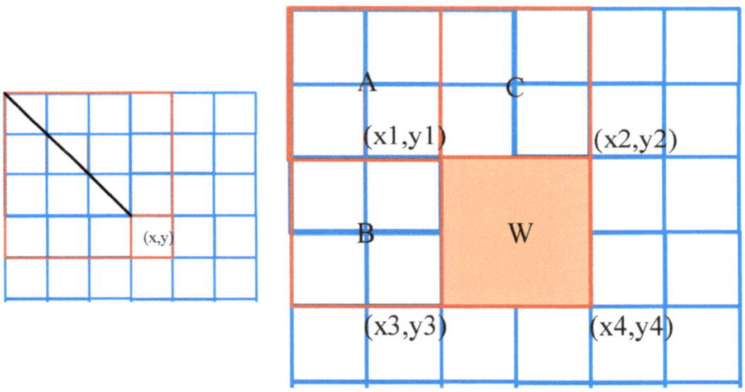

Figure 7. Box filter acceleration.

The value at location $(x1, y1)$ is the sum of the pixels in rectangle A. The value at location $(x2, y2)$ is A + C, at location $(x3, y3)$ is A + C, and at location $(x4, y4)$ is A + B + C + W. The sum within W can be computed as $I(x1, y1) + I(x4, y4) - I(x2, y2) - I(x3, y3)$. The mean gray calculation of four rectangular neighbor sub-windows can be converted to simple additions and subtractions as follows

$$\sum W = \frac{I(x1, y1) + I(x4, y4) - I(x2, y2) - I(x3, y3)}{(x2 - x1)(y4 - y2)} \tag{19}$$

Therefore, the fast local contrast algorithm based on box filter acceleration and structure optimization can extract the local contrast features. Meanwhile, to detect candidate small infrared targets, an adaptive threshold method is used. The candidate small infrared target region $S'(x, y)$ can be expressed as

$$\begin{cases} S'(x, y) = S(x, y) & S(x, y) \geq T \\ S'(x, y) = 0 & S(x, y) < T \\ T = \mu_S + k \bullet \sigma_S \end{cases} \tag{20}$$

where μ_S and σ_S denote the mean and variance of the local contrast map $S(x, y)$, respectively; $k = 3$ is the proportionality coefficient. Further, T is the adaptive threshold value.

As shown in Figure 8, to reduce the false alarm rate, improved circular pipeline filtering is adopted to find candidate target regions using motion characteristics of infrared small targets. The mean gray, target size, and circular region constraints are used in pipeline filtering to extract motion regions [35].

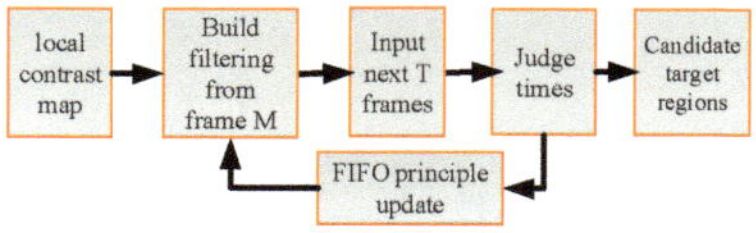

Figure 8. The flowchart of improved circular pipeline filtering.

To find the moving target areas, the local contrast map obtained by the optimal fast local contrast algorithm is regarded as an input of the pipeline as follows:

$$H' = \{H'_1, H'_2, \ldots, H'_l\} \quad (21)$$

where H' denotes the centers of suspected target patches, and l is the number of local contrast patches.

Concentric ring pipeline filtering with an inner radius r_1 and an outer radius r_2 is set, and the step sizes of S_1 and S_2 gradually increase with the length of $T = 5$. The number of times k_M that the center of the suspected target patches in the current frame falls on the corresponding concentric ring pipeline filtering is calculated. Meanwhile, the mean gray and target size have a good match. If k_M exceeds the set threshold $U = 3$, the centers of the suspected target patches are determined as real candidate target regions; otherwise, they are determined as false targets and rejected.

The pipeline filtering is updated according to the principle of First-In-First-Out (FIFO), and the frame that comes in first will be removed first in the update process. Then, the candidate small infrared target regions can be obtained and denoted as R_V.

3.3. Decision Feedback Equalization

For the top-down active attention module, Gaussian shape features are adopted to implement feature extraction, and quaternion cosine transform is used to achieve multi-dimensional fusion of Gaussian shape features. Through numerous experiments, it is found that a large number of Gaussian noises will be retained, affecting the detection of the small infrared targets. The square operation and lower weight can significantly suppress the background Gaussian-like noises. After many experiments, this paper sets the weight to 0.7 for the top-down active attention module. For the bottom-up passive attention module, the local contrast characteristic has been proven to work well in many situations and can achieve target enhancement, and the weight is set to 1. Meanwhile, motion characteristic is also important, and it helps to find small infrared targets in complex scenes. After many experiments, the weight is set to 1.3.

Based on the experimental results, feedback equalization processing is conducted to obtain the optimal weight coefficient and expression:

$$F = 0.7 * R_a + S' + 1.3 * R_V \quad (22)$$

The final candidate regions are regarded as the real target regions.

3.4. Algorithm Analysis of Optimal Fast Local Contrast Algorithm

In the optimal fast local contrast algorithm, the multiscale LOG filter is used to estimate the optimal scale of the infrared small target, and box filter acceleration and structure optimization are used to enhance the computation speed. The primary focus of the algorithm analysis is the detection performance of the optimal fast local contrast algorithm to improve target detection while suppressing the background. Four cases need to be analyzed.

$$d_i(x, y, \sigma_i) = \frac{(m_0^{\sigma_i} - m_{Bii}^{\sigma_i})m_0^{\sigma_i}}{m_{Bii}^{\sigma_i}} (i = 1, 2, \ldots, 4) \quad (23)$$

$$S(x, y, \sigma_i) = \min\{d_1 * d_3, d_2 * d_4\} \tag{24}$$

(1) When the image pixel $f(x, y)$ belongs to the target region, then $m_0^{\sigma_i} > m_{Bii}^{\sigma_i}, \frac{m_0^{\sigma_i}}{m_{Bii}^{\sigma_i}} > 1$, and $d_i > m_0^{\sigma_i} - m_{Bii}^{\sigma_i} > 0$; finally, $S(x, y, \sigma_i) > 0$.

(2) When the image pixel $f(x, y)$ belongs to the background region, then $m_0^{\sigma_i} - m_{Bii}^{\sigma_i} \approx 0$, $\frac{m_0^{\sigma_i}}{m_{Bii}^{\sigma_i}} \approx 1$, and $d_i = m_0^{\sigma_i} - m_{Bii}^{\sigma_i} \approx 0$; finally, $S(x, y, \sigma_i) \approx 0$.

(3) When the image pixel $f(x, y)$ belongs to the strong edge region, for one direction, $m_0^{\sigma_i} > m_{Bii}^{\sigma_i}, \frac{m_0^{\sigma_i}}{m_{Bii}^{\sigma_i}} > 1$, and $d_i > m_0^{\sigma_i} - m_{Bii}^{\sigma_i} > 0$; for the other direction, $m_0^{\sigma_i} - m_{Bii}^{\sigma_i} \approx 0$, $\frac{m_0^{\sigma_i}}{m_{Bii}^{\sigma_i}} \approx 1$, and $d_i = m_0^{\sigma_i} - m_{Bii}^{\sigma_i} \approx 0$; finally, $S(x, y, \sigma_i) \approx 0$.

(4) When the image pixel $f(x, y)$ belongs to the noise region, because it has a smaller size than the target, and its contribution to the mean gray value is limited, then $S(x, y, \sigma_i) > S(x, y, \sigma_i)_{nosie} > 0$.

In summary, the target region and the background clutter region (background, strong edge, and noise region) can be distinguished by defining the contrast by the difference and the ratio, and the back-ground clutter region is suppressed so that the background clutter region tends to be 0, thereby achieving a better effect of infrared target detection.

4. Experimental Results

Experiments were conducted in this section to verify the effectiveness of the proposed method for detecting small targets. The real infrared images are taken by infrared thermography. To investigate the performance of the proposed method in different scenes, four infrared image sequences with complex background clutters were chosen as test sequences. Figure 9 shows the target labeled by a red box in each image, while Table 1 lists the specifics of the four sequences. The background types of these sequences include sea, sky, and ground clutters. In Figure 9(a1), the target is submerged in the cloud, and many noises are randomly distributed in the whole infrared image. In Figure 9(b1,d1), plants and roads have high intensity, and the small infrared target will be blurred and easily confused with the background clutters. In Figure 9(c1), the glint will interfere with the target detection process. The algorithm's ability to detect small infrared targets is evaluated using current metrics both from single-frame and multi-frame methods. The program is executed in Matlab2021a on a personal computer equipped with an Intel® CoreTM i7-11700 CPU @ 2.50 GHz and 32 GB memory.

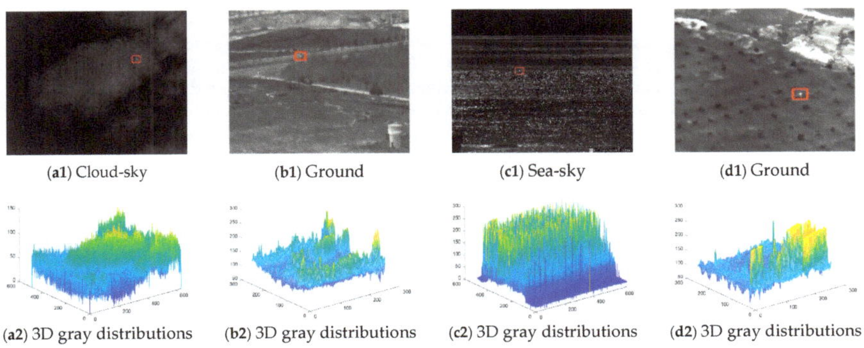

(**a1**) Cloud-sky (**b1**) Ground (**c1**) Sea-sky (**d1**) Ground

(**a2**) 3D gray distributions (**b2**) 3D gray distributions (**c2**) 3D gray distributions (**d2**) 3D gray distributions

Figure 9. Representative infrared images and their 3D gray distributions.

Table 1. The specifics of four real infrared image sequences.

Background Type	Sequence Number	Frame	Target Number	Target-Size Range	Target SCR Range
Cloud-sky	Seq.1	195	1	3×3–5×5	3.16–4.84
Ground	Seq.2	144	1	4×4	1.42–7.74
Sea-sky	Seq.3	108	1	3×3–5×5	2.42–8.77
Ground	Seq.4	399	1	4×4	1.42–30.46

4.1. Evaluation Metrics

Different evaluation metrics are introduced for single-frame and multi-frame small infrared target detection.

The performance of target detection algorithms in single-frame recognition is often evaluated with the SCRG and BSF. The SCRG is an evaluation metric to describe the significance of the enhancement of the target after the process of an algorithm, and BSF reveals the significance of suppression of the background after the process of an algorithm. The SCRG and BSF are defined as

$$SCR = \frac{G_{mt} - G_{mb}}{\sigma_b},\ SCRG = \frac{SCR_{out}}{SCR_{in}},\ BSF = \frac{(\sigma_c)_{in}}{(\sigma_c)_{out}} \tag{25}$$

The average gray of the target area is denoted as G_{mt}, while the average gray and standard variation of the local background area are denoted as G_{mb} and σ_b, respectively. The signal-to-clutter ratio of the input and output images is denoted as SCR_{out} and SCR_{in} respectively, while the standard deviation of the input and output images is represented as $(\sigma_c)_{in}$ and $(\sigma_c)_{out}$ respectively. Additionally, the length ratio of the local background edge to the target region edge is 1.4.

To determine how well multi-frame target recognition performs, the Receiver Operation Characteristic (ROC) curve is utilized. The ROC curve plots the false positive rate (FPR) on the horizontal axis, which is the ratio of the number of false targets detected to the number of real targets, and it is also known as the precision. The true positive rate (TPR) is plotted on the vertical axis, which represents the likelihood that a detected true target is an actual true target, and it is known as the false-alarm ratio. The algorithm achieves optimal detection results when TPR is high and FPR is low, simultaneously indicating a high detection rate and low false-alarm rate. TPR and FPR are mathematically expressed as:

$$\begin{cases} \text{TPR} = \frac{\text{number of true target detected}}{\text{number of real targets}} \\ \text{FPR} = \frac{\text{number of false targets detected}}{\text{number of real targets}} \end{cases} \tag{26}$$

There exists a relationship between the TPR and FPR. Typically, when the FPR is high, the TPR is also high. To assess the TPR and FPR while identifying the true number of targets, the LABELIMAGE2020 software is utilized to label the ground truth of small infrared targets in actual infrared images. To achieve correct detection, the Euclidean distance between the labeled targets and detection results should be less than 16 pixels. However, there could be an error in determining the actual position of the labeled targets. After conducting numerous experiments, a threshold of 16 pixels is set up.

4.2. Effectiveness Analysis for Single-Frame Detection

The effectiveness of single-frame detection is evaluated using SCRG and BSF. When the proposed method achieves high values of both SCRG and BSF, it indicates that the target has been enhanced and the background has been suppressed, making it easier to locate the target. Our proposed method was tested with about 900 infrared images to evaluate its effectiveness. Here, four representative infrared images are taken as an example, as shown in Figure 9(a1–d1). Their 3D gray distributions are shown in Figure 9(a2–d2). After the execution of the proposed method, the target is enhanced, and the clutter is suppressed, making the target easy to detect, as shown in Figure 10(a1–d1). It is noteworthy that the proposed method failed to enhance the target significantly for the image from Seq.2, as shown in Figure 10(b1). However, after using the motion information, the proposed method can enhance the target significantly and suppress the background clutters, and the target is found.

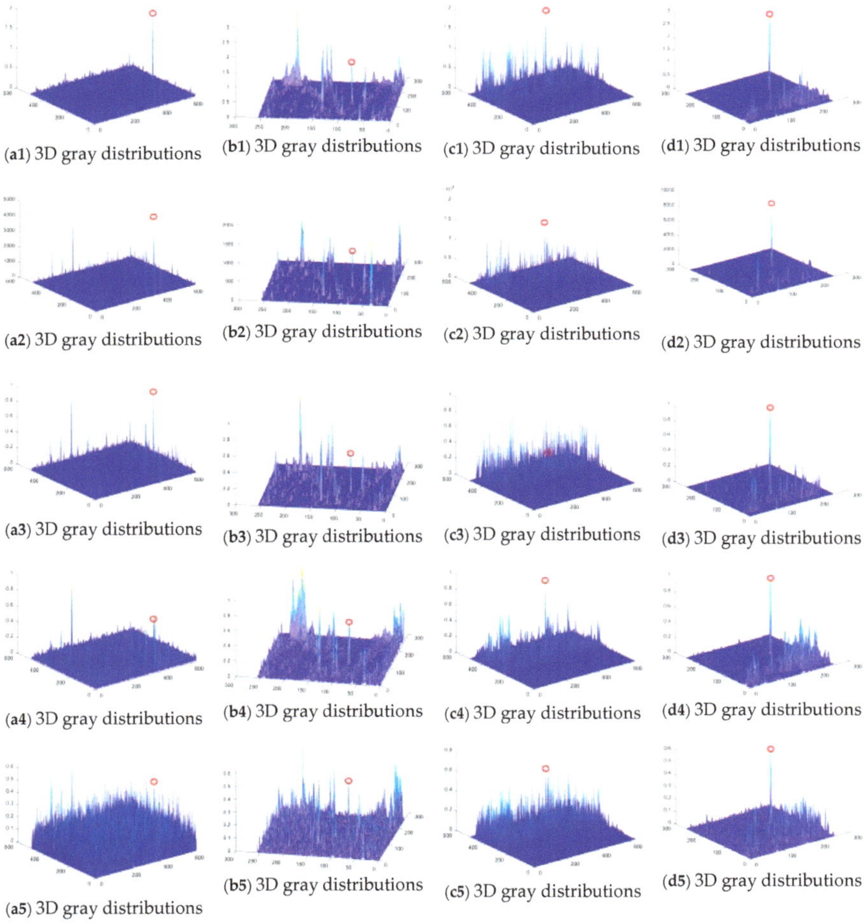

(**a1**) 3D gray distributions (**b1**) 3D gray distributions (**c1**) 3D gray distributions (**d1**) 3D gray distributions

(**a2**) 3D gray distributions (**b2**) 3D gray distributions (**c2**) 3D gray distributions (**d2**) 3D gray distributions

(**a3**) 3D gray distributions (**b3**) 3D gray distributions (**c3**) 3D gray distributions (**d3**) 3D gray distributions

(**a4**) 3D gray distributions (**b4**) 3D gray distributions (**c4**) 3D gray distributions (**d4**) 3D gray distributions

(**a5**) 3D gray distributions (**b5**) 3D gray distributions (**c5**) 3D gray distributions (**d5**) 3D gray distributions

Figure 10. The 3D gray distributions of representative infrared images using five methods (the proposed method, ADMD, MPCM, PQFT, and WASPCM).

4.3. Comparison Analysis for Single-Frame Detection

Figure 10, and Tables 2 and 3 present the comparison of ADMD, MPCM, PQFT, and WASPCM with our proposed method, to demonstrate its efficacy against intricate sea, sky, and ground clutters.

Table 2. EASCRG, EABSF, and the computational cost of the proposed method and four baseline methods for Seq.1, Seq.2, Seq.3, and Seq.4.

	Index / Metrics	Seq.1	Seq.2	Seq.3	Seq.4
Proposed	EASCRG	30.5	2.89	12.5	66.2
	EABSF	2099.5	645	1129.6	1444.4
	Time(s)	0.6027	0.1379	1.1598	0.1316
ADMD [31]	EASCRG	26.9726	83.6801	11.8651	47.0298
	EABSF	0.6574	0.4434	0.0977	0.3810
	Time(s)	0.2083	0.0351	0.9446	0.0298
MPCM [29]	EASCRG	16.7465	2.8000	8.0934	28.8780
	EABSF	3077.1	786.6690	1684.5	2563.3
	Time(s)	0.8113	0.1489	1.7263	0.1361
PQFT [34]	EASCRG	4.3851	0.7913	2.1392	13.6803
	EABSF	1434.04	676.3588	850.3719	1398.2.
	Time(s)	0.3138	0.0360	0.3026	0.3076
WASPCM [36]	EASCRG	4.2918	0.8718	5.35	7.2413
	EABSF	1371.1	663.5638	1667.3	1831.1
	Time(s)	1.3857	0.1674	0.8887	0.1383

Table 3. The ensemble average SCRG with combination of different characteristics.

Each Characteristic	Ensemble Average SCRG
Gaussian	7.5
Gaussian + motion	9.9
Gaussian + contrast	11.7
Gaussian + contrast + motion	12.9

Figure 10 shows the 3D gray distributions obtained when applying ADMD, MPCM, PQFT, and WASPCM. These methods enhance the target while suppressing clutter. Our proposed method outperforms the other methods attributed to its use of Gaussian, contrast, and motion features for target detection, which significantly enhances the target and suppresses the background clutter. For instance, the representative infrared image chosen from Seq.2 has a lot of background clutters, which introduces many false targets. Our proposed method relies on motion characteristics and can obtain real targets. However, ADMD, MPCM, PQFT, and WASPCM cannot accurately find real targets. Obviously, our proposed method surpasses the other four methods, enabling efficient target detection.

To quantitatively compare our proposed method with four other methods in single-frame detection, the ensemble average SCRG (EASCRG) and ensemble average BSF (EABSF) were taken as indicators. Table 2 shows the detection results for Seq.1, Seq.2, Seq.3, and Seq.4. It can be seen that our proposed method achieves the highest ensemble average SCRG in all four sequences, outperforming the other four methods. Although the MPCM method achieves the highest ensemble average BSF, it consumes more time compared to our proposed method. Though the ADMD method has the lowest time consumption in Seq.1, Seq.2, and Seq.4, it obtains worse ensemble average SCRG and ensemble average BSF than our proposed method. Similarly, the PQFT method has the lowest time consumption in Seq.3, but it obtains lower ensemble average SCRG and ensemble average BSF than our proposed method. The WASPCM method has inferior ensemble average SCRG and ensemble average BSF and higher computational cost than our proposed method. Therefore, our proposed method achieves an optimal detection effect in small infrared target detection, especially under the presence of complex sea, sky, and ground clutters.

Meanwhile, an ablation experiment was conducted to demonstrate that the use of each characteristic (Gaussian, contrast, and motion characteristics) in our proposed method helps to enhance the target and suppress the background clutters. The test results of all four sequences are listed in Table 3, and the evaluation metric is ensemble average SCRG. It can be observed that our proposed method with only Gaussian characteristics obtained the lowest performance. Both the contrast and motion characteristics can improve performance. The combination of different characteristics contributes to the best performance with a maximum value of 27.6.

Finally, the sub-window scale of the corresponding image pixel is obtained using the multiscale LOG filter. According to Figure 11, if a simulative small target has a diameter of 7, the multiscale Laplace with a scale-space constant of 3.19 will yield the maximum value. Although calculating multi-scale patches to select the most suitable scale for the target and reduce background clutter is time-consuming, the computation speed can be increased with parallel computations on GPU. To meet the real-time requirement, the algorithm speed can be further enhanced.

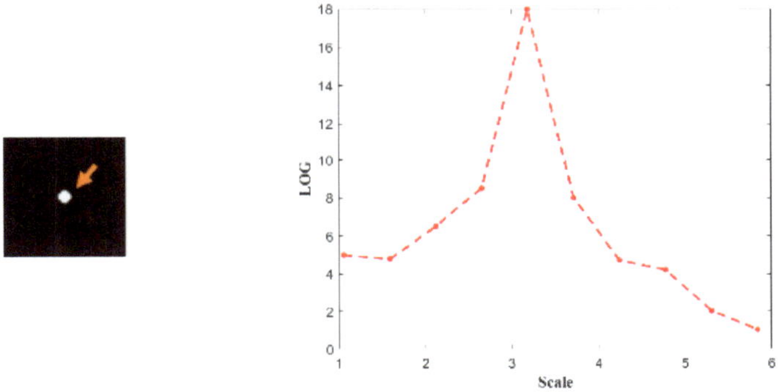

Figure 11. The multiscale Laplace with a scale-space constant 3.19.

4.4. *Effectiveness Analysis for Multi-Frame Detection*

To evaluate the performance of multi-frame target recognition, the TPR and FPR, also known as the precision and false-alarm ratio, are utilized to form the ROC curve. The algorithm's performance is optimal when the precision reaches the maximum under the same FPR. Figure 12 displays the ROC curve of different methods for the entire sequence, including Seq.1, Seq.2, Seq.3, and Seq.4. Our proposed method achieves the highest precision and shows the best detection performance for the entire sequence at the same FPR. The ADMD method has the lowest time consumption in Seq.2 and Seq.4, but it obtains the worst detection performance. The PQFT method and WASPCM method can reach intermediate detection performance to find small infrared targets. The MPCM method, because of its best background suppression effect, can obtain good detection results in most cases. However, its performance is not as good as our proposed method.

Our method was evaluated with qualitative and quantitative analysis methods in this study, and it achieves the highest level of target significance and effectively reduces background clutter, thereby achieving the best detection performance.

It should be noted that, in the capture phase, accurately detecting small infrared targets is important because this is the basis of tracking in military applications. Therefore, it is crucial to study small infrared target detection.

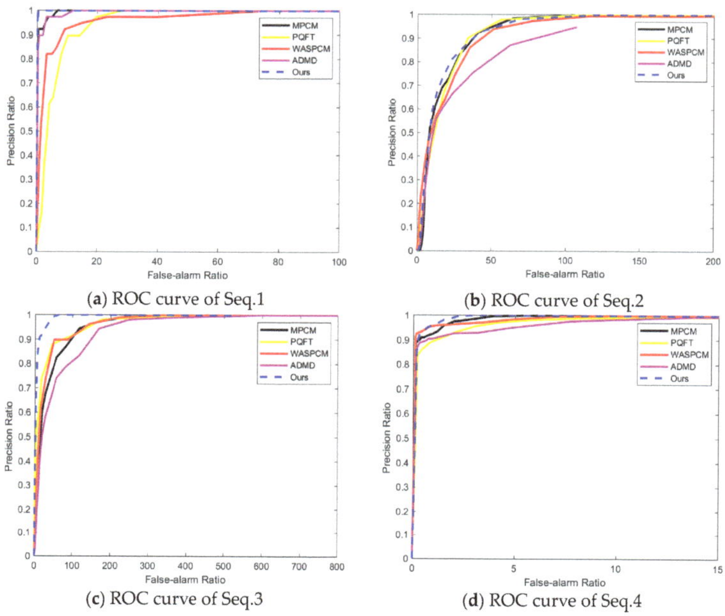

(**a**) ROC curve of Seq.1 (**b**) ROC curve of Seq.2 (**c**) ROC curve of Seq.3 (**d**) ROC curve of Seq.4

Figure 12. The ROC curve of four entire sequences.

5. Conclusions

This paper proposes an infrared small target detection method based on an attention mechanism against complex sea, sky, and ground clutters. Firstly, the top-down active attention module obtains knowledge-experience Gaussian shape features and achieves a complementary fusion of different feature information. Then, an optimal fast local contrast algorithm and improved circular pipeline filtering are adopted to design the bottom-up passive attention module, and the difference in contrast and motion features between the target and the back-ground is found. Finally, decision feedback equalization is adopted to fuse the results of the top-down active attention module and the bottom-up passive attention module, thereby detecting real targets, and this is useful in different scenarios to better adapt to environmental changes. Experiments were conducted on actual infrared images with intricate backgrounds involving sea, sky, and ground clutters, and experimental results reveal that the proposed method can proficiently identify targets and deliver exceptional detection results compared to other methods such as ADMD, MPCM, PQFT, and WASPCM. In particular, the best detection result is obtained in the sea background.

Author Contributions: Conceptualization, X.W. and H.B.; methodology, R.L.; validation, X.W., writing—original draft preparation, Y.L. All authors have read and agreed to the published version of the manuscript.

Funding: The research is supported by the National Natural Science Foundation of China 62106193, the fellowship of China Postdoctoral Science Foundation 2022M712536, the Key Laboratory of Electro-Optical Countermeasures Test & Evaluation Technology Fund GKCP2021002, and the National Key Laboratory of Electromagnetic Space Security foundation JCKY2023230C013.

Institutional Review Board Statement: Not applicable.

Informed Consent Statement: Not applicable.

Data Availability Statement: Not applicable.

Acknowledgments: The authors thank Jie Yan for valuable suggestions and fruitful discussions on this paper.

Conflicts of Interest: The authors declare no conflict of interest.

References

1. Chen, Q.; Wu, C.; Wang, Y. Robust Principal Component Analysis-Based Infrared Small Target Detection. In Proceedings of the AAAI Conference on Artificial Intelligence, Honolulu, HI, USA, 27 January–1 February 2019; Volume 33, pp. 9925–9926.
2. Fan, J.; Wei, J.; Huang, H.; Zhang, D.; Chen, C. IRSDT: A Framework for Infrared Small Target Tracking with Enhanced Detection. *Sensors* **2023**, *23*, 4240. [CrossRef]
3. Xie, F.; Dong, M.; Wang, X.; Yan, J. Infrared Small-Target Detection Using Multiscale Local Average Gray Difference Measure. *Electronics* **2022**, *11*, 1547. [CrossRef]
4. Yao, H.; Liu, L.; Wei, Y.; Chen, D.; Tong, M. Infrared Small-Target Detection Using Multidirectional Local Difference Measure Weighted by Entropy. *Sustainability* **2023**, *15*, 1902. [CrossRef]
5. Dong, L.; Wang, B.; Ming, Z.; Xu, W. Robust Infrared Maritime Target Detection Based on Visual Attention and Spatiotemporal Filtering. *IEEE Trans. Geosci. Remote Sens.* **2017**, *55*, 3037–3050. [CrossRef]
6. Dai, Y.; Wu, Y. Reweighted Infrared Patch-Tensor Model with Both Nonlocal and Local Priors for Single-Frame Small Target Detection. *IEEE J. Sel. Top. Appl. Earth Obs. Remote Sens.* **2017**, *10*, 3752–3767. [CrossRef]
7. Dai, Y.; Wu, Y.; Zhou, F.; Barnard, K. Attentional Local Contrast Networks for Infrared Small Target Detection. *IEEE Trans. Geosci. Remote Sens.* **2021**, *59*, 9813–9824. [CrossRef]
8. Marvasti, F.S.; Mosavi, M.R.; Nasiri, M. Flying small target detection in IR images based on adaptive toggle operator. *IET Comput. Vis.* **2018**, *12*, 527–534. [CrossRef]
9. Wang, X.Y.; Peng, Z.M.; Zhang, P.; He, Y.M. Infrared small target detection via nonnegativity-constrained variational mode decomposition. *IEEE Geosci. Remote Sens. Lett.* **2017**, *14*, 1700–1704. [CrossRef]
10. Achanta, R.; Hemami, S.; Estrada, F.; Susstrunk, S. Frequency-tuned salient region detection. In Proceedings of the IEEE Conference on Computer Vision and Pattern Recognition, Miami, FL, USA, 20–25 June 2009; pp. 1597–1604.
11. Celik, C.; Bilge, H.S. Content based image retrieval with sparse representations and local feature descriptors: A comparative study. *Pattern Recognit.* **2017**, *68*, 1–13. [CrossRef]
12. He, Y.J.; Li, M.; Zhang, J.L.; An, Q. Small infrared target detection based on low-rank and sparse representation. *Infrared Phys. Technol.* **2015**, *68*, 98–109. [CrossRef]
13. Zhang, L.D.; Peng, L.B.; Zhang, T.F.; Cao, S.Y.; Peng, Z.M. Infrared small target detection via nonconvex rank approximation minimization Joint/$_{2,1}$ norm. *Remote Sens.* **2018**, *10*, 1821. [CrossRef]
14. Cui, Z.; Yang, J.L.; Jiang, S.D.; Li, J.B. Infrared Small Target Detect. Algorithm Based High-Speed Local Contrast Method. *Infrared Phys. Technol.* **2016**, *76*, 474–481. [CrossRef]
15. Yi, X.; Wang, B.J.; Zhou, H.X.; Qin, H.L. Dim Small Infrared Target Fast Detect. Guid. By Vis. Saliency. *Infrared Phys. Technol.* **2019**, *97*, 6–14. [CrossRef]
16. Liu, M.; Zhou, Z.; Shang, P.; Xu, D. Fuzzified Image Enhancement for Deep Learning in Iris Recognition. *IEEE Trans. Fuzzy Syst.* **2020**, *28*, 92–99. [CrossRef]
17. Han, J.H.; Liu, S.B.; Qin, G.; Zhao, Q.; Zhang, H.H.; Li, N.N. A local contrast method combined with adaptive background estimation for infrared small target detection. *IEEE Geosci. Remote Sens. Lett.* **2019**, *16*, 1442–1446. [CrossRef]
18. Gao, C.; Meng, D.; Yang, Y.; Wang, Y.; Zhou, X.; Hauptmann, A.G. Infrared Patch-Image Model for Small Target Detection in a Single Image. *IEEE Trans. Image Process.* **2013**, *22*, 4996–5009. [CrossRef]
19. Itti, L.; Niebur, E. A model of saliency-based visual attention for rapid scene analysis. *IEEE Trans. Pattern Anal. Mach. Intell.* **1998**, *20*, 1254–1259. [CrossRef]
20. Wang, X.; Lv, G.; Xu, L. Infrared dim target detection based on visual attention. *Infrared Phys. Technol.* **2012**, *55*, 513–521. [CrossRef]
21. Chen, C.P.; Li, H.; Wei, Y.; Xia, T.; Tang, Y.Y. A Local Contrast Method for Small Infrared Target Detection. *IEEE Trans. Geosci. Remote Sens.* **2013**, *52*, 574–581. [CrossRef]
22. Han, J.; Ma, Y.; Zhou, B.; Fan, F.; Liang, K.; Fang, Y. A Robust Infrared Small Target Detection Algorithm Based on Human Visual System. *IEEE Geosci. Remote Sens. Lett.* **2014**, *11*, 2168–2172.
23. Li, L.; Li, Z.; Li, Y.; Chen, C.; Yu, J.; Zhang, C. Small Infrared Target Detection Based on Local Difference Adaptive Measure. *IEEE Geosci. Remote Sens. Lett.* **2020**, *17*, 1258–1262. [CrossRef]
24. Han, J.; Liang, K.; Zhou, B.; Zhu, X.; Zhao, J.; Zhao, L. Infrared Small Target Detection Utilizing the Multiscale Relative Local Contrast Measure. *IEEE Geosci. Remote Sens. Lett.* **2018**, *15*, 612–616. [CrossRef]
25. Pan, S.D.; Zhang, S.; Zhao, M.; An, B.W. Infrared Small Target Detection Based on Double-layer Local Contrast Measure. *Acta Photonica Sin.* **2020**, *49*, 110003.

26. Nie, J.; Qu, S.; Wei, Y.; Zhang, L.; Deng, L. An Infrared Small Target Detection Method Based on Multiscale Local Homogeneity Measure. *Infrared Phys. Technol.* **2018**, *90*, 186–194. [CrossRef]
27. Zhang, H.; Zhang, L.; Yuan, D.; Chen, H. Infrared small target detection based on local intensity and gradient properties. *Infrared Phys. Technol.* **2017**, *89*, 88–96. [CrossRef]
28. Shi, Y.; Wei, Y.; Yao, H.; Pan, D.; Xiao, G. High-Boost-Based Multiscale Local Contrast Measure for Infrared Small Target Detection. *IEEE Geosci. Remote Sens. Lett.* **2017**, *15*, 33–37. [CrossRef]
29. Wei, Y.; You, X.; Li, H. Multiscale patch-based contrast measure for small infrared target detection. *Pattern Recognit.* **2016**, *58*, 216–226. [CrossRef]
30. Han, J.; Moradi, S.; Faramarzi, I.; Zhang, H.; Zhao, Q.; Zhang, X.; Li, N. Infrared Small Target Detection Based on the Weighted Strengthened Local Contrast Measure. *IEEE Geosci. Remote Sens. Lett.* **2020**, *18*, 1670–1674. [CrossRef]
31. Moradi, S.; Moallem, P.; Sabahi, M. Fast and robust small infrared target detection using absolute directional mean difference algorithm—ScienceDirect. *Signal Process.* **2020**, *177*, 107727. [CrossRef]
32. Yao, S.; Zhu, Q.; Zhang, T.; Cui, W.; Yan, P. Infrared Image Small-Target Detection Based on Improved FCOS and Spatio-Temporal Features. *Electronics* **2022**, *11*, 933. [CrossRef]
33. Ju, M.; Luo, J.; Liu, G.; Luo, H. ISTDet: An efficient end-to-end neural network for infrared small target detection. *Infrared Phys. Technol.* **2021**, *114*, 103659. [CrossRef]
34. Wang, J.; Liu, R.; Wang, H.; Wu, B.; Shi, Y. Quaternion Markov Splicing Detection for Color Images Based on Quaternion Discrete Cosine Transform. *KSII Trans. Internet Inf. Syst.* **2020**, *14*, 2981–2996.
35. Wang, X.; Xie, F.; Liu, W.; Tang, S.; Yan, J. Robust small infrared target detection using multi-scale contrast fuzzy discriminant segmentation. *Expert Syst. Appl.* **2023**, *212*, 118813. [CrossRef]
36. Chen, Y.; Song, B.; Wang, D.; Guo, L. An effective infrared small target detection method based on the human visual attention. *Infrared Phys. Technol.* **2018**, *95*, 128–135. [CrossRef]

sensors

MDPI

Communication

Tool Frame Calibration for Robot-Assisted Ultrasonic Testing

Hanming Zhang [1,*], Jingpin Wang [1] and Canzhi Guo [2]

[1] School of Mechanical and Materials Engineering, North China University of Technology, Beijing 100041, China
[2] School of Mechanical Engineering, Jiangsu University, Zhenjiang 212013, China; guocanzhi@ujs.edu.cn
* Correspondence: zhm@ncut.edu.cn

Abstract: Tool frame calibration has been widely used in robot-assisted printing, welding, and grinding, but it is not suitable for ultrasonic testing because the robot is submerged in water. The purpose of this paper is to present a tool frame calibration method, which is suitable for improving the precision of ultrasonic testing. In uniform mediums, sound travels along a straight line like ray. A reflector is fixed in water to reflect ultrasound, which makes it possible to measure distances between incidence points on a reflector and tool center point (TCP) on an ultrasound transducer. In addition, the positions and poses of the end flange are recorded through a robot controller. Finally, an optimization method is applied to calculate the position and pose errors of the tool frame relative to the end flange according to such records. The presented method was implemented in an ultrasonic testing system. We selected 100 incidence points on the reflector to calculate the assembly errors of the transducer. The pulse amplitude rose obviously after calibration, which verifies that this is an effective method. Considering that ultrasonic transducers can be used as a measuring tool, this paper proposes a tool frame calibration method for ultrasonic testing robots without introducing other measuring devices, which draws the conclusion that tool frame can be calibrated through ultrasound.

Keywords: ultrasonic testing; robot-assisted; calibration; tool frame

Citation: Zhang, H.; Wang, J.; Guo, C. Tool Frame Calibration for Robot-Assisted Ultrasonic Testing. *Sensors* **2023**, *23*, 8820. https://doi.org/10.3390/s23218820

Academic Editors: Zhun Fan and Gang Peng

Received: 25 August 2023
Revised: 25 October 2023
Accepted: 27 October 2023
Published: 30 October 2023

1. Introduction

Ultrasonic testing (UT) is used to detect defects in materials based on the principle of ultrasonic propagation. In the early days, UT was mainly used to ensure product quality. Now, it has been improved for use as an evaluation approach to predict potential risks [1]. The part under test is placed in water and a multi-axial manipulator grasping transducer moves over the part to scan the surface. Frame-type scan machines are widely used as manipulators. However, there are some limitations to such machines when scanning complex structures [2]. Then, general industrial robots were introduced [3]. These are more flexible to achieve complex motions compared to frame-type scan machines.

It is necessary to teach robots paths before running. There are two kinds of teaching methods, i.e., on-line teaching and off-line teaching. Off-line teaching is operated in simulation software, which is more flexible and efficient compared to on-line teaching. However, nominal coordinates inputted in the software may deviate from actual situations due to manufacturing error, assembly error, and so on. These errors cause actual paths to deviate from simulated paths.

Tool assembly error is a cause of path deviation. Some tool frame calibrations have been designed to reduce deviations. For example, Xiaozhi et al. presented a matrix-solving hand-eye calibration method using a laser scanner as a measuring tool to enhance the robotic machining accuracy of complex free form surfaces [4]. Guo et al. calibrated kinematic parameters for an industrial robot using a laser displacement sensor [5]. Lin et al. proposed an enhanced automatic TCP calibration method based on a laser displacement sensor, achieving the purpose of being simple, economical, and time-saving [6]. Similarly, assembly errors in ultrasonic transducers will influence testing precision. It is necessary to design a method to calibrate tool frames for robot-assisted ultrasonic testing.

Many measuring devices with different complexity and cost have been applied in robot calibration. In the early days, some appropriative measuring devices were designed. For example, Renders et al. designed an experimental setup using one magnetostrictive sensor to measure longitudinal displacement and two magnetic transducers to measure transversal displacement [7]. Driels et al. designed a wire potentiometer to measure robot partial pose, which could be used to calibrate PUMA 560 robots [8]. With the development of sensor technology, some universal measuring devices with high precision were applied, including the displacement sensor [9], complementary metal oxide semiconductor (CMOS) camera [10], laser tracker [11], laser interferometer [12], coordinate measuring machine (CMM) [13], etc. On the other hand, some optimization methods for solving non-linear problems have been used for robot calibration. Linear least-squares algorithm was employed for parameter identification to calibrate selective compliance assembly robot arm (SCARA) and a tree-typed modular robot [14]. Omodei et al. used nonlinear optimization, iterative linearization, and extended Kalman filter to identify kinematic parameters of SCARA [15]. Peng et al. used minimal linear combinations to analyze end-effector error and then used linear forecast real-time error compensation to decrease the error [16].

In this research, an ultrasonic transducer grasped by a robot was used as measuring device. A kinematic model was built based on the D–H model. The distance between the transducer and reflector was measured according to time-of-flight (TOF). In error calculation, a constrained non-linear optimization problem, was translated into an unconstrained linear optimization problem. Finally, the assembly errors were identified after several iterations. The experimental results demonstrate that it was feasible to measure and calibrate this tool frame using an ultrasonic robot. The presented method makes it easier to calibrate tool frames for robot-assisted ultrasonic testing system.

2. System Construction

In 2002, Hongwei et al. designed an ultrasonic testing system using a frame-type scan machine as manipulator [17]. It was capable of scanning simple and continuous geometry, but useless for complex curved geometry. A six-degree-of-freedom (DOF)-articulated robot can make up for such shortcomings [18]. The construction of robot-assisted ultrasonic testing system is shown in Figure 1. It consists of three modules: the motion module, the testing module, and the auxiliary module. In the motion module, the industrial robot grasps the transducer scanning over the part. A robot controller is used for motion control and servo drive. The testing module consists of a pulse transmitting–receiving device and transducer. The auxiliary module includes a tank and some other support structures. Positional and signal data are fed into computer for data acquisition, processing, and presentation.

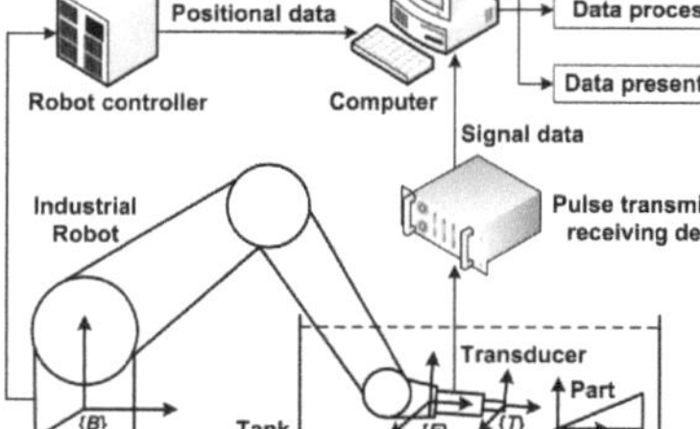

Figure 1. Configuration of robot-assisted ultrasonic testing system.

The robot is fixed on the ground and the part under test is placed in water. The transducer is installed on the end-effector of the robot. Frames are built on components, including the robot base frame {*B*}, end-effector frame {*E*}, tool frame {*T*}, and part frame {*P*}, to describe spatial relationships. All points on the parts are described relative to {*B*} for consistency. The tool frame calibration identifies errors between actual values and nominal values of the tool frame, and errors are due to assembly. Nominal values need to be adjusted during the off-line teaching.

3. Optimization-Based Calibration Method

The kinematic model describes the motion of objects without considering mass and forces. Denavit and Hartenberg presented the D–H model using four parameters associated with a particular convention to describe the position and pose of spatial links. Though many models were developed hereafter, such as the modified D–H (MDH) model, zero reference model, S model, and CPC model, the D–H model remains a popular approach. In this research, the D–H model is used to implement calibration. The values of the tool frame are calculated through an optimization algorithm.

3.1. Robot Kinematic Model

A Staubli-TX90XL industrial robot was used in this research and its structure is shown in Figure 2. The ultrasonic transducer was installed on end-effector of the robot. The frames were built on the base, links, end-effector, and transducer in line with the convention of the DH model.

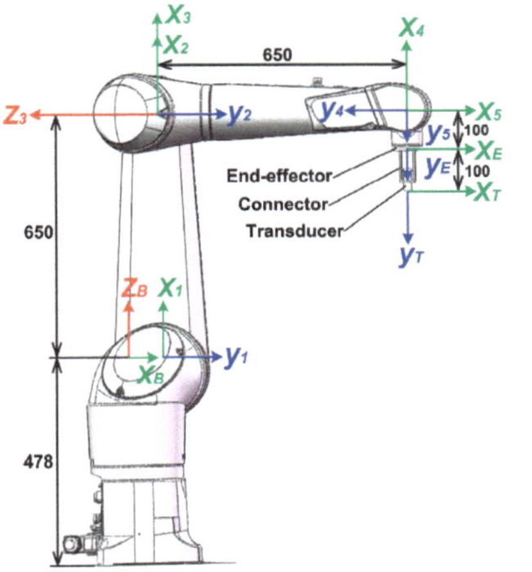

Figure 2. Structure and frames of Staubli-TX90XL industrial robot.

The spatial relationship of two adjacent frames is described by homogeneous transformation matrix $\boldsymbol{T}_{i-1}^{i}$:

$$\boldsymbol{T}_{i-1}^{i} = Rot(Z,\theta_i)Trans(a_i,0,d_i)Rot(X,\alpha_i) = \begin{bmatrix} cos\theta_i & -sin\theta_i cos\alpha_i & sin\theta_i sin\alpha_i & a_i cos\theta_i \\ sin\theta_i & cos\theta_i cos\alpha_i & -cos\theta_i sin\alpha_i & a_i sin\theta_i \\ 0 & sin\alpha_i & cos\alpha_i & d_i \\ 0 & 0 & 0 & 1 \end{bmatrix} \tag{1}$$

where θ_i, a_i, d_i and α_i in Equation (1) are DH parameters: θ_i is the joint angle from the X_{i-1} axis to the X_i axis about the Z_{i-1} axis; a_i is the offset distance from *the* Z_{i-1} axis to *the* Z_i axis along *the* X_i axis; d_i is the joint distance from *the* X_{i-1} axis to *the* X_i axis along xZ_{i-1}

axis; α_i is the offset angle from *the* Z_{i-1} axis to *the* Z_i axis about *the* X_i axis. The values of the D–H parameters are listed in Table 1.

Table 1. DH Parameters of Staubli-TX90XL industrial robot.

Link	θ_i [°]	d_i [mm]	a_i [mm]	α_i [°]
1	θ_1	0	50	−90
2	θ_2−90°	50	650	0
3	θ_3	0	0	90
4	θ_4	−650	0	90
5	θ_5−90°	0	0	90
6	θ_6	100	0	0

Bringing above values into Equation (1), the spatial relationships of two adjacent links were determined as shown in Equation (2), where c_i and s_i are a simplification of $cos\theta_i$ and $sin\theta_i$.

$$
\begin{aligned}
&\boldsymbol{T}_0^1=\begin{bmatrix} c_1 & 0 & -s_1 & a_1c_1 \\ s_1 & 0 & c_1 & a_1s_1 \\ 0 & -1 & 0 & d_1 \\ 0 & 0 & 0 & 1 \end{bmatrix} \boldsymbol{T}_1^2=\begin{bmatrix} c_2 & -s_2 & 0 & a_2c_2 \\ s_2 & c_2 & 0 & a_2s_2 \\ 0 & 0 & 1 & d_2 \\ 0 & 0 & 0 & 1 \end{bmatrix} \\
&\boldsymbol{T}_2^3=\begin{bmatrix} c_3 & 0 & s_3 & 0 \\ s_3 & 0 & -c_3 & 0 \\ 0 & 1 & 0 & 0 \\ 0 & 0 & 0 & 1 \end{bmatrix} \boldsymbol{T}_3^4=\begin{bmatrix} c_4 & 0 & s_4 & 0 \\ s_4 & 0 & -c_4 & 0 \\ 0 & 1 & 0 & d_4 \\ 0 & 0 & 0 & 1 \end{bmatrix} \\
&\boldsymbol{T}_4^5=\begin{bmatrix} c_5 & 0 & s_5 & 0 \\ s_5 & 0 & -c_5 & 0 \\ 0 & 1 & 0 & 0 \\ 0 & 0 & 0 & 1 \end{bmatrix} \boldsymbol{T}_5^6=\begin{bmatrix} c_6 & -s_6 & 0 & 0 \\ s_6 & c_6 & 0 & 0 \\ 0 & 0 & 1 & d_6 \\ 0 & 0 & 0 & 1 \end{bmatrix}
\end{aligned} \tag{2}
$$

Furthermore, the spatial relationships of two arbitrary links were calculated through matrix multiplication. For example, the relationship between frames {E} and {B} in Figure 2 is expressed as

$$
\boldsymbol{T}_B^E=\boldsymbol{T}_B^1\boldsymbol{T}_1^2\boldsymbol{T}_2^3\boldsymbol{T}_3^4\boldsymbol{T}_4^5\boldsymbol{T}_5^E=\begin{bmatrix} n_x^E & o_x^E & a_x^E & p_x^E \\ n_y^E & o_y^E & a_y^E & p_y^E \\ n_z^E & o_z^E & a_z^E & p_z^E \\ 0 & 0 & 0 & 1 \end{bmatrix} \tag{3}
$$

Equation (3) is simplified as

$$
\begin{bmatrix} n_x^E & o_x^E & a_x^E \\ n_y^E & o_y^E & a_y^E \\ n_z^E & o_z^E & a_z^E \end{bmatrix}=\boldsymbol{R}_B^E,\ \begin{bmatrix} p_x^E \\ p_y^E \\ p_y^E \end{bmatrix}=\boldsymbol{P}_B^E\ \Rightarrow\ \boldsymbol{T}_B^E=\begin{bmatrix} \boldsymbol{R}_B^E & \boldsymbol{P}_B^E \\ \boldsymbol{0} & 1 \end{bmatrix} \tag{4}
$$

where $\boldsymbol{R}_B^E$ and $\boldsymbol{P}_B^E$ describe the pose and position of {E} relative to {B}. Similarly, the relationship between {T} and {B} is expressed as

$$
\boldsymbol{T}_B^T=\boldsymbol{T}_B^1\boldsymbol{T}_1^2\boldsymbol{T}_2^3\boldsymbol{T}_3^4\boldsymbol{T}_4^5\boldsymbol{T}_5^E\boldsymbol{T}_E^T=\begin{bmatrix} \boldsymbol{R}_B^T & \boldsymbol{P}_B^T \\ \boldsymbol{0} & 1 \end{bmatrix} \tag{5}
$$

where $\boldsymbol{R}_B^T$ and $\boldsymbol{P}_B^T$ describe the pose and position of {T} relative to {B}; $\boldsymbol{T}_E^T$ describes spatial relationship between {T} and {E}. The tool frame calibration in this research was to correct values in $\boldsymbol{T}_E^T$.

3.2. *Error Calibration*

Laser trackers, CMM, and other measuring devices have been widely used to calibrate the tool frame in robotic systems. In this research, an ultrasonic transducer was introduced to calibrate the tool frame, which is uncommon in previous reports.

Ultrasonic transducers are sound–electricity converting components based on the piezoelectric effect in semiconductor elements. In the transmitting state, the pulse transmitting–receiving device transmits a narrow-pulse to drive the transducers. Then, the device turns into the receiving state and a series of echoes is received and converted into electrical signal. The driving pulse and subsequent echoes received with different incidence distances and angles are shown in Figure 3. Comparing Figure 3a,b, the TOF between the driving pulse and surface echo pulse changes linearly with the distance between transducer and reflector; comparing Figure 3a,c, the amplitude of the echo pulse decreases when the incident beam deflects a certain angle from the normal direction. Consequently, the distance between the transducer and the reflector can be measured based on the principle of ultrasonic propagation when it is conducted in homogeneous mediums like water.

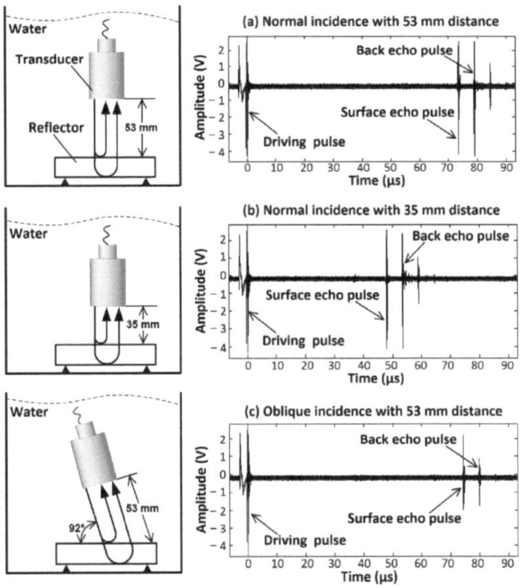

Figure 3. Waveform of ultrasonic with different distance and angle.

In Figure 4, there is a fixed spatial relationship between the robot and the reflector. The ultrasonic beam transmits Δl along Z axis of $\{T\}$ and intersects at point N on the reflector. A frame $\{N\}$ is fixed at point N to describe the position and pose of N in space. It can be inferred from Equation (5) and coordinate conversion that the position and pose of N relative to $\{B\}$ is

$$\begin{gathered} \boldsymbol{T}_B^N = \boldsymbol{T}_B^E \boldsymbol{T}_E^T \Delta \boldsymbol{L} \\ \Rightarrow \begin{bmatrix} \boldsymbol{R}_B^N & \boldsymbol{P}_B^N \\ \boldsymbol{0} & 1 \end{bmatrix} = \begin{bmatrix} \boldsymbol{R}_B^E & \boldsymbol{P}_B^E \\ \boldsymbol{0} & 1 \end{bmatrix} \begin{bmatrix} \boldsymbol{R}_E^T & \boldsymbol{P}_E^T \\ \boldsymbol{0} & 1 \end{bmatrix} \Delta \boldsymbol{L} \end{gathered} \tag{6}$$

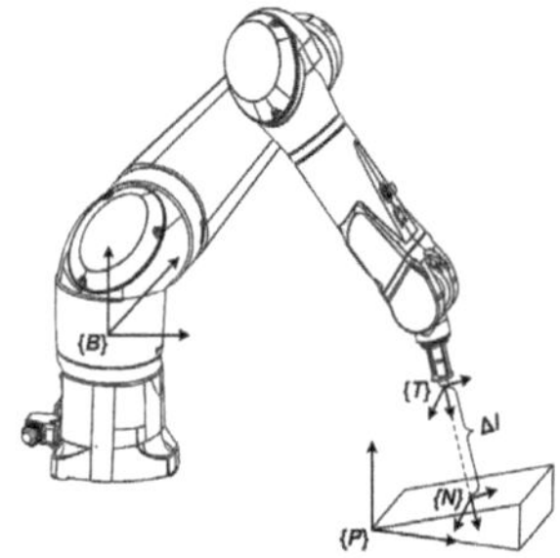

Figure 4. Measuring distance between transducer and reflector.

The elements of the matrixes in Equation (6) are

$$T_B^N = \begin{bmatrix} n_x^N & o_x^N & a_x^N & p_x^N \\ n_y^N & o_y^N & a_y^N & p_y^N \\ n_z^N & o_z^N & a_z^N & p_z^N \\ 0 & 0 & 0 & 1 \end{bmatrix} \quad T_B^E = \begin{bmatrix} n_x^E & o_x^E & a_x^E & p_x^E \\ n_y^E & o_y^E & a_y^E & p_y^E \\ n_z^E & o_z^E & a_z^E & p_z^E \\ 0 & 0 & 0 & 1 \end{bmatrix}$$

$$T_E^T = \begin{bmatrix} n_x^T & o_x^T & a_x^T & p_x^T \\ n_y^T & o_y^T & a_y^T & p_y^T \\ n_z^T & o_z^T & a_z^T & p_z^T \\ 0 & 0 & 0 & 1 \end{bmatrix} \quad \Delta L = \begin{bmatrix} 1 & 0 & 0 & 0 \\ 0 & 1 & 0 & 0 \\ 0 & 0 & 1 & \Delta l \\ 0 & 0 & 0 & 1 \end{bmatrix}$$

The elements above have following relationship:

$$\begin{pmatrix} p_x^N \\ p_y^N \\ p_z^N \end{pmatrix} = \begin{pmatrix} p_x^E \\ p_y^E \\ p_z^E \end{pmatrix} + \begin{pmatrix} n_x^E & o_x^E & a_x^E \\ n_y^E & o_y^E & a_y^E \\ n_z^E & o_z^E & a_z^E \end{pmatrix} \begin{pmatrix} \Delta l a_x^T + p_x^T \\ \Delta l a_y^T + p_y^T \\ \Delta l a_z^T + p_z^T \end{pmatrix} \tag{7}$$

Among which, (n_x^E, n_y^E, n_z^E), (o_x^E, o_y^E, o_z^E), (a_x^E, a_y^E, a_z^E), and (p_x^E, p_y^E, p_z^E) are calculated according to the position feedback from the robot controller and Δl is measured by the transducer. (a_x^T, a_y^T, a_z^T) is a unit vector that satisfies a constrain:

$$\left(a_x^T\right)^2 + \left(a_y^T\right)^2 + \left(a_z^T\right)^2 = 1 \tag{8}$$

All incidence points on the reflector, such as N, are in line with a plane equation. Assuming that, there are three points on the reflector, noted as A (x_A, y_A, z_A), B (x_B, y_B, z_B), and C (x_C, y_C, z_C). The normal vector of the reflector is described as

$$\begin{aligned} \boldsymbol{n} &= (n_x, n_y, n_z) \\ &= [(y_B - y_A)(z_C - z_A) - (z_B - z_A)(y_C - y_A)]\boldsymbol{i} \\ &+ [(z_B - z_A)(x_C - x_A) - (x_B - x_A)(z_C - z_A)]\boldsymbol{j} \\ &+ [(x_B - x_A)(y_C - y_A) - (y_B - y_A)(x_C - x_A)]\boldsymbol{k} \end{aligned} \tag{9}$$

The plane equation is determined through the coordinates of A, B, and C, which are described as

$$n_x(x - x_A) + n_y(y - y_A) + n_z(z - z_A) = 0 \tag{10}$$

Point $\boldsymbol{N}$ (p_x^N, p_y^N, p_z^N) shown in Figure 4 is in line with Equation (10):

$$n_x\left(p_x^N - x_A\right) + n_y(p_y^N - y_A) + n_z(p_z^N - z_A) = 0 \tag{11}$$

The values of $\boldsymbol{t}$ (p_x^T, p_y^T, p_z^T, a_x^T, a_y^T, a_z^T) are unknown when taking Equation (7) into Equation (11). A measure and calculation process was designed to acquire the values.

The measuring process is shown in Figure 4. m incidence points were selected on the reflector. The transducer grasped by the robot was moved to the selected points with different poses. The spatial data of $\{E\}$ relative to $\{B\}$ were recorded in each point, and then the data were taken into Equation (11) and an optimization problem was established as

$$Objective\ function : f(\boldsymbol{t}) = \sum_{i=1}^{m} \left[n_x \left(p_x^i - x_A \right) + n_y (p_y^i - y_A) + n_z (p_z^i - z_A) \right]^2 \tag{12}$$

$$Constrain\ function : c(\boldsymbol{t}) = \left(a_x^T \right)^2 + \left(a_y^T \right)^2 + \left(a_z^T \right)^2 - 1 = 0 \tag{13}$$

Equations (12) and (13) constitute a non-linear optimization problem, which is difficult to solve directly. They were transformed into a linear optimization problem without constraint, which can be solved according to essential and sufficient conditions.

According to first-order essential condition, the objective and constrain function were rewritten together making up a Lagrange function:

$$L(\boldsymbol{t}, \lambda) = f(\boldsymbol{t}) + \lambda \cdot c(\boldsymbol{t}) \tag{14}$$

among which λ is noted as a Lagrange multiplier. Assuming that $\boldsymbol{t}^*$ is an optimal solution, $f(\boldsymbol{t})$ and $c(\boldsymbol{t})$ are continuous and differentiable in the neighbourhood of $\boldsymbol{t}^*$, then there is a λ^* satisfying:

$$\nabla L(\boldsymbol{t}^*, \lambda^*) = \nabla f(\boldsymbol{t}^*) + \lambda^* \cdot \nabla c(\boldsymbol{t}^*) = 0 \tag{15}$$

Equation (15) transforms a non-linear optimization problem with equality constraint into a linear optimization problem that is solved by iteration. While $\boldsymbol{t}^*$ acquired from iterations could be an extremal point, it needed to be further confirmed through second-order sufficient condition. The Hesse matrix relative to the Lagrange function was used. If it was a positive definite matrix when taking $\boldsymbol{t}^*$ into Hesse matrix, then $\boldsymbol{t}^*$ was confirmed as an extremal point. Consequently, $\boldsymbol{t}^*$ was the corrected coordinates of $\{T\}$ relative to $\{E\}$, which are the calibration results.

4. Experiment Results

An experimental platform in line with Section 2 was established, as shown in Figure 5, to verify the proposed calibration. The industrial robot used in the experiment was produced by Staubli International AG (Pfäffikon, Switzerland), model TX90XL. The pulse transmitting–receiving device is an ultrasonic pulser/receiver produced by Olympus Corporation (Tokyo, Japan), model 5077PR. An ultrasonic transducer with a frequency of 25 MHz and an element size of 6 mm was selected. The model of the transducer is V324-SU-F5.25IN, which was also produced by Olympus Corporation.

Figure 5. Experimental platform based on Staubli robot.

It was demonstrated in Section 3 that all the incidence points on the reflector in line with a plane equation and the equation can be determined through three points on the reflector. Contact measuring was implemented, as shown in Figure 6, to acquire the coordinates of the three points. A standard tool was installed on the end-effector and a board (20 mm × 30 mm) was fixed in water.

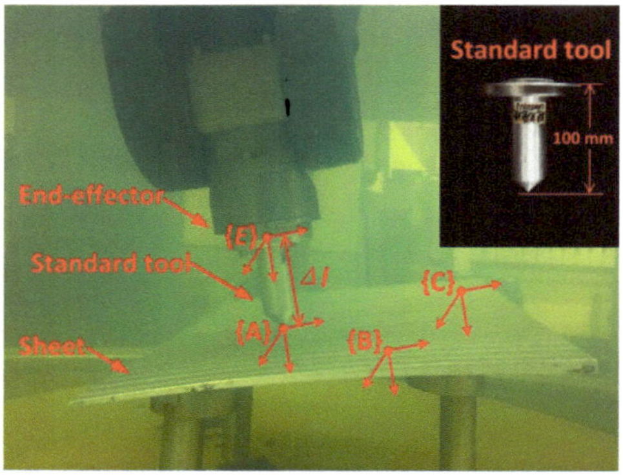

Figure 6. Contact measuring through standard tool to acquire plane equation.

The specific measuring steps were as follows:

Step 1 The robot is operated by making the standard tool contact the board with an arbitrary pose;

Step 2 The position coordinates of the contact point (such as A) relative to $\{B\}$ are calculated through equation (6);

Step 3 Step 1 and 2 are repeated three times to acquire the position coordinates of point A, B, and C.

Step 4 The plane equation is calculated according to Equation (10).

Measuring steps 1–4 are repeated three times to reduce the influence of deformation when contacting the board. In addition, the standard tool can be replaced with a laser sensor to avoid deformation, but the measuring steps need to be implemented before pouring water into the tank. The coordinates of points are shown in Table 2. The plane equation calculated according to the points is

$$-0.0088x - 0.13y + 0.99z + 864.81 = 0 \tag{16}$$

Table 2. Coordinates of nine points.

i	x_i (mm)	y_i (mm)	z_i (mm)
1	1035.06	366.72	−814.92
2	1035.04	519.10	−794.92
3	992.91	509.89	−796.71
4	992.96	341.27	−819.36
5	956.59	361.46	−817.28
6	956.56	536.33	−793.95
7	836.25	536.20	−795.29
8	844.11	371.18	−817.68
9	726.88	384.30	−816.16

Then, the standard tool was replaced with an ultrasonic transducer and the board was used as a reflector as shown in Figure 7. The robot was operated by moving over the reflector with a different pose. During the operation 100 points and corresponding TOF were recorded.

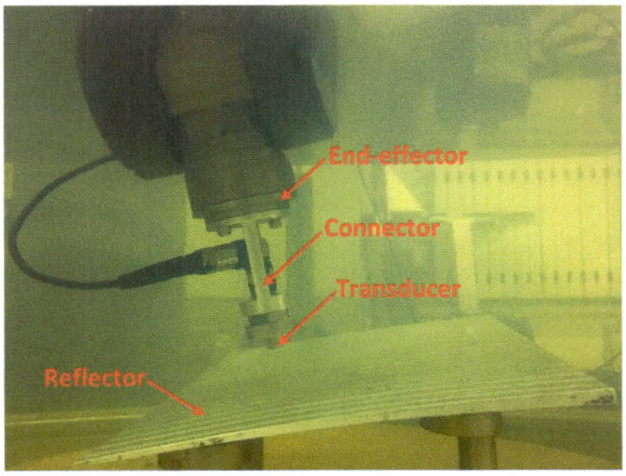

Figure 7. Non-contact measuring through ultrasonic transducer to acquire distance.

Several points were selected from 100 incidence points to calculate the coordinates of the tool frame based on the optimization algorithm presented in Section 3. The number of selected points was increased by ten each time from 1 to 100. The position errors along each axis and the angle errors around each axis between the calculated values and nominal values are shown in Figure 8a,b. The trend in the curves in Figure 8 demonstrate that the position and angle error decreased with the number of selected points. Consequently, fewer points can not obtain stable results and the calculated results' trend become stable when the selected points are greater than 70. The calibration results are shown in Table 3.

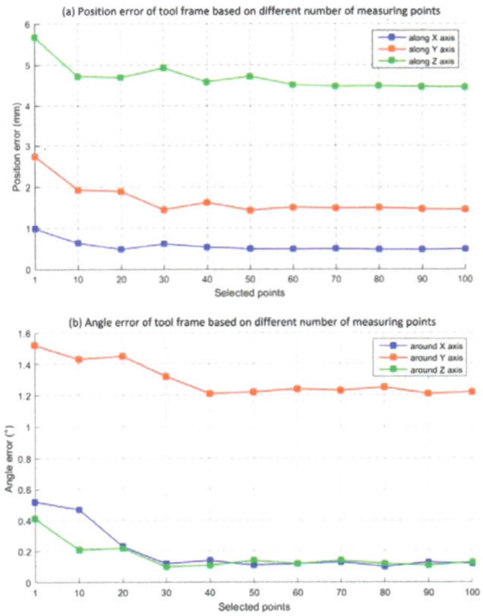

Figure 8. Position and angle errors of tool frame with different number of measuring points.

Table 3. Tool frame calibration results.

Axis	Position Error (mm)	Angle Error (mm)
X	0.48	0.12
Y	1.48	1.22
Z	4.45	0.13

If there is a slight offset between the beam and the normal direction of the reflector, pulse amplitude will descend as shown in Figure 3c. This principle can be used to verify the effectiveness of the calibration method. The pulse amplitudes of the 100 incidence points before and after calibration are shown in Figure 9. The pulse amplitude of each measuring point after calibration is higher than that before calibration, which demonstrates that this is an effective calibration method.

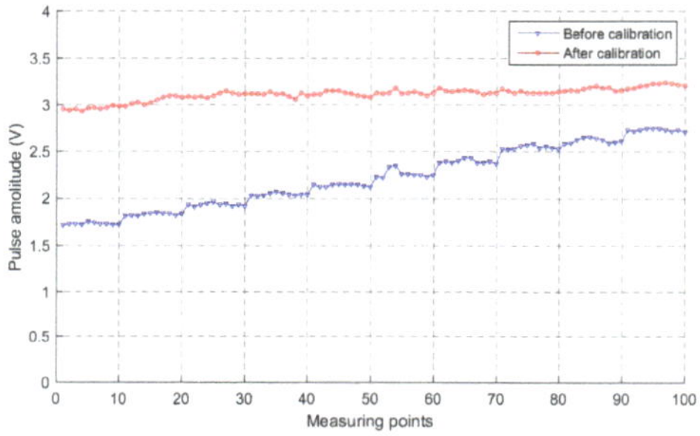

Figure 9. Comparison of pulse amplitude before and after calibration.

5. Conclusions

Assembly errors in the transducer installed on a robot will influence ultrasonic testing precision. In this work, a tool frame calibration method based on ultrasonic principles is presented. The tool frame can be calibrated using a piece of board, which is economical and easily operated. Four conclusions are drawn as follows:

(1) A robot-assisted system which is more suitable to test complex curved parts compared with frame-type scan machines is proposed.
(2) The DH model is used to describe the statical relationship of each component. The tool frame calibration is modelled and translated into a mathematical problem to calculate the values in T_E^T.
(3) The calculation is described as a constrained non-linear optimization problem, then it is transformed into a linear optimization problem that can be solved by iteration.
(4) In this experiment, we find that the incidence points should be scattered on the reflector; the propagation distance should be different on each incidence point; and the transducer pose should be distributed in different angles.

The ultrasonic transducer can only measure distance in a single direction. Assembly errors are calculated by iteration. Though the calibration accuracy of this method is lower than that of universal measuring devices, such as laser trackers or coordinate measuring machines, it is suitable for underwater applications. Our future work will aim at improve the iteration accuracy.

Author Contributions: C.G. designed and established the experimental platform; all measurements and data processing were performed by J.W.; H.Z. wrote the draft and edited the paper. All authors have read and agreed to the published version of the manuscript.

Funding: This research was funded by [Research Start-up Fund of North China University of Technology] grant number [110051360023XN224-31] And The APC was funded by [Innovation and entrepreneurship training program for college students].

Institutional Review Board Statement: The study did not require ethical approval.

Informed Consent Statement: The study did not involve humans.

Data Availability Statement: The data are contained within the article.

Conflicts of Interest: The authors declare no conflict of interest.

References

1. Liu, S.; Liu, F.; Yang, Y.; Li, L.; Li, Z. Nondestructive Evaluation 4.0: Ultrasonic Intelligent Nondestructive Testing and Evaluation for Composites. *Res. Nondestruct. Eval.* **2020**, *31*, 370–388.
2. Mineo, C.; MacLeod, C.; Morozov, M.; Pierce, S.G.; Lardner, T.; Summan, R.; Powell, J.; McCubbin, P.; McCubbin, C.; Munro, G.; et al. Fast ultrasonic phased array inspection of complex geometries delivered through robotic manipulators and high speed data acquisition instrumentation. In Proceedings of the IEEE International Ultrasonics Symposium (IUS), Tours, France, 18–21 September 2016.
3. Guo, C.; Xu, C.; Hao, J.; Xiao, D.; Yang, W. Ultrasonic Non-Destructive Testing System of Semi-Enclosed Workpiece with Dual-Robot Testing System. *Sensors* **2019**, *19*, 3359. [CrossRef]
4. Feng, X.; Tian, D.; Wu, H.; Qian, C.; Zhu, D. A matrix-solving hand-eye calibration method considering robot kinematic errors. *J. Manuf. Process* **2023**, *99*, 618–635.
5. Guo, Y.; Song, B.; Tang, X.Q.; Zhou, X.D.; Xie, Y.L.; Jin, J. Calibration for kinematic parameters of industrial robot by a laser displacement sensor. In Proceedings of the International Conference on Control, Automation, Robotics and Vision (ICARCV), Singapore, 18–21 November 2018.
6. Lin, C.J.; Wang, H.C.; Wang, C.C. Automatic Calibration of Tool Center Point for Six Degree of Freedom Robot. *Actuators* **2023**, *12*, 107. [CrossRef]
7. Renders, J.M.; Rossignol, E.; Becquet, M.; Hanus, R. Kinematic calibration and geometrical parameter identification for robots. *IEEE T. Robotic. Autom.* **1991**, *7*, 721–732. [CrossRef]
8. Driels, M.R.; Swayze, W.E. Automated partial pose measurement system for manipulator calibration experiments. *IEEE Trans. Robot. Autom.* **1994**, *10*, 430–440. [CrossRef]
9. Liu, Y.; Zhuang, Z.; Li, Y. Closed-Loop Kinematic calibration of robots using a six-point measuring device. *IEEE Trans. Instrum. Meas.* **2022**, *71*, 1–12.
10. Tang, X.; Zhou, H.; Xu, T. A geometric errors identification method for the rotating axis of five-axis welding equipment. *Int. J. Precis. Eng. Manuf.* **2023**, *24*, 1355–1367. [CrossRef]
11. Xiao, P.; Ju, H.; Li, Q.; Meng, J.; Chen, F. A new fixed axis-invariant based calibration approach to improve absolute positioning accuracy of manipulators. *IEEE Access* **2020**, *8*, 134224–134232.
12. Soichi, I.; Ryota, U. A novel error mapping of bi-directional angular positioning deviation of rotary axes in a SCARA-type robot by "open-loop" tracking interferometer measurement. *Precis. Eng.* **2022**, *74*, 60–68.
13. Gharaaty, S.; Shu, T.; Joubair, A.; Xie, W.; Bonev, I.A. Online pose correction of an industrial robot using an optical coordinate measure machine system. *Int. J. Adv. Robot. Syst.* **2018**, *15*, 1729881418787915.
14. Chen, I.-M.; Yang, G.; Tan, C.T.; Yeo, S.H. Local POE model for robot kinematic calibration. *Mech. Mach. Theory* **2001**, *36*, 1215–1239.
15. Omodei, A.; Legnani, G.; Adamini, R. Three Methodologies for the Calibration of Industrial Manipulators: Experimental Results on a SCARA Robot. *J. Robot. Syst.* **2000**, *17*, 291–307.
16. Chang, P.; Li, C.; Li, T. Kinematic calibration and forecast error compensation of a 2-DOF planar parallel manipulator. *Chin. J. Mech. Eng.* **2011**, *24*, 992–998.
17. Ma, H.W.; Zhang, X.H.; Wei, J. Research on an ultrasonic NDT system for complex surface parts. *J. Mater. Process. Technol.* **2002**, *129*, 667–670.
18. Aparicio Secanellas, S.; Gauna León, I.; Parrilla, M.; Acebes, M.; Ibáñez, A.; de Matías Jiménez, H.; Martínez-Graullera, Ó.; Álvarez de Pablos, A.; González Hernández, M.; Anaya Velayos, J.J. Methodology for the Generation of High-Quality Ultrasonic Images of Complex Geometry Pieces Using Industrial Robots. *Sensors* **2023**, *23*, 2684.

sensors

MDPI

Article

Reprojection Error Analysis and Algorithm Optimization of Hand–Eye Calibration for Manipulator System

Gang Peng [1,2], Zhenyu Ren [1,2], Qiang Gao [1,2,*] and Zhun Fan [2,3,*]

1 School of Artificial Intelligence and Automation, Huazhong University of Science and Technology, Wuhan 430074, China; penggang@hust.edu.cn (G.P.); 13781366835@163.com (Z.R.)
2 Key Laboratory of Image Processing and Intelligent Control, Ministry of Education, Wuhan 430074, China
3 College of Engineering, Shantou University, Shantou 515063, China
* Correspondence: gao_qiang@hust.edu.cn (Q.G.); zfan@stu.edu.cn (Z.F.)

Abstract: The Euclidean distance error of calibration results cannot be calculated during the hand–eye calibration process of a manipulator because the true values of the hand–eye conversion matrix cannot be obtained. In this study, a new method for error analysis and algorithm optimization is presented. An error analysis of the method is carried out using a priori knowledge that the location of the augmented reality markers is fixed during the calibration process. The coordinates of the AR marker center point are reprojected onto the pixel coordinate system and then compared with the true pixel coordinates of the AR marker center point obtained by corner detection or manual labeling to obtain the Euclidean distance between the two coordinates as the basis for the error analysis. We then fine-tune the results of the hand–eye calibration algorithm to obtain the smallest reprojection error, thereby obtaining higher-precision calibration results. The experimental results show that, compared with the Tsai–Lenz algorithm, the optimized algorithm in this study reduces the average reprojection error by 44.43% and the average visual positioning error by 50.63%. Therefore, the proposed optimization method can significantly improve the accuracy of hand–eye calibration results.

Keywords: hand–eye calibration; reprojection error analysis; manipulator object grasping

Citation: Peng, G.; Ren, Z.; Gao, Q.; Fan, Z. Reprojection Error Analysis and Algorithm Optimization of Hand–Eye Calibration for Manipulator System. *Sensors* **2024**, *24*, 113. https://doi.org/10.3390/s24010113

Academic Editor: Andrey V. Savkin

Received: 27 October 2023
Revised: 19 December 2023
Accepted: 22 December 2023
Published: 25 December 2023

1. Introduction

With the development of robotics and sensor technology [1–4], robotic grasping has gradually become a significant function for robots. It involves identifying and locating a target object through a visual sensor. Therefore, many sensor calibration methods [5,6] have emerged, including hand–eye calibration algorithms [7]. To ensure the accuracy of visual information and to achieve coordinated hand–eye motion, it is essential to analyze the manipulator's hand–eye visual calibration problem and to improve its accuracy. Hand–eye calibration establishes a transformation matrix between the pixel coordinate system of the camera and spatial coordinate system of the manipulator. By transforming the pixel coordinates into the coordinate system of the manipulator, the robot can calculate the motor movements necessary to reach the target position and to control the manipulator. Hand–eye calibration can be divided into eye-to-hand and eye-in-hand calibrations, depending on the camera installation position. This study focuses on experiments using an eye-in-hand installation.

Traditional calibration methods build calibration models based on preset imaging scenes and select optimal algorithms, including reference-based, active vision, and self-calibration methods, to calculate camera parameters based on scene geometric constraints. Based on the reference object camera calibration method, the corner points of the target image were extracted as control points, a system of equations corresponding to the pixels and spatial coordinates was established, and an optimization algorithm was used to calculate the parameters in which the shape and size of the reference object and other information were known. The reference object is divided into one-dimensional straight lines [8], two-dimensional plane calibration plates, and three-dimensional solid blocks, according to the spatial dimension. Because of their simple production and controllable accuracy, flat calibration plates are often used as targets in industrial applications instead of calibration blocks. Commonly used calibration-plate patterns include checkerboards, solid circles [9], and concentric rings. Recently, various templates have been proposed for this purpose [10]. The Zhengyou calibration method [11], based on a chessboard calibration board, is a classic representative of this type of method. This method has strong imaging constraints, a simple calibration process, and a high algorithm robustness. This type of method has strong imaging constraints, a simple calibration process, and high algorithm robustness; however, the production and maintenance cost of high-precision reference objects is high, and it is lost in situations where it is impossible to carry the significance of the reference objects. Camera calibration methods based on active vision obtain multiple images by accurately controlling special movements, such as the pure rotation and translation of the camera or target, and use controllable quantitative motion constraints to determine the internal and external information of the camera. This is an important aspect of self-calibration methods. Typical methods include calibrations based on pure rotational motion [12], three orthogonal translation motions, plane orthogonal motion, an infinite plane homography matrix, hand–eye calibration [13], and self-calibration based on projection reconstruction. Active visual calibration technology can linearly solve internal camera parameters with strong algorithm robustness. However, strict requirements for the control equipment limit their use and promotion. The camera self-calibration method does not require the setting up of a reference object or controlling its precise movement. It only uses the geometric consistency constraints of corresponding points in multiple image frames [14,15] to solve the basic camera matrix and does not rely on scene structure and motion information, including directly solving the Kruppa equation, methods based on absolute quadratic curves and absolute quadric surfaces [16,17], Pollefeys module-constrained calibration [18,19], and hierarchical stepwise calibration methods under variable internal parameters [20].

The focus of this study was a hand–eye calibration algorithm. Since 1989, Shiu [21] and Tscai [7] presented the hand–eye calibration problem. Domestic and foreign scholars have conducted considerable research. For example, Chen et al. proposed a noise-tolerant algorithm for robot sensor calibration using a planar disk in any three-dimensional direction [22]. Li et al. proposed a hand–eye calibration method for linear laser sensors based on three-dimensional reconstruction [23]. Zhang et al. proposed a calibration method of a hand–eye system with rotation and translation coupling [24]. The solutions to hand–eye calibration problems can be divided into two categories according to the solution order of the calibration matrix. The first solves the rotation and translation vectors of the matrix simultaneously. A typical algorithm is Andreff 's closed-loop solution of the hand–eye calibration equation based on the matrix direct product for the calibration of small-scale moving measurement scenarios [25]. Tabb proposed a hand–eye calibration algorithm based on an iterative optimization method and solved it using a nonlinear optimizer [26]. Jiang et al. proposed a method to calibrate the hand–eye of the EOD robot by solving the AXB = YCZD problem [27]. The second solves the rotation matrix of the calibration matrix first and then the translation vector of the calibration matrix. The most common method is the method proposed by Tsai and Lenz [7], which introduces the rotation axis–rotation angle system to describe the rotation motion. Liu et al. proposed a hand–eye calibration method for a robot vision measurement system [28]. Zou et al. performed a hand–eye calibration on arc welding robots and laser vision sensors through semi-definite programming [29]. Deng et al. proposed a hand–eye calibration method based on a monocular robot [30]; however, the above algorithm still has some space for optimization.

In the hand–eye calibration process of a real manipulator, the actual error of the calibration result cannot be calculated because the true value of the hand–eye transformation matrix cannot be obtained. Therefore, a new error analysis method is required. A reprojection error analysis method was proposed. The contribution of this study is to determine the reason for certain errors in each calibration algorithm; they ignore the calculation errors in the transformation matrix of the augmented reality (AR) marker coordinate system relative to the camera coordinate system, ${}^{c}_{m}T$, making it difficult to obtain high-precision calibration results. The method proposed in this study innovatively utilizes the prior knowledge that the position marked by the AR is fixed in the calibration process to conduct error analysis and algorithm optimization. First, the coordinates of the center point of the AR mark are reprojected to the pixel coordinate system and are then compared with the real pixel coordinates of the center point of the AR mark obtained by corner detection or manual labeling to obtain the Euclidean distance between the two coordinates, which is the basis for error analysis. Finally, we fine-tuned the results of the hand–eye calibration algorithm to obtain the smallest reprojection error, thereby obtaining high-precision calibration results.

2. Coordinate System Definitions and Hand–Eye Calibration Equation

2.1. Coordinate System Definitions

In the robot arm hand–eye vision grasping system, the visual information obtained through the camera is described in the camera coordinate system. However, the reference coordinate system for robot arm motion planning is the robot arm base coordinate system. Therefore, to realize the vision of a robot arm, grasping usually requires converting the visual information into a robot arm base coordinate system for description. Let ${}^{b}_{e}T$ be the transformation matrix of the robot arm end coordinate system relative to the base coordinate system and ${}^{e}_{c}T$ be the transformation matrix of the camera coordinate system relative to the robot arm end coordinate system. Then, the transformation matrix of the camera coordinate system relative to the base coordinate system of the robotic arm is ${}^{b}_{e}T = {}^{b}_{e}T \cdot {}^{e}_{c}T$. Among them, ${}^{b}_{e}T$ can be calculated using the joint angles of the robotic arm and the forward kinematics equation, whereas the matrix ${}^{e}_{c}T$ is solved through the hand–eye calibration algorithm.

To solve the above hand–eye calibration problem, it is necessary to use a calibration object, which requires the attitude of the calibration object coordinate system relative to the camera coordinate system to be calculated in real time. AR markers were used in this study. The camera for the robotic arm visual grasping system was installed in an eye-in-hand manner, and the origins of the manipulator base, manipulator end, camera, and AR marker coordinate systems are defined as O_b, O_e, O_c, and O_m, respectively. The coordinate systems and their transformation relationships during the hand–eye calibration process are shown in Figure 1.

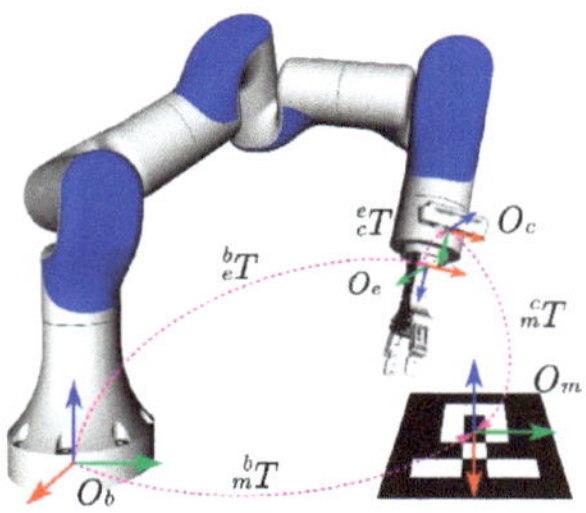

Figure 1. Coordinate system and transformation relationship diagram in the process of hand–eye calibration. The coordinate systems for the manipulator base, manipulator end, camera, and AR marker are defined as O_b, O_e, O_c, and O_m. In the coordinate system in the figure, the red arrow represents the x-axis, the green arrow represents the y-axis, and the blue arrow represents the z-axis.

Let ${}^{c}_{m}T$ be the transformation matrix of the AR marker coordinate system relative to the camera coordinate system. According to Figure 1, the transformation relationship between the AR marker coordinate system and base coordinate system of the manipulator ${}^{b}_{m}T$ is

$$ {}^{b}_{m}T = {}^{b}_{e}T \cdot {}^{e}_{c}T \cdot {}^{c}_{m}T \quad (1) $$

2.2. Hand–Eye Calibration Equation

In Equation (1), ${}^{e}_{c}T$ is fixed, which is the hand–eye transformation matrix to be solved. If the position of the AR marker relative to the base coordinate system of the manipulator is constant, ${}^{b}_{m}T$ is fixed. For a certain state $S_i | i \in \mathbb{N}$ of the manipulator, ${}^{c}_{m}T_i$ can be calculated using the size of the AR marker, the corner coordinates, and the internal parameters of the camera. Therefore, for a certain state S_i of the manipulator, the above equation can be expressed as:

$$ {}^{b}_{m}T = {}^{b}_{e}T \cdot {}^{e}_{c}T \cdot {}^{c}_{m}T_i\,,\ i \in N \tag{2} $$

There are two fixed unknown matrices, ${}^{b}_{m}T$ and ${}^{e}_{c}T$, in the above equation: to solve these two unknown matrices, it is necessary to control the manipulator to move in two different states, and the position of the AR marker should remain unchanged during movement. Using these two states, the following equations were obtained:

$$ \begin{cases} {}^{b}_{m}T = {}^{b}_{e}T \cdot {}^{e}_{c}T \cdot {}^{c}_{m}T_1 \\ {}^{b}_{m}T = {}^{b}_{e}T \cdot {}^{e}_{c}T \cdot {}^{c}_{m}T_2 \end{cases} \tag{3} $$

According to the above Equation, the following can be obtained

$$ {}^{b}_{e}T \cdot {}^{e}_{c}T \cdot {}^{c}_{m}T_1 = {}^{b}_{m}T = {}^{b}_{e}T \cdot {}^{e}_{c}T \cdot {}^{c}_{m}T_2 \tag{4} $$

The above Equation can be converted to

$$ \underbrace{{}^{b}_{e}T_2^{-1} \cdot {}^{b}_{e}T_1}_{A} \cdot \underbrace{{}^{e}_{c}T}_{X} = \underbrace{{}^{e}_{c}T}_{X} \cdot \underbrace{{}^{c}_{m}T_2 \cdot {}^{c}_{m}T_1^{-1}}_{B} \tag{5} $$

Furthermore, the problem of solving the hand–eye transformation matrix ${}^{e}_{c}T$ is transformed into a problem of solving the homogeneous Equation AX = XB, where A, X, and B are 4×4 homogeneous transformation matrices.

To further solve the homogeneous equation AX = XB, the homogeneous transformation matrix in Equation (5) is expressed in the form of a rotation matrix and translation vector as follows:

$$ \underbrace{\begin{bmatrix} R_A & t_A \\ 0 & 1 \end{bmatrix}}_{A} \underbrace{\begin{bmatrix} R & t \\ 0 & 1 \end{bmatrix}}_{X} = \underbrace{\begin{bmatrix} R & t \\ 0 & 1 \end{bmatrix}}_{X} \underbrace{\begin{bmatrix} R_B & t_B \\ 0 & 1 \end{bmatrix}}_{B} \tag{6} $$

By expanding the above formula, the equations to be solved can be obtained as

$$ \begin{cases} R_A \cdot R = R \cdot R_B \\ (R_A - I)t = R \cdot t_B - t_A \end{cases} \tag{7} $$

In the above Equation, R_A, R_B, t_A, and t_B can be measured, and I is the unit matrix. There are many solutions for obtaining rotation matrix R and translation vector t from this equation set, such as the Tsai–Lenz algorithm [7] and the Horaud [31], Andreff [25], and Daniilidis [32] algorithms. Subsequently, we conducted simulation experiments and developed an error analysis method to analyze and optimize the above four hand–eye calibration algorithms.

3. Reprojection Error Analysis Method of Calibration Algorithms

3.1. Hand–Eye Calibration Algorithm Simulation Experiments

In this study, the ROS and Gazebo simulation platforms were used to build the simulation environment shown in Figure 2 to test the performance of the above four hand–eye calibration algorithms [7,25,31,32]. With reference to the manipulator employed in the practical experiment, the simulation employed the standard DH parameter method to conduct the kinematic modeling of the 7-degree-of-freedom manipulator. The link coordinate system of the manipulator was established as illustrated in Figure 3, and the respective DH parameters are listed in Table 1. The resolution of the RGB camera in the test environment was 640 × 480 pixels, and Gaussian noise with a mean value of 0 and a variance of 0.07 was added to the collected images. The internal parameters are Equations (19) and (20), which are the internal camera parameters used in the real experiment. In the simulation environment, the size of the AR marks was 10 × 10 cm. The true values of ${}^{e}_{c}T$ and ${}^{b}_{m}T$ for the simulation experiments are listed in Table 2.

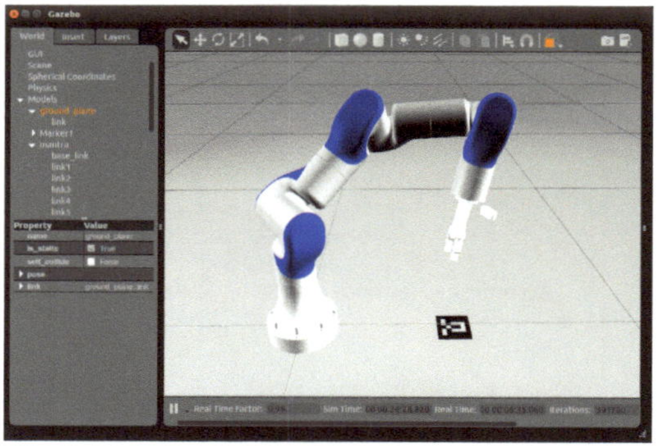

Figure 2. Hand–eye calibration simulation experimental environment built in ROS and Gazebo simulation platforms. The resolution of the RGB camera in the test environment is 640 × 480 pixels and the size of the AR mark is 10 cm × 10 cm.

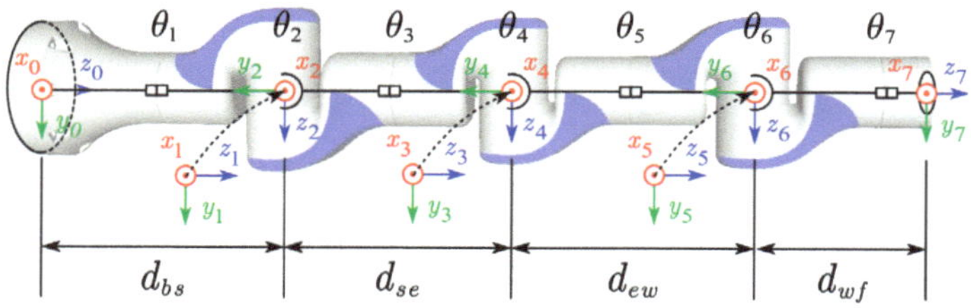

Figure 3. Establishment of the coordinate system of the 7-DOF manipulator in this article. In the coordinate system in the figure, the red arrow represents the x-axis, the green arrow represents the y-axis, and the blue arrow represents the z-axis.

Table 1. DH parameter table of 7-DOF robot arm in this article.

i	$\alpha_{i-1}(deg)$	$a_{i-1}(m)$	$d_i(m)$	$\theta_i(deg)$	$\theta_{imin}(deg)$	$\theta_{imax}(deg)$
1	−90	0	d_{bs}	θ_1	−180	180
2	90	0	0	θ_2	−110	110
3	90	0	d_{se}	θ_3	−180	180
4	−90	0	0	θ_4	−107	107
5	−90	0	d_{ew}	θ_5	−180	180
6	90	0	0	θ_6	−110	110
7	0	0	d_{wf}	θ_7	−180	180

Table 2. True values of parameters in a simulation environment.

	Translation Matrix t_r (m)	Rotation Matrix R_r	Eulerian Angle E_r (rad)
${}^{e}_{c}T$	$\begin{bmatrix} -0.0700 \\ 0.00000 \\ 0.03450 \end{bmatrix}$	$\begin{bmatrix} 0 & 1 & 0 \\ -1 & 0 & 0 \\ 0 & 0 & 1 \end{bmatrix}$	$\begin{bmatrix} 0 \\ 0 \\ -\pi/2 \end{bmatrix}$
${}^{b}_{m}T$	$\begin{bmatrix} 0.45000 \\ 0.00000 \\ 0.00100 \end{bmatrix}$	$\begin{bmatrix} 0 & 1 & 0 \\ -1 & 0 & 0 \\ 0 & 0 & 1 \end{bmatrix}$	$\begin{bmatrix} \pi/2 \\ 0 \\ \pi/2 \end{bmatrix}$

In the simulation experiment, 22 different states of the robotic arm were manually set to ensure that the center point information of the AR mark could be detected in each state and that the joint angles of the robotic arm changed significantly between each state. Some of the collected images and AR marker center point detection results are shown in Figure 4. The calculation results for each calibration algorithm are listed in Table 3.

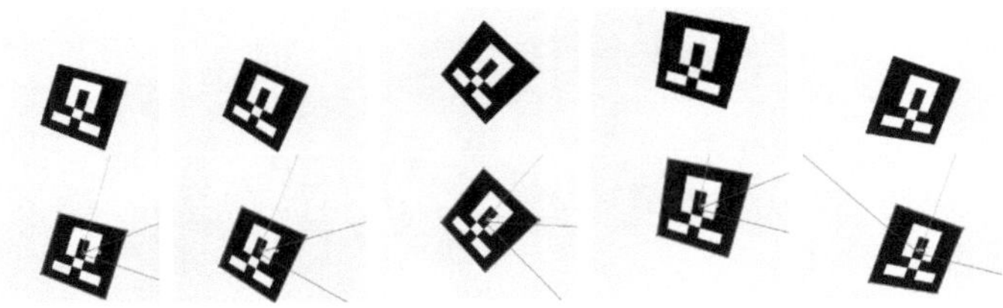

Figure 4. AR marker images and corner detection results collected in the simulation environment.

Table 3. Hand–eye transformation matrix calculated by calibration algorithms in a simulation environment.

Calibration Algorithm	${}^{e}_{c}T$ Computation	
	Translation Matrix	Rotation Matrix
TsaiLenz [7]	$\begin{bmatrix} -0.07058 \\ 0.00085 \\ 0.00325 \end{bmatrix}$	$\begin{bmatrix} 0.00203 & 0.99999 & 0.00337 \\ -0.99999 & 0.00204 & -0.00317 \\ -0.00318 & -0.00336 & 0.99999 \end{bmatrix}$
Horaud [31]	$\begin{bmatrix} -0.07160 \\ -0.00116 \\ 0.00315 \end{bmatrix}$	$\begin{bmatrix} 0.00190 & 0.99999 & 0.00326 \\ -0.99999 & 0.00191 & -0.00272 \\ -0.00273 & -0.00326 & 0.99999 \end{bmatrix}$
Andreff [25]	$\begin{bmatrix} -0.07205 \\ -0.00032 \\ 0.00343 \end{bmatrix}$	$\begin{bmatrix} 0.00185 & 0.99999 & 0.00323 \\ -0.99999 & 0.00186 & -0.00250 \\ -0.00251 & -0.00322 & 0.99999 \end{bmatrix}$
Daniilidi [32]	$\begin{bmatrix} -0.07198 \\ -0.00116 \\ 0.00315 \end{bmatrix}$	$\begin{bmatrix} 0.00161 & 0.99999 & 0.00371 \\ -0.99999 & 0.00162 & -0.00215 \\ -0.00216 & -0.00371 & 0.99999 \end{bmatrix}$

To quantitatively evaluate the performance of each calibration algorithm, the translation error err_t of the hand–eye transformation matrix was defined as the two norms of the difference between the calculated value t_c of the translation vector and the true value t_r. The translation error was measured using the following Euclidean distance:

$$err_t = \|t_c - t_r\|_2 \tag{8}$$

Similarly, the rotation matrix R is first converted to a Euler angle E, and the Euler angle is expressed in vector form as $E = [roll, pitch, yaw]^T$, and then the rotation error err_R of the hand–eye conversion matrix can be defined as:

$$err_R = \|E_c - E_r\|_2 \tag{9}$$

According to the real values of the parameters in the simulation environment, the statistical results of the translation and rotation errors of the hand–eye transformation matrix calculated by each calibration algorithm are shown in Figure 5. It can be observed from the statistical figure that the translation error of the hand–eye conversion matrix calculated by the Tsai–Lenz and Andreff algorithms is significantly lower than that of the other two algorithms; however, the rotation error of the hand–eye conversion matrix calculated by the Tsai–Lenz algorithm is slightly higher than that of the other algorithms. Overall, the calibration accuracies of the Tsai–Lenz and Andreff algorithms are relatively high in the simulation environment.

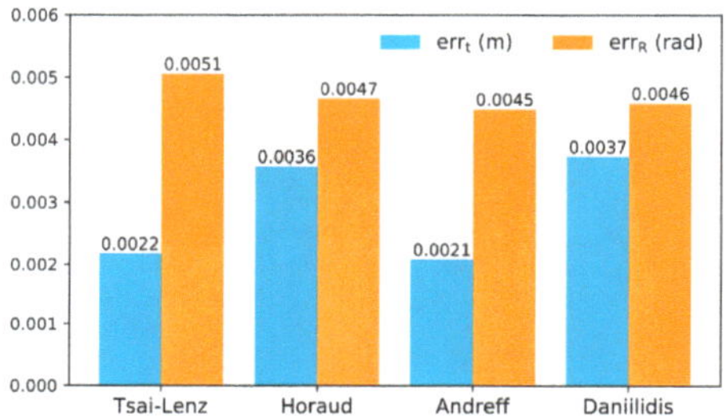

Figure 5. Euclidean distance error statistics of each calibration algorithm (according to the real values of the parameters in the simulation environment in Table 2). The statistical figure reveals that the translation error in the hand–eye conversion matrix, as computed by the Tsai–Lenz and Andreff algorithms, is notably lower compared to the other two algorithms. However, it is noteworthy that the rotation error in the hand–eye conversion matrix, calculated using the Tsai–Lenz algorithm, is marginally higher than in the other algorithms.

3.2. Heuristic Error Analysis

As shown in Figure 5, each algorithm has a certain optimization space. The following presents a heuristic error analysis of the simulation results. During the eye-in-hand calibration process, the position of the calibration object is fixed relative to the base coordinate system of the manipulator. Therefore, in theory, ${}^b_mT$ should be a fixed value; however, according to the coordinate transformation relationship shown in Figure 1, after calculating the hand–eye transformation matrix ${}^e_cT$, ${}^b_mT$ can be calculated by the following equation

$${}^b_mT = {}^b_eT \cdot {}^e_cT \cdot {}^c_mT \tag{10}$$

In the process of hand–eye calibration, it is necessary to collect multiple datasets; hence, the set $C = \left\{{}^b_mT_i\right\}, i \in \mathbb{N}$ can be calculated. Because ${}^b_mT$, calculated by the above formula, will be affected by the parameter error in ${}^e_cT$, a simple conjecture can be established. If the fluctuation degree of the data in set C is smaller, the errors of ${}^e_cT$ are smaller.

To facilitate the analysis of the degree of data fluctuation, box plots were used to visualize the fluctuation range of each parameter of the translation matrix t_i in ${}^b_mT_i$, and the results are shown in Figure 6. The red dotted line in the figure represents the real value, whereas the orange solid and green dotted lines represent the median and average of the data calculated using each algorithm, respectively. The left/right sides of the rectangle are the lower/upper quartiles, respectively; the size of the I-shaped area reflects the fluctuation range of the data; and the open circles represent the outliers.

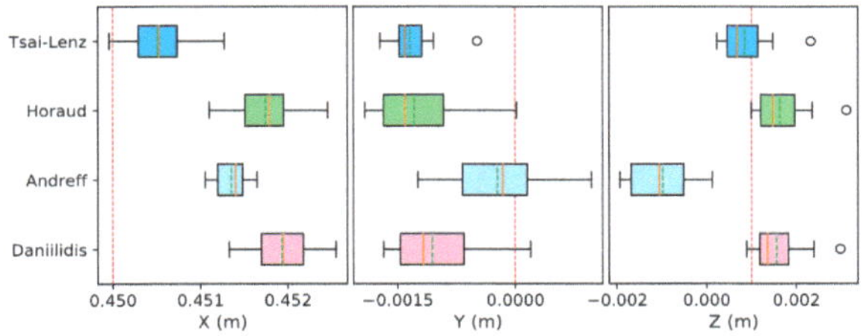

Figure 6. Visualization of fluctuation range of parameters in the translation matrix of ${}^{b}_{m}T_i$. In the illustration, the authentic data are represented by the red dotted line, whereas the orange solid line and green dotted line correspond to the median and average values derived from each algorithm, respectively. The left and right sides of the rectangle signify the lower and upper quartiles, respectively. The extent of the I-shaped area conveys the fluctuation range of the data, and the outliers are denoted by open circles.

The corresponding data fluctuation range of the Tsai–Lenz algorithm is small, which is in line with expectations. However, the Andreff algorithm has a large fluctuation range for the corresponding data, which is inconsistent with expectations. Therefore, it is unreasonable to determine the error of the hand–eye conversion matrix ${}^{e}_{c}T$ by the fluctuation degree of the data in set C because the above conjecture ignores the influence of the error of ${}^{c}_{m}T$ on the calculation result of ${}^{b}_{m}T$.

3.3. Reprojection Error Analysis

From the results of the heuristic error analysis, it can be observed that ${}^{e}_{c}T$ and ${}^{c}_{m}T$ may have errors; therefore, it is unreasonable to calculate ${}^{b}_{m}T$ using Equation (10). Because the position of the AR marker is fixed during the calibration process, the following error analysis process assumes that ${}^{b}_{m}T$ is known and fixed.

According to the coordinate transformation relationship shown in Figure 1, the pose representation of the AR marker in the camera coordinate system can be obtained as follows:

$$ {}^{c}_{m}T = {}^{c}_{e}T \cdot {}^{e}_{b}T \cdot {}^{b}_{m}T \tag{11} $$

If the AR marker coordinate system $\{O_{marker}\}$ is defined as the world coordinate system, then ${}^{c}_{w}T = {}^{c}_{m}T$. According to the pinhole camera imaging model, the conversion relationship between the coordinates (X_w, Y_w, Z_w), the pixel coordinates (u, v) in the AR marker coordinate system and the z-axis coordinate Z_c in the camera coordinate system can be obtained as follows:

$$ Z_c \begin{bmatrix} u \\ v \\ 1 \end{bmatrix} = M {}^{c}_{w}T \begin{bmatrix} X_w \\ Y_w \\ Z_w \\ 1 \end{bmatrix} = M {}^{c}_{m}T \begin{bmatrix} X_m \\ Y_m \\ Z_m \\ 1 \end{bmatrix} = M {}^{c}_{e}T {}^{e}_{b}T {}^{b}_{m}T \begin{bmatrix} X_m \\ Y_m \\ Z_m \\ 1 \end{bmatrix} \tag{12} $$

Based on the above definition, the coordinates of the origin O_{marker} of the AR marker coordinate system in the world coordinate system $(X_w, Y_w, Z_w) = (X_m, Y_m, Z_m) = (0,0,0)$, and the pixel coordinates (u, v) of the AR marker center point can be calculated using the following formula:

$$Z_c \begin{bmatrix} u \\ v \\ 1 \end{bmatrix} = M_e^cT_b^eT_m^bT \begin{bmatrix} 0 \\ 0 \\ 0 \\ 1 \end{bmatrix} \tag{13}$$

In the above formula, M is the inherent property of the camera, ${}_e^cT$ is the hand–eye conversion matrix to be calibrated, ${}_b^eT$ can be calculated by the forward kinematics equation of the manipulator, and ${}_m^bT$ is known and fixed.

Because the translation part of the homogeneous transformation matrix ${}_m^bT$ reflects the coordinates of the AR marker center point in the base coordinate system of the manipulator, the function of Equation (13) is to remap the coordinates of the AR marker center point into the pixel coordinate system. For a certain position $P_i, i \in \mathbb{N}$, the manipulator moves during the calibration process, the AR mark image captured by the camera is denoted as $\mathrm{img_I}$, and the pixel coordinate after the reprojection of the AR marker center point in img_i is denoted as $q_i = [u_i, v_i]^T$, and then:

$$q_i = [u_i, v_i]^T = proj\left({}_e^cT, {}_b^eT_i, {}_m^bT\right), i \in \mathbb{N} \tag{14}$$

Because the real pixel coordinates $Q_i = [U_i, V_i]^T$ of the AR marker center point in $\mathrm{img_I}$ can be obtained by corner detection or manual labelling, the reprojection error err_{proj_i} corresponding to img_i can be defined as the Euclidean distance between the real pixel coordinates of the AR marker center point and the reprojection coordinates:

$$err_{proj_i} = \|Q_i - q_i\|_2 = \left\|[U_i, V_i]^T - proj\left({}_e^cT, {}_b^eT_i, {}_m^bT\right)\right\|_2, i \in N \tag{15}$$

If the manipulator moves to N positions during the hand–eye calibration process, the average reprojection error can be defined as

$$err\_proj_{avg} = \sum_{i=1}^{N} err\_proj_i / N \tag{16}$$

According to Equations (15) and (16), the reprojection error of each group of simulation experimental data was calculated. The results are shown in Figure 7. The horizontal line in the figure represents the average reprojection error of each calibration algorithm. The average reprojection error corresponding to the calculation results of the Tsai–Lenz and Andreff algorithms was small, and the fluctuation in the reprojection error of each group of data was relatively small. In addition, from the previous analysis results, the Euclidean distance errors of these two algorithms were relatively small, proving that the reprojection error reflects the accuracy of the calibration results to a certain extent. Generally, the smaller the reprojection error, the higher the accuracy of the calibration results. Because the real value of the hand–eye transformation matrix cannot be obtained in the process of the hand–eye calibration of a real manipulator, the Euclidean distance error of the calibration results cannot be calculated, and the reprojection error can be used as the evaluation standard for calibration accuracy.

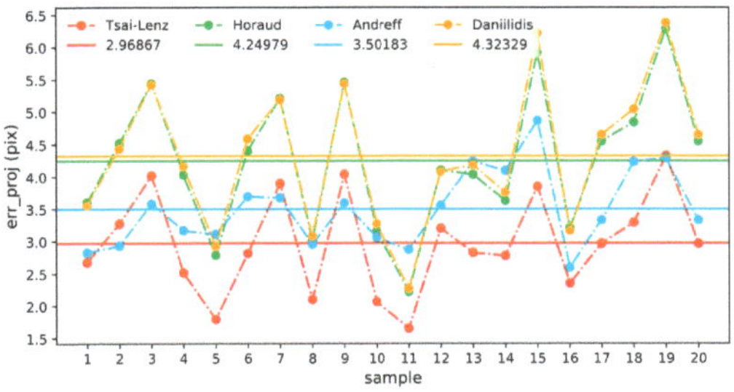

Figure 7. Reprojection error comparison (simulation environment). The average reprojection error associated with the computations from the Tsai–Lenz and Andreff algorithms is minimal, and the variation in the reprojection error for each dataset is relatively modest. This observation aligns with the earlier analyses depicted in Figure 4.

3.4. Optimizing Calibration Algorithm by Minimizing Reprojection Error Analysis

From the statistical results of the Euclidean distance error of each algorithm in Figure 1, even if the hand–eye calibration is carried out in the simulation environment, the translation error of the hand–eye conversion matrix calculated by different calibration algorithms is also quite different. Both are greater than 2 mm, which indicates that each calibration algorithm still has a large optimization space.

In the process of hand–eye calibration in a simulation environment, the only error source is the pose calculation error of the AR marker. However, when the above four common algorithms are used for calibration, the calculated ${}^{c}_{m}T$ is considered error-free, resulting in a certain error in the calibration results of each algorithm. In other words, the conventional hand–eye calibration algorithm pays more attention to versatility and does not use prior knowledge that the position of the AR marker is fixed in the calibration process; therefore, it is difficult to obtain high-precision calibration results. The definition of the reprojection error of the hand–eye calibration results makes full use of this prior knowledge. According to the previous analysis, the smaller the average reprojection error, the higher the accuracy of the calibration results by minimizing the reprojection error.

Based on the above analysis, the following exploratory experiments were carried out by controlling the variables to test whether smaller average reprojection errors can be obtained by adjusting the parameters in the calibrated hand–eye transformation matrix. In the experiment, the ${}^{b}_{m}T$ used to calculate the reprojection error takes the real values in Table 2, and the translation parameters calibrated by the Tsai–Lenz algorithm in Table 3 are taken as the initial values. The three translation parameters x, y, and z were adjusted with a step length of 0.001 m, and the adjusted hand–eye transformation matrix was substituted into Equations (15) and (16) to calculate the average reprojection error for each group of samples in the simulation experiment. The step size is selected based on the order of magnitude of the translation matrix. If the step size is too large, the search may be too fast, and the optimal solution may be missed. When the step size is too small, it may cause increased computational overhead, especially in high-dimensional parameter spaces. The experimental results are shown in Figure 8. The purple dotted line represents the parameter value of the minimum point, whereas the black dotted line represents the average reprojection error of the minimum point.

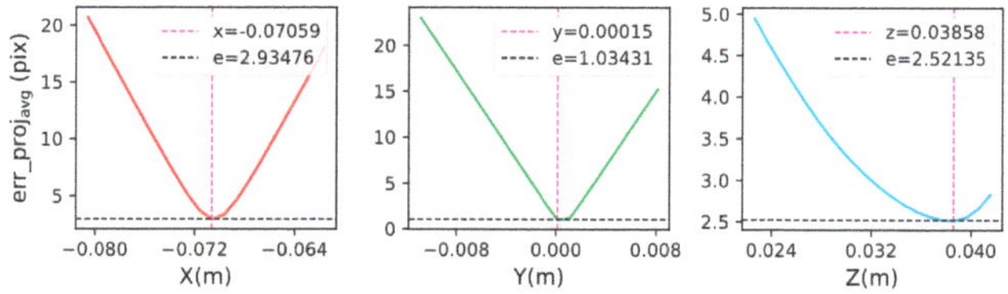

Figure 8. Change curves of reprojection error when adjusting x, y, and z, respectively.

As shown in Figure 8, the average reprojection error can be reduced to a certain extent by adjusting the translation parameters separately. The translation parameters calibrated by the Tsai–Lenz algorithm are used as the initial values, which can reduce the search space of the parameters and can help determine the translation parameters corresponding to the lowest point of the reprojection error. However, when z = 0.03858 m, the average reprojection error was minimized; however, this value deviated from the real value $z_r = 0.0345$ m (Table 3). Therefore, translation parameters with higher accuracy cannot be guaranteed by adjusting x, y, and z alone.

Next, we used 0.001 m as the step length and simultaneously adjusted the x, y, and z parameters. The change rule for the reprojection error is shown in Figure 9. The colors of the data points in the figure reflect the size of the reprojection errors. The translation matrix that minimizes the average reprojection error is $t_m = [-0.07059, 0.00015, 0.03558]^T$, and the corresponding average reprojection error $err_{proj_{avg}} = 0.69861$. The corresponding translation error, $err_t = 0.00165$, was calculated using Equation (8). Figures 5 and 7 show that the translation error, $err_t = 0.0022$, and the average reprojection, error $err_{proj_{avg}} = 2.96867$, of the calibration results were calculated using the Tsai–Lenz algorithm. There is a set of translation parameters, $t_b = (x_b, y_b, z_b)$, that can minimize the average reprojection error, and the translation error of this set of parameters, err_t, is less than that of the parameters calibrated using the Tsai–Lenz algorithm. In other words, the accuracy of the hand–eye calibration results can be improved by simultaneously adjusting the three parameters x, y, and z to minimize the reprojection error.

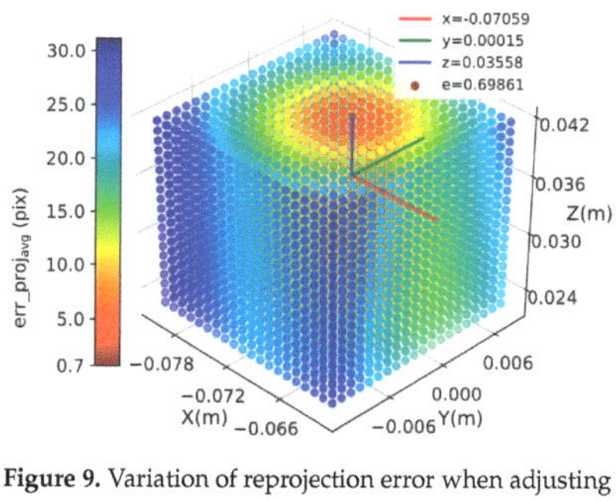

Figure 9. Variation of reprojection error when adjusting x, y, and z parameters simultaneously.

During the calibration process, after a certain position P_i of the manipulator is determined, the four parameters U_i, V_i, ${}^e_bT_i$, ${}^b_mT$ in Equation (15) are determined accordingly. Therefore, ${}^c_eT = {}^e_cT^{-1}$ determines the size of the reprojection error err_{proj_i}. According to Equation (16), if N positions are determined, then the magnitude of the average reprojection error, $err_{proj_{avg}}$, is uniquely determined by ${}_{c}^{e}T$, and the mapping function of the hand–eye transformation matrix, ${}_{c}^{e}T$, to the average reprojection error, $err_{proj_{avg}}$, can be defined as

$$err\_proj_{avg} = f({}_{c}^{e}T) \tag{17}$$

Furthermore, ${}_{c}^{e}T$ can be uniquely determined by the translation parameters $t = (x, y, z)$ and the rotation parameter $r = (roll, pitch, yaw)$; hence, the mapping function F of the translation and rotation parameters to the average reprojection error, $err_{proj_{avg}}$, can be defined as follows:

$$err\_proj_{avg} = F(t, r) = F(x, y, z, roll, pitch, yaw) \tag{18}$$

Based on the above definition, this study transforms the optimization problem of the hand–eye conversion matrix into the problem of finding the minimum value of the objective function F and optimizes the hand–eye conversion matrix from the perspective of minimizing the reprojection error.

Next, with a step size of 0.0001 m, the three parameters x,y, and z were adjusted to search for the translation parameter that minimizes the function F. The optimal translation parameter $t_b = (-0.07058, 0.00039, 0.03483)$ and the corresponding average reprojection error, $err_{proj_{avg}}$, is 0.36084, which is 87.845% lower than that of the Tsai–Lenz algorithm. The translation error, err_t, is 0.0007836 m, which is 64.382% lower than that of the Tsai–Lenz algorithm.

In the above optimization process, only the translation parameters in the hand–eye conversion matrix were adjusted. After numerous simulation calibration experiments, it was deduced that the translation parameters had a greater influence on the accuracy of the calibration results than the rotation parameters, and the number of translation parameters was lower. Generally, adjusting only the translation parameters yields ideal calibration results. If the translation and rotation parameters are adjusted simultaneously, higher-precision calibration results can be obtained theoretically, but this process is highly time consuming.

4. Hand–Eye Calibration Algorithm Experiment

4.1. Calibration Process and Results

To realize the real-time calculation of the spatial posture of the end of the robotic arm, the standard DH parameter method is used to carry out the kinematic modeling of the 7-degree-of-freedom robotic arm in this paper, and the robotic arm link coordinate system is established as shown in Figure 3. The corresponding DH parameters are listed in Table 4. In the table, i represents the joint index, α_{i-1} denotes the rotation angle of link i relative to link $i-1$, a_{i-1} is the length of the previous link, d_i indicates the offset distance of the i-th joint, and θ_i represents the rotational angle of each joint. Specifically, $d_{bs} = 0.3705\ m$, $d_{se} = 0.3300\ m$, $d_{ew} = 0.3200\ m$, and $d_{wf} = 0.2255\ m$. In the experimental environment, the size of the AR mark was 10 cm × 10 cm. The RGB camera internal parameters are shown below:

$$[f_x, f_y, u_0, v_0] = [1373.72, 1374.10, 965.286, 554.510] \tag{19}$$

$$[k_1, k_2, p_1, p_2] = [0.142, -0.288,\ 0.002,\ 0.000] \tag{20}$$

Table 4. Hand–eye transformation matrix from calibration algorithms in a real environment.

Calibration Algorithm	${}^{e}_{c}T$ Computation	
	Translation Matrix	**Rotation Matrix**
TsaiLenz [7]	$\begin{bmatrix} -0.03740 \\ -0.09599 \\ 0.04845 \end{bmatrix}$	$\begin{bmatrix} 0.99981 & -0.01313 & -0.01423 \\ 0.01346 & 0.99963 & 0.02330 \\ 0.01392 & -0.02348 & 0.99962 \end{bmatrix}$
Horaud [31]	$\begin{bmatrix} -0.03740 \\ -0.09599 \\ 0.04845 \end{bmatrix}$	$\begin{bmatrix} 0.99981 & -0.01315 & -0.01422 \\ 0.01348 & 0.99963 & 0.02330 \\ 0.01391 & -0.02349 & 0.99962 \end{bmatrix}$
Andreff [25]	$\begin{bmatrix} -0.03658 \\ -0.09808 \\ 0.06191 \end{bmatrix}$	$\begin{bmatrix} 0.99981 & -0.01258 & -0.01464 \\ 0.01292 & 0.99965 & 0.02307 \\ 0.01434 & -0.02325 & 0.99962 \end{bmatrix}$
Daniilidi [32]	$\begin{bmatrix} -0.03746 \\ -0.09580 \\ 0.04834 \end{bmatrix}$	$\begin{bmatrix} 0.99981 & -0.01179 & -0.01497 \\ 0.01214 & 0.99965 & 0.02313 \\ 0.01469 & -0.02330 & 0.99962 \end{bmatrix}$
Optimized	$\begin{bmatrix} -0.03540 \\ -0.09349 \\ 0.05795 \end{bmatrix}$	$\begin{bmatrix} 0.99981 & -0.01313 & -0.01423 \\ 0.01346 & 0.99963 & 0.02330 \\ 0.01392 & -0.02348 & 0.99962 \end{bmatrix}$

Under the premise that the internal parameters of the camera have been calibrated, the hand–eye calibration experiment is performed in a real environment. The experimental configuration is shown in Figure 10a, and the AR mark is placed on the workbench in front of the robotic arm. To obtain the real position of the AR marker, an auxiliary calibration tool is installed at the end of the manipulator, and then the manipulator is manually controlled to align the top of the calibration tool with the center point of the AR marker (as shown in Figure 10b). Finally, the forward kinematics equation of the manipulator is used to calculate the translation matrix of the center point of the AR marker relative to the base coordinate system of the manipulator. The translation matrix is $t_r = \begin{bmatrix} 0.53514\,m & 0.00406\,m & 0.25409\,m \end{bmatrix}^T$.

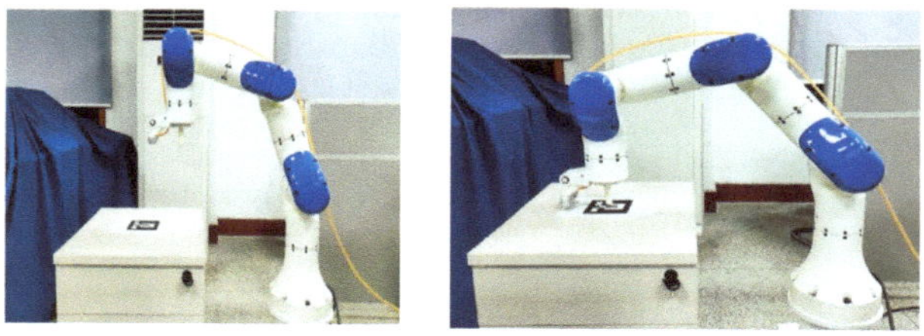

(a) (b)

Figure 10. Hand–eye calibration experiment configuration and AR marker center point acquisition. (**a**) Hand–eye calibration experiment configuration. (**b**) AR marker center point acquisition.

According to the experimental and error analysis results in the simulation environment, the optimized hand–eye calibration process is as follows.

- The real translation matrix t_r of the AR marker coordinate system relative to the base coordinate system of the manipulator was obtained using the auxiliary calibration tool, and the position of the AR marker remained unchanged.
- The manipulator was controlled to move to 20 different states where the corner information of the AR marker could be detected, and the corresponding 20 groups of coordinate system transformation data were collected and recorded.
- The mean value ${}^{b}_{m}T_{avg}$ of ${}^{b}_{m}T$ is calculated using the coordinate transformation data of each group, and the translation matrix in ${}^{b}_{m}T_{avg}$ is replaced by t_r to obtain ${}^{b}_{m}T_{proj}$ for calculating the reprojection error.

- The Tsai–Lenz algorithm is used to calculate the initial value of the hand–eye transformation matrix ${}^{e}_{c}T_{init}$, and its translation parameters are automatically adjusted to minimize the average reprojection error. The optimized hand–eye transformation matrix is ${}^{e}_{c}T_{optimized}$.

Because it is difficult to obtain the true value of the rotation parameter in ${}^{b}_{m}T$ in a real environment, the average value was used when calculating the reprojection error in the above calibration process. According to the above process, a hand–eye calibration experiment was carried out in a real environment, and the hand–eye conversion matrix ${}^{e}_{c}T$ was calculated using the four traditional algorithms and the optimized algorithms mentioned above. The results are summarized in Table 4. In addition, in terms of algorithm calibration efficiency, because the calibration formula includes four common arithmetic and matrix operations, there is no requirement for computing power. With the current mainstream CPU (i7 10750H), the calibration time does not exceed 1 ms.

4.2. Reprojection Error Analysis

To evaluate the performance difference between the traditional algorithm and the optimized algorithm in a real environment, the reprojection errors corresponding to each hand–eye transformation matrix in Table 4 were calculated using the coordinate transformation data of each group. The results are shown in Figure 11, where the horizontal line indicates the average reprojection error for each method. Because the difference between the hand and eye conversion matrices calculated by the Tsai–Lenz and Horaud algorithms is very small, only the reprojection error corresponding to the Tsai–Lenz algorithm is shown in the figure. In a real environment, except for the Andreff algorithm, the performances of the other traditional algorithms are similar. It is worth mentioning that the average reprojection error of the optimized algorithm is reduced by 44.43% compared with the Tsai–Lenz algorithm. Furthermore, the variability observed in the connecting lines between discrete points in the graph served as an indicator of the robustness of the optimization algorithm. An observation of the figure readily reveals that the fluctuation level in the yellow connecting line, representing the optimization algorithm, is notably reduced compared with the other algorithms. This distinct pattern underscores the heightened robustness of the optimization algorithm compared with its counterparts.

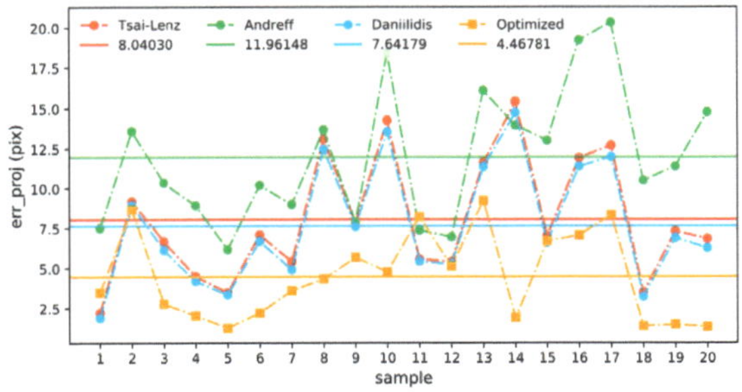

Figure 11. Comparison of reprojection errors. In the real-world context, excluding the Andreff algorithm, traditional algorithms demonstrate similar performance levels. Notably, the optimized algorithm stands out with a significant 44.43% reduction in the average reprojection error compared to the Tsai–Lenz algorithm, showcasing its enhanced efficacy in practical scenarios.

4.3. Visual Positioning Error Analysis

In a real scene, the calibration results of each algorithm are used to test the positioning accuracy of the manipulator's visual positioning. During the test, the AR marker was moved several times, and the state of the manipulator was adjusted to ensure that the corner information of the AR marker could be detected. Position P_c of the center point of the AR marker in the base coordinate system of the manipulator was then calculated using Equation (2). Finally, the manipulator was manually controlled, and an auxiliary calibration tool was used to obtain the real position P_r of the AR marker center point.

The positioning accuracy of the manipulator in a real scene was quantitatively evaluated, and the visual positioning error was defined as the two norms of the difference between the calculated value P_c and the real value P_r of the AR marker center point position.

In the actual test process, 10 datasets were collected, and the hand–eye conversion matrix calculated by each algorithm was used for the visual positioning of the manipulator. The error statistics are shown in Figure 12. The horizontal line represents the average visual positioning error for each algorithm. The optimized hand–eye calibration method can significantly reduce the visual positioning error of the manipulator. Compared with the traditional Tsai–Lenz algorithm, the average visual positioning error was reduced by 50.63%.

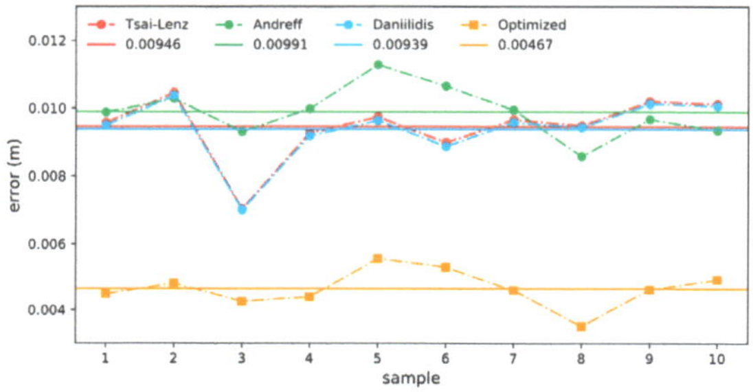

Figure 12. Visual positioning error comparison. The optimized hand–eye calibration method demonstrated a significant reduction in visual positioning error compared to the traditional Tsai–Lenz algorithm, achieving an impressive 50.63% average error reduction. This highlights the efficacy of the optimized approach in substantially enhancing the manipulator's visual positioning accuracy in real-world scenarios.

5. Summary

To ensure the precision of visual information obtained by a robot, this study delves into the hand–eye calibration algorithm of the manipulator. Commonly used hand–eye calibration algorithms are tested in a simulated environment, and an error analysis is conducted on the results. Subsequently, the reprojection error of the hand–eye calibration results are defined, and an optimization method for hand–eye calibration is proposed based on the eye-in-hand manipulator. Experimental validation is performed in a real environment using the manipulator visual grasping system. The results indicate a significant enhancement in calibration accuracy. Specifically, compared to the Tsai–Lenz algorithm, the proposed optimization algorithm reduces the average reprojection error by 44.43% and the average visual positioning error by 50.63%. Thus, the proposed optimization method markedly enhances the precision of hand–eye calibration results, elevating the overall performance of hand–eye calibration algorithms.

Author Contributions: Methodology, G.P. and Z.R.; software, Q.G.; validation, Q.G.; investigation, Z.R. and Q.G.; data curation, Z.R. and Q.G.; project administration, G.P.; structural optimization and writing improvement, Z.F. All authors have read and agreed to the published version of the manuscript.

Funding: This work was supported by the Hubei Province Core Technology for Bridging Development Gaps Project (HBSNYT202213), the Hubei Province Unveiling Science and Technology Project (2021BEC008), and the Hubei Province Natural Science Foundation of China (No. 2019CFB526).

Institutional Review Board Statement: Not applicable.

Informed Consent Statement: Not applicable.

Data Availability Statement: Data sharing is not applicable to this article.

Conflicts of Interest: The authors declare no conflict of interest.

References

1. Macenski, S.; Foote, T.; Gerkey, B.; Lalancette, C.; Woodall, W. Robot Operating System 2: Design, architecture, and uses in the wild. *Sci. Robot.* **2022**, *7*, eabm6074. [CrossRef] [PubMed]
2. Yang, L.; Yu, J.; Guo, Y.; Chen, S.; Tan, K.; Li, S. An electrode-grounded droplet-based electricity generator (EG-DEG) for liquid motion monitoring. *Adv. Funct. Mater.* **2023**, *33*, 202302147. [CrossRef]
3. Markku, S.; Karayiannidis, Y.; Kyrki, V. A survey of robot manipulation in contact. *Robot. Auton. Syst.* **2022**, *156*, 104224.
4. Lin, W.; Liang, P.; Luo, G.; Zhao, Z.; Zhang, C. Research of Online Hand–Eye Calibration Method Based on ChArUco Board. *Sensors* **2022**, *22*, 3805. [CrossRef] [PubMed]
5. Wang, D.; Wang, L.; Wu, J.; Ye, H. An experimental study on the dynamics calibration of a 3-DOF parallel tool head. *IEEE/ASME Trans. Mechatron.* **2019**, *24*, 2931–2941. [CrossRef]
6. Zhang, J.; Li, X.; Dai, X. Camera calibration method based on 3D board. *J. Southeast Univ. (Nat. Sci. Ed.)* **2011**, *43*, 543–548. [CrossRef]
7. Tsai, R.Y.; Lenz, R.K. A new technique for fully autonomous and efficient 3 d robotics hand/eye calibration. *IEEE Trans. Robot. Autom.* **1989**, *5*, 345–358. [CrossRef]
8. Ge, P.; Wang, Y.; Wang, H.; Li, G.; Zhang, M. Multivision Sensor Extrinsic Calibration Method with Non-Overlapping Fields of View Using Encoded 1D Target. *IEEE Sens. J.* **2022**, *22*, 13519–13528. [CrossRef]
9. Bu, L.; Huo, H.; Liu, X.; Bu, F. Concentric circle grids for camera calibration with considering lens distortion. *Opt. Lasers Eng.* **2021**, *140*, 106527. [CrossRef]
10. Chen, X.; Song, X.; Wu, J.; Xiao, Y.; Wang, Y.; Wang, Y. Camera calibration with global LBP-coded phase-shifting wedge grating arrays. *Opt. Lasers Eng.* **2021**, *136*, 106314. [CrossRef]
11. Zhang, Z. A flexible new technique for camera calibration. *IEEE Trans. Pattern Anal. Mach. Intell.* **2000**, *22*, 1330–1334. [CrossRef]
12. Hartley, R.I. Self-calibration of stationary cameras. *Int. J. Comput. Vis.* **1997**, *22*, 5–23. [CrossRef]
13. Li, M. Camera calibration of a head-eye system for active vision. In Proceedings of the Computer Vision—ECCV'94: Third European Conference on Computer Vision, Stockholm, Sweden, 2–6 May 1994; Springer: Berlin/Heidelberg, Germany, 1994; Volume I, p. 3.
14. Hartley, R.I. Projective reconstruction and invariants from multiple images. *IEEE Trans. Pattern Anal. Mach. Intell.* **1994**, *16*, 1036–1041. [CrossRef]
15. Pollefeys, M.; Van Gool, L.; Oosterlinck, A. The modulus constraint: A new constraint self-calibration. In Proceedings of the 13th International Conference on Pattern Recognition, Vienna, Austria, 25–29 August 1996; Volume 1.
16. Triggs, B. Autocalibration and the absolute quadric. In Proceedings of the IEEE Computer Society Conference on Computer Vision and Pattern Recognition, San Juan, PR, USA, 17–19 June 1997.
17. Heyden, A.; Astrom, K. Flexible calibration: Minimal cases for auto-calibration. In Proceedings of the Seventh IEEE International Conference on Computer Vision, Kerkyra, Greece, 20–27 September 1999; Volume 1.
18. Pollefeys, M.; Koch, R.; Van Gool, L. Self-calibration and metric reconstruction inspite of varying and unknown intrinsic camera parameters. *Int. J. Comput. Vis.* **1999**, *32*, 7–25. [CrossRef]
19. Pollefeys, M. Self-Calibration and Metric 3D Reconstruction from Uncalibrated Image Sequences. Ph.D. Thesis, Katholieke Universiteit Leuven, Leuven, Belgium, 1999.
20. Hartley, R.I.; Hayman, E.; de Agapito, L.; Reid, I. Camera calibration and the search for infinity. In Proceedings of the Seventh IEEE International Conference on Computer Vision, Kerkyra, Greece, 20–27 September 1999; Volume 1.
21. Shiu, Y.C.; Ahmad, S. Calibration of wrist mounted robotic sensors by solving homogeneous transform equations of the form AX = XB. *IEEE J. Robot. Autom.* **1989**, *5*, 16–29. [CrossRef]
22. Chen, W.; Du, J.; Xiong, W.; Wang, Y.; Chia, S.; Liu, B.; Cheng, J.; Gu, Y. A Noise-Tolerant Algorithm for Robot-Sensor Calibration Using a Planar Disk of Arbitrary 3-D Orientation. *IEEE Trans. Autom. Sci. Eng.* **2018**, *15*, 251–263. [CrossRef]
23. Li, M.; Du, Z.; Ma, X.; Dong, W.; Gao, Y. A robot hand-eye calibration method of line laser sensor based on 3D reconstruction. *Robot. Comput. -Integr. Manuf.* **2021**, *71*, 102136. [CrossRef]
24. Zhang, Y.; Qiu, Z.; Zhang, X. Calibration method for hand-eye system with rotation and translation couplings. *Appl. Opt.* **2019**, *58*, 5375–5387. [CrossRef]
25. Andreff; Nicolas; Horaud, R.; Espiau, B. On-line hand-eye calibration. In Proceedings of the Second International Conference on 3-D Digital Imaging and Modeling (Cat. No. PR00062), Ottawa, ON, Canada, 8 October 1999.
26. Tabb, A.; Yousef, K.M.A. Solving the robot-world hand-eye (s) calibration problem with iterative methods. *Mach. Vis. Appl.* **2017**, *28*, 569–590. [CrossRef]
27. Jiang, J.; Luo, X.; Xu, S.; Luo, Q.; Li, M. Hand-Eye Calibration of EOD Robot by Solving the AXB = YCZD Problem. *IEEE Access* **2022**, *10*, 3415–3429. [CrossRef]

28. Liu, Y.; Wang, Q.; Li, Y. A method for hand-eye calibration of a robot vision measuring system. In Proceedings of the 2015 10th Asian Control Conference (ASCC), Kota Kinabalu, Malaysia, 31 May–3 June 2015; pp. 1–6. [CrossRef]
29. Zou, Y.; Chen, X. Hand–eye calibration of arc welding robot and laser vision sensor through semidefinite programming. *Ind. Robot* **2019**, *45*, 597–610. [CrossRef]
30. Deng, S.; Mei, F.; Yang, L.; Liang, C.; Jiang, Y.; Yu, G.; Chen, Y. Research on the Hand-eye calibration Method Based on Monocular Robot. In Proceedings of the 2021 International Conference on Mechanical Engineering, Intelligent Manufacturing and Automation Technology (MEMAT), Guilin, China, 15–17 January 2021. [CrossRef]
31. Horaud, R.; Dornaika, F. Hand-eye calibration. *Int. J. Robot. Res.* **1995**, *14*, 195–210. [CrossRef]
32. Daniilidis, K. Hand-eye calibration using dual quaternions. *Int. J. Robot. Res.* **1999**, *18*, 286–298. [CrossRef]

MDPI AG
Grosspeteranlage 5
4052 Basel
Switzerland
Tel.: +41 61 683 77 34

Sensors Editorial Office
E-mail: sensors@mdpi.com
www.mdpi.com/journal/sensors

www.ingramcontent.com/pod-product-compliance
Lightning Source LLC
Chambersburg PA
CBHW041412210926
51850CB00013B/3254
9783725828630